Preserving the Legacy

Basics of Industrial Hygiene

Debra Nims

JOHN WILEY & SONS, INC.

New York / Chichester / Weinheim / Brisbane / Singapore / Toronto

Technical Illustration: Richard J. Washichek, Graphic Dimensions, Inc.
Photo Research: Beatrice Hohenegger

This book is printed on acid-free paper. ∞

This material is based on work supported by the National Science Foundation under Grant No. DUE94-54521. Any opinions, findings, and conclusions or recommendations expressed in this material are those of the author(s) and do not necessarily reflect those of the National Science Foundation.

This publication is designed to provide accurate and authoritative information in regard to the subject matter covered. It is sold with the understanding that the publisher is not engaged in rendering legal, accounting, or other professional services. If legal advice or other expert assistance is required, the services of a competent professional person should be sought.

Library of Congress Cataloging-in-Publication Data:

0471 29983-9

Printed in the United States of America

10 9 8 7 6 5 4

PRESERVING

THE LEGACY

Table of Contents

Preface

Basics of Industrial Hygiene is the fifth volume in the **Preserving the Legacy** series developed by INTELECOM Intelligent Telecommunications in association with the Partnership for Environmental Technology Education (PETE). This is an introductory textbook designed to present the basic concepts of industrial hygiene at the technician level. When these concepts are supported by a program of study including biology, chemistry, physics, and English, the student should gain sufficient knowledge to be able to effectively perform the job duties and functions typically required of an industrial hygiene technician. It is also hoped that the information presented will spark the interest of students and provide a sound basis for moving on to further studies in the field of occupational safety and health.

This book is intended to introduce students to the areas typically encompassed by industrial hygiene, and the importance of the industrial hygienist in protecting the safety and health of workers. Following a brief history of industrial hygiene, its growth and representation by different professional organizations is discussed. Although familiarity with basic biology and toxicology is assumed, reviews of the principles and terminology that are foundational to the understanding of industrial hygiene are included.

A series of chapters address airborne contaminants, their measurement and control as well as the underlying OSHA regulations, and the occurrence and symptoms of occupational diseases associated with airborne particles. Solved examples involving the calculation of time-weighted averages, PELs, and the mixture rule help students understand the importance of sampling, analytical techniques, and exposure levels. Sick building syndrome, radon, and their associated health problems are considered, along with the instruments and calculations that can be used to determine the air exchange rate for the HVAC system.

After a brief review of the function and anatomy of the skin, various types of skin-damaging agents and methods for preventing exposures are presented. Occupational noise as well as its physics, measurement, control, and methods used to protect the hearing of the worker are discussed. Exposure of workers to ionizing and nonionizing radiation is presented, along with its measurement and control as well as electromagnetic and ionizing radiation. Ergonomics and temperature extremes focus on the physical comfort and safety of employees.

When engineering and administrative controls fail to provide the desired employee safety, consideration turns to the use of personal protective equipment (PPE) as either a temporary measure or a last resort. The use and maintenance of different types of respirators and chemical protective clothing are presented.

Student learning aids have been incorporated throughout the book. Each chapter starts with measurable learning objectives to help students focus on the broad concepts to be mastered. At the end of each section, questions are provided so students may check their understanding of the material as they move through the chapter. Critical thinking questions and sample problems are included at the end of each chapter to provide students an opportunity to apply their skills and knowledge in real-world situations. Figures and tables are liberally used throughout to illustrate and supplement the text. Each technical word that is considered critical to the understanding of the topic is bolded on first usage. The glossary provides definitions and additional information about each of these words, in the context of industrial hygiene usage. The bibliography directs students to other sources that contain information on the subject. A carefully developed index will assist students in the location of a topic of immediate interest.

The completion of this book would not have been possible without the partial financial support of the National Science Foundation's Department of Undergraduate Education (DUE) and the dedication and hard work of the many people at INTELECOM. Many thanks to the members of the *Preserving the Legacy's* National Academic Council (NAC) who, as PETE regional representatives, acted as designers and reviewers of the manuscript. They are:

—David Y. Boon, Front Range Community College

—Ann Boyce, Bakersfield College

— Eldon Enger, Delta College

— William T. Engle, Jr., PhD, University of Florida, TREEO Center

— Douglas A. Feil, Kirkwood Community College

— Steven R. Onstot, Esq., Fullerton College

— Douglas Nelson, SUNY, Morrisville, New York

— Ray Seitz, PhD, Columbia Basin College (Retired)

— Andrew J. Silva, South Dakota School of Mines

Their comments and suggestions added both quality and clarity to the writing. A special thanks to Ryan Wilke for his input on the chapter on indoor air quality, and to Cheryl Floreen for her reviews of the early chapters. Thanks also to Howard Guyer, Academic Team Leader, for his tenacity and input throughout the entire process; to Sally V. Beaty, INTELECOM President, for her vision and support; and to Beatrice Hohenegger, INTELECOM's Senior Editor, whose skills and attention to detail added immeasurably to the quality of the final product.

Thanks to Dick and Mary Ann Washichek of Graphic Dimensions, Inc. for their creativity, proofing, and polishing the manuscript into its educationally sound format.

A special thank you to my husband, friend, and colleague, Mark Langlois, for his contributions to the chapter on ionizing and nonionizing radiation, and also for his unwavering support throughout this project.

Debra Nims, Author

To my parents, JoAnn and Myron Nims, who taught all of us daughters to always strive for the best and who supported our careers in fields traditionally dominated by men.

Introduction to Industrial Hygiene

Chapter Objectives

Upon completing this chapter, the student will be able to:

1 **Identify** several persons who have made significant contributions to industrial hygiene in the past.

2 **Define** the terms industrial hygienist and industrial hygiene.

3 **Name** five responsibilities or tasks that an industrial hygienist might perform.

4 **Identify** five areas of applied science that are used in industrial hygiene practice.

5 **List**, in preferred order of application, the three basic methods used for controlling health hazards.

6 **Describe** the key elements of an industrial hygiene program and explain its relationship to other aspects of a company's overall safety program.

Chapter Sections

1-1 Introduction

Industrial hygiene has been defined as part science and part art. To put it simply, **industrial hygiene** is the application of scientific principles in the workplace to prevent the development of occupational disease or injury. It requires knowledge of chemistry and physics, anatomy and physiology as well as mathematics. Successful application of industrial hygiene principles also requires curiosity, creativity, and the ability to communicate effectively. This text is designed to cover the basics of industrial hygiene using a pyramid-like approach, presenting concepts and then building upon them to explore the many facets of industrial hygiene practice as they exist in today's changing world.

The present chapter includes some historical highlights in the development of industrial hygiene and introduces the role of the industrial hygienist in assuring worker health protection. Some of the common duties and responsibilities of an industrial hygienist are presented, and the importance of an industrial hygiene program as part of an organization's overall worker health and safety policy is discussed.

1-2 Early Industrial Hygiene

Diseases resulting from exposure to chemical and physical agents have existed as long as people have found it necessary or useful to handle materials that have toxic potential. Today, we face a dizzying array of chemicals at work and at home, from industrial solvents to cleaning preparations, paints, and deodorizers. Although in the past causes were not always recognized and correctly associated with their effects, some descriptions of occupational diseases that were recorded hundreds of years ago are remarkably accurate and insightful for their time. Some of them even allow us to put present-day names on the described diseases.

Among the earliest recordings of work-related diseases are observations of lead poisoning among miners by Hippocrates in the fourth century BCE (Before Common Era). A half-century later Pliny the Elder described dangers to workers exposed to zinc and sulfur. More records of occupational disease appeared during the Middle Ages in Europe. One of the most famous of these was written by a Saxon named Georg Bauer, better known by the Latin version of his name, Georgius Agricola (ah GRICK o la), who lived from 1494 to1555. He was the town physician in Joachimstal, where silver mining was one of the primary occupations.

> The many mines along the border between Germany and Czechoslovakia produce silver, nickel, and cobalt; they were, in the past, one of the world's chief sources of uranium. The uranium mines are now closed due to high levels of radioactivity and chemicals inside the tunnels and nearby tailing ponds.

Agricola's work, a 12-volume set called *De Re Metallica*, was published in 1556. Among the topics included in *De Re Metallica* are: mining geology, environmental contamination, management techniques (including scheduling of shift work and layoffs), mine ventilation, ergonomics, and the illnesses suffered by miners.

De Re Metallica represented what was, at the time, a comprehensive, state-of-the art text on mining, smelting, and refining operations. Agricola described the diseases and ailments suffered by the miners, including lung, joint, and eye afflictions. The symptoms and effects were described in enough detail that we can deduce that the miners suffered

from diseases such as silicosis, tuberculosis, and lung cancer: "If the dust has corrosive qualities, it eats away at the lungs, and implants consumption in the

Figure 1-1 (A and B): The hazards associated with airborne contaminants were already recognized four hundred years ago. These woodcuts from Agricola's *De Re Metallica* (1556) show some of the methods used to provide ventilation in hazardous atmospheres.

body...*" Workers also exhibited symptoms that we now recognize as various manifestations of toxicity of arsenic and cadmium: "...there is found in the mines black pompholyx, which eats wounds and ulcers to the bone; this also corrodes iron... there is a certain kind of cadmia which eats away at the feet of workmen when they have become wet, and similarly their hands, and injures their lungs and eyes." Woodcut illustrations excerpted from Agricola's books also show smelting processes, ventilation systems, and mechanical lifting machines. Butter is the recommended antidote for lead toxicity; and a goat's bladder is the featured respiratory protection for an iron furnace worker. Other woodcuts illustrate principles of ventilation for removal of stagnant air and poisonous vapors, and the use of mechanical devices for lifting heavy loads.

An American mining engineer named Herbert C. Hoover and his wife, L. H. Hoover, translated Agricola's work into English. The translation was published in London in *The Mining Magazine* in 1912. Mr. Hoover went on to pursue a career in politics.

Another early work describing the health problems of miners was published in 1567. The author, Theophrastus Bombastus von Hohenheim (1493-1591), also called Paracelsus, is known among toxicologists for uttering the phrase "All substances are poisons; there is none which is not a poison. The right dose differentiates a poison and a remedy." This statement provides the basis for the concept of the **dose-response relationship**; that is, the toxicity of a substance depends not only on its toxic properties, but also on the amount of exposure, or the dose. (More on toxicology in Chapter 2.) His work, titled *On the Miners' Sickness and Other Diseases of Miners*, was specific to the diseases of miners and smelter workers. Paracelsus' descriptions of the worker's conditions differentiated between chronic (low-level, long-term) and acute (high-level, short-term) poisonings; as in Agricola's work, his descriptions of symptoms were fairly detailed. In fact, Paracelsus' description of the physical and behavioral effects on mercury-exposed workers closely resembles current descriptions of mercury poisoning.

An important contributor to the record of occupational disease was Bernardino Ramazzini (1633-1714), who wrote a 40-chapter book, *De Morbis Artificum* (Diseases of Workers) that earned him the distinction of being generally credited with spawning the field of occupational medicine. Ramazzini urged physicians to include questions about their patient's occupation ("Of what trade are you?") as part of the medical examination. Ramazzini's book included descriptions of diseases associated with most of the occupations of his time, including trades that were considered to be lower-class, such as corpse carriers and laundresses, which would be recognized today as occupations presenting many unseen health hazards to workers. Physicians in other countries also recognized the association between exposure and worker disease, and papers dealing with mining and worker disease were published in Europe and Asia in the late eighteenth century. In the late 1770s, Sir George Baker correctly linked "Devonshire colic" to lead in cider; another English physician, Percival Pott, made the connection between soot exposure and the development of scrotal cancer among London chimney sweeps.

The Mad Hatter in Lewis Carroll's *Alice in Wonderland* also exhibited symptoms of mercury poisoning, such as mental and personality changes marked by depression and a tendency to withdraw. Mercury was an ingredient used in processing the animal hides that were made into hats; the bars on the windows of hat factories were likely installed to prevent the mercury-affected workers from leaping from the windows to the street below in a bout of depression.

Throughout the nineteenth century, scattered reports of work-related disease were reported; by 1900 physicians were experimenting with laboratory animals in an effort to anticipate work-related health effects. The need for such type of work became more obvious as the industrialization increased along with the number of occupational injuries and diseases. Great Britain led the European community by passing the English Factory Act in 1833, establishing a means for workers injured on the job to receive compensation. Later, in 1878, revisions to the Factory Act created a Factory Inspectorate. The Factory Act did not require employers to prevent the unhealthy conditions, but it was not long before business owners recognized it was more cost-effective to take steps to reduce incidence of worker injury rather than incur the associated costs of compensation. Worker's compensation and other worker protection legislation

Figure 1-2 (A and B): These woodcuts (also from *De Re Metallica*) illustrate examples of protective equipment used by mine and smelter workers. Workers protected themselves from inhalation hazards of mines and smelting operations by using what would be considered - by present-day standards - a crude respirator (A and B), sometimes made from a goat's bladder. One of the workers in (B) appears to be using a hand-held shield against the heat.

in the United States did not follow until the early 1900s, and it was 1970 before Congress passed the **Occupational Safety and Health Act (OSHAct)**. The majority of the **Occupational Safety and Health Administration (OSHA)** regulations were adopted at that time, or shortly thereafter, and addressed aspects of worker safety from recordkeeping and chemical exposures to welding in confined spaces, and more. Health and safety regulations will be addressed in more detail in Chapter 3.

Checking Your Understanding

1. What was the significance of *De Re Metallica*?

2. Who said "All substances are poisons; there is none which is not a poison. The right dose differentiates a poison and a remedy"?

3. Name a cancer that was correctly linked to the occupation of chimney sweep.

4. When did Congress pass the OSHAct?

1-3 Industrial Hygiene as a Recognized Profession

In the 1900s, there were few practicing industrial hygienists in the United States and the profession was generally unknown and unrecognized by the general public – a situation that has changed somewhat in time, although there still are misconceptions about industrial hygiene. Physicians often saw the industrial hygienist as a threat or an invader of the doctor's realm of expertise. One physician who did not share this view was Alice Hamilton, who is credited with sparking the growth of occupational medicine in the United States. Dr. Hamilton, a champion of worker health and safety through social responsibility, directly related worker illnesses to the toxic materials that were the cause, and she offered specific recommendations for preventing the illness from occurring. Gradually, the public became aware of the relationship between exposure and disease; trade unions found worker safety a better cause than hazard pay; and state and federal governments were pressed to pass worker compensation and worker protection legislation. In 1939, the American Industrial Hygiene Association was formed by a group of professionals whose occupations were aimed at protecting worker health. Members of the group included individuals from industry, universities, and government agencies, all with the common interest of protecting worker health through application of preventive measures.

Since that time, the profession of industrial hygiene has continued to grow and expand. In other countries, professional organizations have been established, and other terms – such as occupational hygiene, environmental hygiene, and environmental health – have come into use to describe the industrial hygiene function. Industrial hygiene will probably continue to evolve as it becomes more widely recognized throughout the world.

Professional Organizations

Three large professional industrial hygiene organizations exist in the United States; they are the American Industrial Hygiene Association, the American Academy of Industrial Hygiene, and the American Conference of Governmental Industrial Hygienists. Each organization grew to meet different needs of industrial hygienists. In addition, representatives from each of the three groups make up a fourth organization, which is concerned with the certification process. Recently the three groups have collaborated to produce a single code of ethics for the professional practice of industrial hygiene, and the formation of a single, unified professional organization is currently being considered by memberships of all three. Consideration of each group's mission and goals may clarify both the differences and similarities between the groups.

The American Industrial Hygiene Association

The **American Industrial Hygiene Association (AIHA)** is an organization whose membership includes professional industrial hygienists, students, health care professionals, and others with an interest in industrial hygiene. The AIHA's purposes are to promote the field of industrial hygiene; provide education and training; provide a forum for the exchange of ideas and information; and represent the interests of industrial hygienists and those they serve.

AIHA provides services to its members through government affairs representation, and commenting and providing input to Congress and Congressional committees on proposed health and safety regulations. AIHA members serve on technical committees, sharing knowledge and experience in areas such as aerosol technology, biological monitoring, occupational medicine, law, and toxicology. The AIHA also administers several laboratory accreditation and/or proficiency programs, aimed at ensuring analytical accuracy for lead, asbestos, metals, silica, and organic solvents. Other services of the AIHA include continuing education programs, a job-search service, and multimedia publications and materials addressing nearly every aspect of industrial hygiene. The AIHA also publishes the peer-reviewed *American Industrial Hygiene Association Journal*. Smaller groups of professionals may form a Local Section, and most major industrialized areas of the United States have at least one such organization. The AIHA has also been a joint sponsor (with the ACGIH) of the American Industrial Hygiene Conference and Exposition, which has grown into one of the largest international forums on industrial hygiene.

The American Conference of Governmental Industrial Hygienists

The **American Conference of Governmental Industrial Hygienists**, or **ACGIH**, is an organization whose mission is to promote excellence in occupational and environmental health. It began as an organization of industrial hygienists employed by federal, state, and local government health and safety agencies. Originally called the National Conference of Governmental Industrial Hygienists, in 1938 the organization created nine standing committees to develop standard practices that could be used by all industrial hygiene professionals. In 1946 the organization changed its name to the American Conference of Governmental Industrial Hygienists and offered membership to all industrial hygiene personnel working for governmental agencies, including those in countries outside the United States.

The ACGIH has made major contributions to industrial hygiene, especially in the area of informational products and services. Its technical committees now number 14 and remain focused on providing other industrial hygienists with information in the form of texts, reports, conferences, seminars, and other avenues of learning. ACGIH publications include some considered to be the authoritative work in their respective area of industrial hygiene, among them: *Industrial Ventilation: A Manual of Recommended Practice* (published first in 1951, and with regular reviews, updates, and re-issues since then, it is currently in its 23rd edition); and *Documentation of Threshold Limit Values and Biological Exposure Indices*. The ACGIH also reviews and publishes annually a booklet of exposure guidelines called Threshold Limit Values for Chemical Substances and Physical Agents and Biological Exposure Indices, commonly referred to among industrial hygienists as the "TLV Booklet." The 1968 edition of this booklet provided OSHA with the basis for the original permissible exposure limits set in the 1970 OSHAct. The ACGIH also publishes a peer-reviewed journal, *Applied Occupational and Environmental Hygiene* (formerly *Applied Industrial Hygiene*). Other functions of the ACGIH include co-sponsorship of the American Industrial Hygiene Conference and Exposition, as well as other seminars that focus on specific areas of interest to occupational health professionals.

The American Academy of Industrial Hygiene

The **American Academy of Industrial Hygiene (AAIH)** is a non-profit professional organization whose members are individuals who have successfully met the certification requirements for industrial hygiene as set forth by the American Board of Industrial Hygiene (ABIH). (The ABIH was formed initially by representatives from the AIHA and the ACGIH, when, in 1959, members of these two organizations agreed upon a mechanism for voluntary certification of industrial hygienists.) The purpose and goals of the AAIH are: 1) to recruit and train industrial hygienists from among graduates of scientific and engineering disciplines; 2) to promote recognition of industrial hygiene practices; 3) to promote the ABIH Certification as a basic qualification for employment as an industrial hygiene professional; and 4) to establish guidelines for ethical conduct in the practice of industrial hygiene.

The American Board of Industrial Hygiene

The **American Board of Industrial Hygiene (ABIH)** is comprised of six representatives from each of the above three organizations, making a total of 18 individuals. The ABIH serves as an independent organization that administers certification programs for industrial hygiene professionals. The certification process involves a review of the applicant's education and experience by the Board. Individuals whose applications are approved must then pass a written examination that consists of two parts, a core examination followed by a second exam. Professionals may obtain certifications in the following aspects: acoustical; air pollution; chemical; comprehensive; engineering; radiological; and toxicological. Successful completion of the Core exam earns one the designation Industrial Hygienist in Training (IHIT). The designation Certified Industrial Hygienist (CIH) is awarded upon passing both the core and the second examination, which may be in one of the specialty areas. Besides administering the certifications, the ABIH is also responsible for the certification maintenance process, whereby those certified in industrial hygiene practice must earn points to keep their certification in good standing. Certification maintenance points are awarded for such things as full-time practice of industrial hygiene, participation in seminars, attending short courses, publication of research papers, and lecturing or teaching on industrial hygiene or a related topic. Certificate holders keep track of points earned and submit them every six years to the ABIH for review. Failure to earn enough maintenance points results in having to take the examinations again in order to maintain CIH status.

In Canada, the certification process is administered by the Canadian Registration Board of Occupational Hygiene, and successful passing of the examination earns the right to use the designation of Registered Occupational Hygienist, or ROH.

Checking Your Understanding

1. What are the four industrial hygiene professional organizations?

2. Name two other terms that are used to refer to the industrial hygiene function.

1-4 Definition of Industrial Hygiene

There are many ways to define an **industrial hygienist (IH)** and what the industrial hygienist does. Since these definitions are similar in content, one representative definition was selected for this book. The following is paraphrased from the AIHA definition:

An industrial hygienist is a person having a college or university degree(s) in engineering, chemistry, physics, medicine, or related physical and biological sciences, who has also received specialized training in recognition, evaluation, and control of workplace stressors and therefore achieved competence in industrial hygiene. The specialized studies and training must be sufficient so that the individual is able to: 1) anticipate and recognize the environmental factors and understand their effects on people and their well-being; 2) evaluate, on the basis of experience and with the aid of quantitative measurement techniques, the magnitude of these stresses in terms of the stressor's ability to impair human health and well-being; and 3) prescribe methods to eliminate, control, or reduce such stresses when necessary to diminish their effects.

Box 1-1 ■ A European View of Industrial Hygiene

In 1991 a European regional conference on Occupational Hygiene met in Copenhagen. The proceedings of the conference included a profile of the profession. The conference, which included participation by U.S. representatives, adopted a document listing eleven items that should be within the capability of a professional occupational hygienist. These eleven items are:

1. Anticipate the health hazards that may result from work processes, operations, and equipment, and accordingly advise on their planning and design.

2. Recognize and understand, in the work environment, the occurrence (real or potential) of chemical, physical, and biological agents, as well as other stresses, including their interactions, which may affect the health and well-being of workers.

3. Understand the possible routes of agent entry into the human body, and the effects that such agents and other factors may have on health.

4. Assess worker exposure to potentially harmful agents and factors and evaluate the results.

5. Evaluate work processes and methods, with regard to the possible generation and release/propagation of potentially harmful agents and other factors, with views to eliminating exposures, or reducing them to acceptable levels.

6. Design, recommend for adoption, and evaluate the effectiveness of control strategies, alone or in collaboration with other professionals, to ensure effective and economical control.

7. Participate in overall risk analysis and management of an agent, process, or workplace, and contribute to the establishment of priorities for risk management.

8. Understand the legal framework for occupational hygiene practice in his/her country.

9. Educate, train, inform, and advise persons at all levels, in all aspects of hazard communication.

10. Work effectively in a multidisciplinary team involving other professionals.

11. Recognize agents and factors that may have environmental impact and understand the need to integrate occupational hygiene practice with environmental protection.

From: Occupational Hygiene in Europe: Development of the Profession. World Health Organization Regional Office for Europe, European Occupational Health Series No. 3, Copenhagen, 1991; document EUR/RC41/Inf.doc./ 1Rev. 1, pp 19-22.

A 1973 publication of the U.S. Department of Health, Education and Welfare, titled *The Industrial Environment – Its Evaluation and Control*, is considered a classic text on industrial hygiene. This book describes the scope of industrial hygiene as containing three elements: recognition, evaluation, and control. Recognition involves identification of health problems that are created or exist in a workplace. Causes of these health problems include **chemical agents**, such as dusts, mists, fumes, vapors and gases; **physical agents** in the forms of ionizing and nonionizing radiation, noise, vibration, and temperature extremes; **biological agents** such as insects, molds, yeasts, fungi, bacteria, and viruses; and finally, **ergonomic agents**, such as monotony, fatigue, and repetitive motion. After some experience in the field, an industrial hygienist may actually recognize the potential for hazards to develop, rather than seeing the hazards after they occur; this is called anticipation.

The second aspect, workplace evaluation, is usually accomplished through observations and quantitative measurement of the agent(s) of concern in the work environment coupled with the experience and knowledge of the industrial hygiene professional.

The third basic tenet of industrial hygiene, control of hazards, is aimed at eliminating existing problems and preventing potential hazards from developing. Traditionally, the methods for achieving these goals are, in order of preference: 1) to engineer out the hazard by, for example, changing the process so that workers are not exposed to a chemical, or substituting with a non-hazardous material; 2) to reduce the number of people who are exposed through implementing **administrative controls** such as procedures and work area access restrictions; and 3) to provide proper work clothing and/or protective equipment to reduce hazards to the worker when other means for reducing exposure have been employed to the extent possible. Often, the controls that are used for worker protection involve some combination of all three of the above methods.

Also included in *The Industrial Environment – Its Evaluation and Control* are the responsibilities of the industrial hygienist as described by the editor of the book, George D. Clayton, who has contributed much to the profession of the industrial hygienist. Although Mr. Clayton prepared this list in the 1970s, it continues to present a good model for most present-day industrial hygiene professionals. Mr. Clayton lists the responsibilities of the industrial hygienist as follows:

1. Examination of the industrial environment.

Box 1-2 ■ Code of Ethics for the Industrial Hygiene Profession

In 1995, the following code of ethics for conduct of industrial hygiene was adopted by all three professional industrial hygiene organizations in the United States (AIHA, ACGIH, and AAIH).

Objective: These canons provide standards of ethical conduct for industrial hygienists as they practice their profession and exercise their primary mission, to protect the health and well-being of working people and the public from chemical, microbiological, and physical health hazards present at, or emanating from, the workplace.

Canons of Ethical Conduct

Industrial hygienists shall:

1. Practice their profession following recognized scientific principles with the realization that the lives, health, and well-being of people may depend upon their professional judgment and that they are obligated to protect the health and well-being of people.

2. Counsel affected parties factually regarding potential health risks and precautions necessary to avoid adverse health effects.

3. Keep confidential personal and business information obtained during the exercise of industrial hygiene activities, except when required by law or overriding health and safety considerations.

4. Avoid circumstances where a compromise of professional judgement or conflict may arise.

5. Perform services only in the area of their competence.

6. Act responsibly to uphold the integrity of the profession.

2. Interpretation of gathered data from studies made in the industrial environment.

3. Preparation of control measures and proper implementation of these control measures.

Disciplines Involved	Applications in Industrial Hygiene
Physics, mathematics, human anatomy, and physiology	Hazard evaluations of: noise illumination, lasers, nonionizing radiation, ionizing radiation, and ergonomics.
Chemistry, anatomy and physiology, toxicology	Toxic chemical exposure evaluations of: carcinogen hazard assessments and reproductive hazard assessments.
Physics, chemistry, statistics	Measuring exposures to chemical and physical agents. Interpreting laboratory analytical reports. Use of direct-reading instruments.
Statistics, epidemiology, physics, chemistry, anatomy and physiology, toxicology, language skills	Interpreting study and laboratory results; critical review of research; performing research.
Language skills	Interactions with workers, management, and clients; report writing; preparing manuscripts of original research studies for publication; design and delivery of employee education programs.

Table 1-1: Industrial hygienists draw from knowledge and experience in multiple areas of applied science. The importance of language skills should not go unnoticed as communication of hazards and worker health protection is among the most important roles of the industrial hygienist.

4. Creation of regulatory standards for work conditions.

5. Presentation of competent, meaningful testimony when called upon to do so by boards, commissions, agencies, courts, or investigative bodies.

6. Preparation of adequate warnings and precautions where dangers exist.

7. Education of the working community in the field of industrial hygiene.

8. Conduct of epidemiological studies to uncover the presence of occupation-related diseases.

As we can see, the field of industrial hygiene encompasses aspects of chemistry, engineering, biology, physics, and mathematics, as well as toxicology, physiology, and biochemistry, which is why a training program in industrial hygiene often contains courses in such diverse areas. Table 1-1 illustrates how multiple disciplines may be employed in the evaluation of hazards and various other applications in the workplace.

Checking Your Understanding

1. What are the primary tenets of the practice of industrial hygiene?

2. Name several disciplines or areas of study in which an industrial hygienist must be conversant.

3. What are some responsibilities that are typically assigned to an industrial hygienist?

1-5 Industrial Hygiene as Part of Worker Safety Programs

An effective industrial hygiene program is an important part of an organization's overall program for worker health protection and safety. It complements the other aspects that may be included in a worker health program: wellness/fitness programs, substance abuse prevention and treatment, periodic medical examinations, and routine screening for health problems. The industrial hygiene program is a source of information for the physician or nurse relative to the employee's working conditions, including the possible causes of, or factors contributing to, the employee's symptoms. The industrial hygiene program also provides information that is necessary for an effective **medical surveillance program**, which is a periodic evaluation of an employee by a health professional in order to assure that health problems associated with chemical exposures or physical agents are detected early, when there is time to prevent permanent or debilitating injury. Examples of applications of a medical surveillance program include periodic hearing tests to detect noise-induced hearing loss; another is examination of the respiratory system including chest X-ray and lung capacity measurements to detect scarring of the lungs due to asbestos exposure.

An industrial hygiene program consists of the following basic elements (as a minimum):

1. Anticipation/recognition of health hazards;

2. Evaluation of health hazards;

3. Control of health hazards;

4. Recordkeeping;

5. Employee training;

6. Periodic program review, changes, and updates.

The complexity and size of an industrial hygiene program will depend upon the complexity and size of the organization of which it is a part. It may fill several three-ring binders and require a large staff for effective implementation; it may fill only a slim volume and be administered entirely by one or two individuals. Whatever the situation, the goal of the industrial hygienist remains the same: to protect the health of the workers. The industrial hygiene program that meets this need will ideally be a written plan that clearly assigns roles and responsibilities; describes the methods that will be used in meeting its stated goals and requirements; provides for accurate and detailed recordkeeping, and records retention; and assures that employees will receive benefit of the program through education and training about the hazards in their workplace, including means by which they may protect their health. The written program should be a living document that is periodically reviewed, updated, and changed as necessary.

Elements of an Industrial Hygiene Program

Anticipation/Recognition of Health Hazards

The industrial hygienist is expected to anticipate health hazards before they develop, and recognize health-threatening conditions as they arise, by becoming familiar with the operations and processes at the site. One method for accomplishing this is to perform a **walk-through survey** of the process, if possible with a knowledgeable individual (such as an operator), who can explain the process in detail. This provides an opportunity for the industrial hygienist to become familiar with the process through watching, asking questions, observing work practices, and reviewing chemical inventories; through this process the industrial hygienist will attempt to identify all of the chemical, physical, biological, and ergonomic hazards that are present. This walk-through survey should be performed on a regular basis, for example once or twice each year, or more frequently if there are significant changes in the process. This will ensure that identified hazards remain under control, and that new hazards are identified and evaluated in order to protect worker health. The frequency with which surveys are performed as well as the individuals participating in the surveys should be specified. Records of the surveys, including identified hazards and recommended evaluation or control actions, as well as follow-ups to assure that controls are implemented will need to be maintained; the mechanism for records should also be included in the written program.

Although they can be conducted according to a regular and convenient schedule, investigations may also be conducted as part of an emergency response action, as a result of employee complaints, or at the request of workers or management who are concerned about a potential health hazard in their work area. In these situations, the industrial hygienist generally employs the same methodical approach outlined above for reviewing the overall process to identify the real or potential source of workplace stresses.

Anticipation of hazards before they exist is perhaps what prompts defining industrial hygiene as both science and art. As the industrial hygienist becomes more experienced, anticipation will become an important tool in the professional's repertoire of skills. Each walk-through, each evaluation, each challenge to find and implement an effective control increases the industrial hygienist's ability to anticipate hazards. The beginning industrial hygienist will see this skill grow and develop in himself or herself as he/she continues in the practice of industrial hygiene.

One aspect of hazard recognition that may be overlooked is the inclusion of the industrial hygiene professional in the design review process, both for new processes and for proposed changes to existing processes. Including the industrial hygienist in the planning stages helps assure the provision of health and safety features in the process, rather than requiring retrofitted features that can delay startup and add significant cost to the project. Typical health and safety reviews include:

— Discussions with engineers during the initial planning stages;

— Review of design specifications, blueprints, and process flow to identify potential health hazards such as chemical exposures, noise, and ergonomic issues;

— Review of material safety data sheets for chemicals that will be used in the process to ensure that any necessary controls, such as ventilation systems, are included in the design;

— Visits to the production facility to examine the process machinery prior to shipment to the plant;

— A final check once the equipment is installed in the plant to ensure that health and safety safeguards are functional;

— Follow-up during the initial weeks of production to verify that safe operating conditions exist;

— Any new or modified process should also be added to the schedule of periodic walk-through surveys.

Evaluation of Hazards

Evaluation of hazards identified in the walk-through survey is accomplished most often through obtaining objective data on the level of chemical or physical agent(s) present. This is usually done through the use of some form of measurement: air sampling for hazards such as organic vapors or metal fumes; the use of special instruments, such as noise meters, to evaluate exposure to industrial noise; evaluation of workstation design to identify ergonomic stresses; or some other method, depending on the hazard being evaluated. The data obtained by the industrial hygienist are used to assess the risk posed to employees by the hazard; the method of measurement must therefore take into account the toxic or harmful properties of the health hazard as well as the employee's overall **dose**, that is, the level or amount of the exposure, and the length of time or the duration of the exposure. Instruments used in measuring exposures must be maintained and calibrated and, depending on the agent involved, there may be specific regulatory requirements for sampling intervals and allowable levels of employee exposure. For example, federal regulations for lead specify how often the employer must monitor employee exposures, and what exposures are allowed.

The method used for gathering and evaluation of the data is also important. In some cases, the method to be used is dictated by OSHA as a regulatory requirement, and the entire sampling process including the instruments, sample time and flow rate, as well as the method of analysis will be spelled out clearly. In other cases, some professional discretion is allowed and may even be required. For example, the industrial hygienist might have to select one method from among several in order to detect the contaminant of concern, especially if there are other possible interfering compounds also present in the work area. This requires that the industrial hygienist have a good working knowledge of the process, including some expectations about the measured concentration of the contaminant.

The **National Institute of Occupational Safety and Health**, or **NIOSH**, is a source of sampling methods for a large number of chemical and physical hazards. The NIOSH method for a particular substance addresses the entire sampling process from obtaining the sample (use of the correct sampling pump, flow rate, and collection medium) to analysis (how to set up the laboratory instrument so that accurate results are obtained). OSHA also has published a set of analytical methods that address sampling methods and analytical techniques. Sampling and evaluation of hazards is addressed in more detail in later chapters.

As with other aspects of the industrial hygiene program, keeping accurate records of walk-through surveys, inspections, exposure evaluations, monitoring methods and results, as well as recommendations for control measures, is a must. Exposure monitoring records become part of the employee's permanent file, and OSHA requires these records to be maintained for a period of 30 years. The industrial hygiene program should describe the recordkeeping system and assign responsibilities for each aspect; for example, the medical department may have a role in maintaining copies of employee exposure data, while the industrial hygiene department will retain records of the evaluation process, including methods, instrumentation calibration and readings, laboratory results, workplace observations, and the like.

Control of Hazards

Determining appropriate and effective control measures is sometimes one of the industrial hygienist's most challenging tasks. The reader should remember from earlier in the chapter that there is a preferred hierarchy for implementation of hazard controls: engineering controls; administrative controls; and finally, personal protective clothing or equipment. Examples of engineering controls include replacement of toxic materials with those that pose a lesser hazard, or use of a local exhaust ventilation system to control a dusty or gaseous contaminant. The best-known type of administrative control is the use of worker rotation to reduce exposure. When it is impossible or infeasible to utilize one of the other methods, personal protective equipment

may be necessary to reduce or eliminate worker exposure; an example of this would be use of respirators during asbestos removal inside of a temporary work enclosure. Whatever the mechanism, recordkeeping is again important, as is follow-up to evaluate the effectiveness of the controls in eliminating or ameliorating the hazard.

Recordkeeping

Previous sections have already mentioned the importance of keeping accurate and complete industrial hygiene records, yet this is often the weak spot of industrial hygiene programs. Records are important for a multitude of reasons: regulations require them; they are a source of valuable information for trending exposures and identifying and evaluating workplace hazards; they may be used for developing and defending an exposure monitoring strategy; and they often become a legal document that must be relied upon to defend a particular allegation, or to prove an exposure relative to a regulatory limit. As the primary recorder and generator of this important information, the industrial hygienist – in many cases acting as the company's representative – must assure that all forms and records are completed accurately, distributed appropriately, and maintained in a secure manner for historical purposes. The industrial hygienist working as a consultant must be just as diligent in recordkeeping, and perhaps more so.

Recordkeeping is addressed in the Occupational Safety and Health Administration regulations found in 29 CFR 1910.20, "Access to Employee Exposure Medical Records." This regulation defines what constitutes an "employee exposure record"; it may surprise you to learn that OSHA considers an MSDS indicating that a material may pose a hazard to human health to be such a record. OSHA also includes all sample collection information, such as calibration records and sampling methodology, calculations, and other background data. The regulation includes definitions of other key terms, and stipulates requirements for:

— Preservation of records (a 30-year retention time beyond the last date of employment is required).

— Access to records by employees and other authorized personnel (employees must give consent for the release of certain medical records).

Box 1-3 ■ Recordkeeping

The following are some basic ground rules for good recordkeeping:

1. Complete all industrial hygiene forms accurately with indelible (permanent, waterproof) black ink.

2. Leave no blank spaces; include all relevant data where indicated and appropriate. For spaces or lines where data do not exist (such as a flow rate for a noise meter), mark "N/A" or use a similar method so that no data will be missed.

3. Do not scribble out errors or changes; strike out with a single line, initial, and date.

4. Write (or print) legibly.

5. If desirable, keep a personal log or record book. Use a bound book with numbered pages and follow the above rules.

6. Include a provision in your recordkeeping procedure for review of records by another individual who can double-check your work (flow rates and sample volumes, for example) as a safeguard to make sure your notes are complete and understandable.

7. Maintain duplicate copies of all records that you generate and store them in a secure location separate from the primary records storage location.

—Ensuring the confidentiality of trade secrets.

—Transferring records to a safe repository (NIOSH) if the 30-year retention requirement cannot be met.

Employee Training

With the implementation of the OSHA Right-to-Know regulation, employees have become very aware of the chemical and physical hazards that surround them. The industrial hygiene professional is often a key player in the hazard communication program, not only providing training in formal sessions but also through new employee orientations, responding to specific questions, and evaluating new materials proposed for use by employees or in a process.

Training needs may be evident through observing performance on the part of affected employees, for example they may not be wearing their respirators or hearing protectors in the correct manner. Often training is done to meet a regulatory requirement, as is the case in hazard communication training, which is required by OSHA for all employees prior to working with hazardous materials. Training may also be required to familiarize employees with process changes, how to operate engineering controls, or how to wear personal protective equipment. Training methods may include films or videos, workbooks, handouts, overhead slides, group discussions and short talks; hands-on training is a useful mechanism for very specific training such as respirator use. An entire industry has developed to meet health and safety training needs; available materials include everything from short videos to entire programs complete with manuals, films, workbooks, and student tests.

Training should be conducted in a setting that allows the participants to focus on the learning experience; it should be comfortable in terms of temperature, lighting, seating, and viewing, and it should be appropriate for the methods being used in the session. Finally, the participants should be allowed to evaluate their experience and provide feedback to the training provider. Course evaluations can be a source of suggestions for improvement, reveal strengths and weaknesses of the program, and may also be useful in identifying additional training needs.

Records of training sessions should include the date, time and place, topics covered, the name of the person(s) who led the training, and the names of the attendees. This record is most conveniently obtained through a sign-in sheet, which then becomes part of the training record.

Program Review

Periodic review of a written industrial hygiene program as well as its implementation are as important as any other element. Depending on the facility and the processes that the industrial hygiene program supports, this could be the most important aspect of the entire program. Changes in regulatory requirements, new information about the toxic properties of a chemical or physical agent, and changes in the process itself are circumstances that can invalidate parts of a written program or procedure, or render it obsolete. For example, the OSHA permissible exposure limit for an eight-hour exposure to cadmium was reduced when new regulations were issued and went into effect in 1992. Some employers were forced to make changes in order to ensure compliance with the new limit for worker exposure and health protection; often, these changes included additional exposure monitoring in order to evaluate worker exposure relative to allowable limits. Assuming that a particular industrial hygiene program addressed monitoring and control of cadmium hazards, changes to the regulations would necessarily have been reflected by changes in the industrial hygiene program. If there is no mechanism for reviewing and updating a written program, it can easily become outdated and useless. Some mechanisms that might be used to evaluate the industrial hygiene program and its effective implementation are:

—Audits of a particular program or element.

—Performance of internal OSHA-type inspections.

—Review of procedures following changes in regulations.

—Self-evaluations performed by users or participants in the program.

Audits can be useful for evaluating specific areas such as hazard communication or respiratory protection. The simplest criteria for these evaluations would be adherence to regulatory requirements and effective implementation of procedures. For example: are the respirators stored in a clean and sanitary location; are the material safety data sheets available for employee review? Examples of program elements that could be audited include recordkeeping, equipment calibration procedures, and chemical hazard evaluations, to name only a few.

A mock OSHA inspection may be performed by a single person or a team, and encompass the entire facility or concentrate on one process; this can provide a detailed and comprehensive evaluation. The OSHA regulations that apply to the process serve as the criteria; violations can be self-identified and corrected. This method is preferable over a bona fide inspection, which can bring fines and other penalties.

Upon release of new or altered regulations, existing procedures need to be examined to ensure that the minimum requirements as set forth by OSHA are being met. In some cases, internal or corporate (self-imposed) standards may exist, and existing procedures should also be checked against these for compliance.

Another approach is to allow users of the program to evaluate it; for example, employees who participate in the hearing conservation program might uncover weaknesses that would go unnoticed by the industrial hygiene or medical staff.

Checking Your Understanding

1. What is the purpose of a medical surveillance program?

2. How might an industrial hygiene program be useful to the plant physician or nurse?

3. When should the industrial hygiene program be changed?

4. Name four elements of an industrial hygiene program.

5. What aspect of plant operations is most commonly omitted from the review process?

6. What kinds of records must be maintained as part of an industrial hygiene program?

7. List four good rules of practice to follow in recordkeeping.

Summary

This chapter describes the earliest beginnings of industrial hygiene, starting with Hippocrates in the fourth century BCE, through the Middle Ages. Among the first to recognize the link between occupation and disease was a physician named Agricola, whose extensive accounts of the maladies suffered by miners was sufficiently detailed to allow us to deduce the exposures and resulting diseases that workers of that time suffered from. Later, physicians such as Percival Pott and Alice Hamilton furthered our knowledge of the relationship between work and disease. It is from these beginnings that industrial hygiene has evolved to a multi-disciplinary science with its primary focus on preventing occupational health problems. The passage of the Factory Act in England in 1833 was one of the first workers compensation laws; it provided compensation for those injured on the job. It was not until 1970 that the Occupational Safety and Health Act was passed by the United States Congress. This landmark legislation had as its focus the protection of worker health and safety; it also established standards and limits for healthful work conditions. As the OSHA standards have gone on to become greater in number and complexity, the demand for worker health and safety specialists has increased. One of these professionals is the industrial hygienist.

The present-day industrial hygienist may perform many functions associated with worker safety and environmental health. However, the primary mission of the industrial hygiene professional can be described as the anticipation and recognition of health hazards, followed by their evaluation and control. Implementing health hazard controls is generally done in the preferred order of: 1) engineering out the hazard; 2) implementing administrative controls; and 3) use of personal protective clothing and equipment. The industrial hygienist accomplishes their mission of hazard anticipation, evaluation, and control through a combination of experience, talent, and technology, applied in a systematic manner.

Potentially hazardous environmental contaminants and physical agents are identified and quantified using sampling and measurement techniques that involve the use of specialized equipment. Measured levels are compared against allowable limits, and appropriate controls are recommended by the industrial hygienist for implementation. Workers may require training to ensure that they can apply the necessary hazard controls effectively; the industrial hygienist has an important role in such training programs. Regulatory requirements dictate some activities of the industrial hygiene professional, including keeping accurate records for everything from air sampling to worker training, as well as written programs and procedures that describe the methods used for ensuring worker protection. A written industrial hygiene program provides guidance and structure for ensuring that worker health is protected while meeting regulatory and other requirements. Written industrial hygiene programs should be living documents that are reviewed and revised to reflect changes in conditions and regulatory requirements.

Critical Thinking Questions

1. OSHA regulations for controlling worker exposure to health hazards generally require that employers utilize engineering controls to the extent feasible before other control methods are employed. Why is engineering out the hazard the preferred option?

2. A group of concerned employees stops you outside your office and asks you many questions about a new chemical being used in their work area and the new protective gear they are now required to wear. Based on their concerns you decide to do a follow-up investigation and assess the situation. Explain how you might conduct your investigation, including any preliminary research and follow-up activities that might be necessary.

3. Industrial hygiene is a multi-disciplinary science. Explain what this means.

Toxicology Review

Chapter Objectives

Upon completing this chapter, the student will be able to:

1. **Explain** the dose-response relationship and the concept of threshold dose.

2. **Explain** how toxins enter the body and are transported to different organs and tissues.

3. **Describe** the types of responses or toxic effects that can result from exposure to a substance.

4. **Describe** how toxins are altered, detoxified, and eliminated from the human body.

5. **List** several classes of toxins that are important in occupational health protection.

Chapter Sections

2-1 Introduction

In the previous chapter we learned that people have long recognized the link between hazardous exposures and illness or injury. We also briefly traced the development of the industrial hygiene profession and the role of the industrial hygienist as a professional charged with the recognition, evaluation, and control of chemical, physical, biological, and ergonomic hazards in the workplace. In this chapter, we will look at the harmful interactions between chemicals and living organisms and, specifically, at how these interactions occur in humans. The study of harmful chemical interactions is called **toxicology**.

In the centuries since Agricola documented symptoms of metal poisoning among miners, we have produced more than 10 million chemicals, about 60,000 of which have a reasonable potential for human exposure. These chemicals present an assortment of health hazards that the industrial hygiene professional is expected to recognize, evaluate, and control; to be able to do this, an understanding of some basic toxicological principles is essential. This chapter presents a basic review of toxicological principles that are important in industrial hygiene; more detailed information can be obtained from any of the references listed at the end of this book.

2-2 Toxicology in the Occupational Setting

All of us have probably experienced at least one episode of toxicity in our lifetime or know of someone who has, whether a friend, a coworker, or an incident reported in the news. Injuries and deaths occur each year as a result of accidental poisonings, smoke inhalation, entry into confined spaces, and many other situations. These events occur at construction sites, in factories and hospitals, and even in homes. Sometimes the events pass without being recognized for what they are, since exposure to some hazardous materials may cause symptoms similar to other illnesses, like colds or the flu. In other cases, the symptoms are mild and pass without concern, or they are non-specific, like nausea or a headache, and are attributed to other causes.

The OSHA Hazard Communication Standard went into effect in 1986; it requires employers to inform employees about health hazards specific to the materials they handle or use at work and about ways to protect themselves against those hazards. As a result, workers are now better informed about the hazardous materials present in their workplaces. Workers receive specific training about the kinds of health problems that can result from exposure to the hazardous materials; they know what steps to take to protect themselves from these hazards, such as the use of protective respirators or gloves. However, each year overexposures or accidental releases of hazardous materials occur and cause injury, despite the training and information available to workers. The industrial hygiene professional plays a key role in recognizing potential hazards and implementing controls to protect worker health.

The Dose-Response Relationship

The dose-response relationship (see Figure 2-1) is one of the most important concepts in toxicology. Its premise is that a dose, or a time of exposure (to a chemical, drug, or toxic substance), will cause an effect (response) on the exposed organism. If the amount or intensity of the exposure increases, there will be a proportional increase in the effect. When this principle is applied to occupational exposures, we use the term dose to refer to the amount of the substance that is absorbed. The response is the mani-

festation of the effects of that absorbed dose; the reaction may be mild or severe. For example, administration of a therapeutic drug at the prescribed dosage is necessary to achieve the desired result, say, for relief of pain. However, at higher dosages the drug may produce additional, and sometimes undesirable, effects such as drowsiness or even unconsciousness. You may remember from Chapter 1 that this relationship was first formally recognized by the physician Paracelsus (1493-1541). Paracelsus used the phrase to explain how a substance that was a locally recognized poison could actually be used as a medicine, providing the dose was correct (too much, the patient would die from the poison's toxic effect; too little, the patient would continue to suffer from disease). The practical side of this argument should be apparent: there is, theoretically, a dose or exposure level, below which the adverse effects of a substance are not expressed by the exposed population. This dose is called the **threshold dose**.

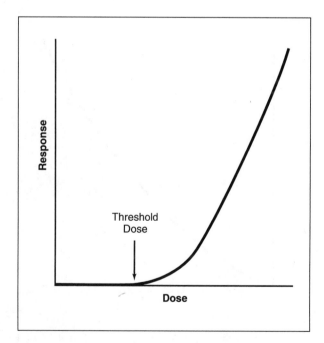

Figure 2-1: A dose-response curve illustrating the threshold dose. For some substances, there is an exposure below which there is no apparent response. This level is called the threshold dose, and it may represent a "safe" level of exposure. There are many substances – for example carcinogens – for which no threshold dose has been established.

The threshold dose is also referred to as the **no observed adverse effect level (NOAEL)**, or the **no effect level (NEL)**.

> Dose: The amount of a substance administered (or absorbed), usually expressed in milligrams of substance per kilograms of the exposed organism (mg/kg); more aptly described as dosage.
>
> Response: Term used to refer to the effect or effects of a substance; these may be positive effects as in the therapeutic dosage of a drug, or negative effects such as, for example, severe irritation of the respiratory tract.

While the response to a substance may be positive, such as curing a disease or relieving pain, in toxicology we generally are more concerned with the harmful or unwanted effects associated with exposure. The expression of a toxic effect may be a single event or outcome, such as the development of a specific disease or even death. There may also be symptoms of varying degrees of intensity ranging from mild irritation to sensory impairment to incapacitation or permanent injury, with death as the ultimate negative response.

The dose-response relationship is useful to occupational health and safety professionals because it helps define acceptable levels of exposure. The ideal acceptable exposure is one that has no negative effects on any of the individuals that are exposed. In some instances, the negative effects we are being protected against are simple ones that will disappear when the cause is removed: symptoms of irritation such as burning, watering eyes, or coughing are examples. For some materials, however, there may be no acceptable level of exposure, usually because any dose can have potentially harmful effects. Phosphine gas is an example of a material that causes tremendous irritation at any level of exposure, even though it has an established exposure level. Another example is the case of many **carcinogens**; for some of these, there is arguably no safe level of exposure,

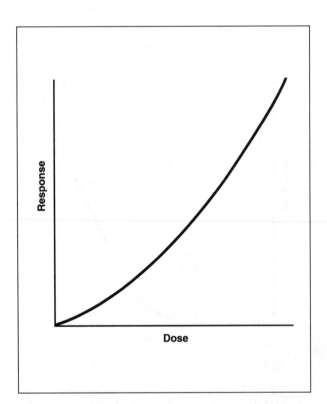

Figure 2-2: Dose-response curve showing simple dose-response relationship. This graph illustrates the observation by Paracelsus regarding the relationship between dose and the response of the exposed organism.

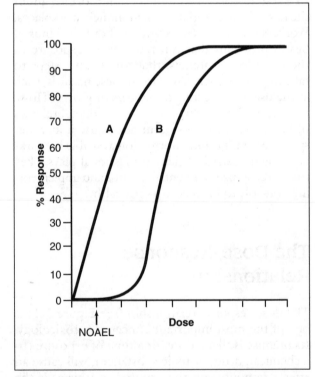

Figure 2-3: Dose-response curves contrasting two different materials with different toxicities. Comparison of the dose-response relationship for different materials can be an aid in evaluating toxicity. The slope of the dose-response curve is generally related to toxicity, with more toxic materials exhibiting a steeper slope.

since any exposure could result in cancer. This means that for some carcinogens, there is no threshold dose, since there appears to be a chance that cancer will develop as a result of exposure, no matter how low the dose. (We will address carcinogens in more detail later in the chapter.)

Another possible complication of the dose-response relationship is the consideration of each individual's susceptibility. There is a great deal of variation in response to a toxic material among individuals. What might be an acceptable level of exposure for one individual may be enough to cause severe irritation, or trigger disease development, in another.

The dose-response relationship is often expressed graphically, showing the dose on the x-axis and the response on the y-axis. Sometimes the logarithm of the dose value is used, to allow illustration across a wider range of doses. This is a convenient method for expression of data but it poses a dilemma for those who rely on the graphs as a rule of toxicity; since there is usually a lack of data for very low (or high) levels of exposure, extrapolation of the dose-response relationship at the extreme ends of the scale should be done with caution.

Agent	LD_{50} (mg/kg)
Ethyl alcohol	10,000
Sodium chloride	4,000
Morphine sulfate	900
Strychnine sulfate	2
Nicotine	1
Hemicholinium-3	0.2
Dioxin (TCDD)	0.001
Botulinum toxin	0.00001

Table 2-1: The LD_{50} of a material provides some additional information compared to a dose-response curve. In order to evaluate this value and compare it to that of other toxic substances, the route of administration, the species involved, and the units for the dose must be known.

Indicators of Relative Toxicity

The **toxicity** of a substance is its ability to cause harm or adversely affect an organism. As we have seen, the toxic properties of a particular substance are a function of more than just a single chemical or physical characteristic. At this point we shall look at some terms used to describe the toxicity of various materials. We have already said that there is a spectrum of effects that can be elicited by exposure to a toxic substance; correspondingly, there is a spectrum of dosages needed to effect these responses. One term used to describe such a dosage is the LD_{50}, which is the amount of a substance needed to produce death in 50 per cent of the treated (exposed) population. Often the literature will indicate the route of administration as well as the species; this information will be listed after the LD_{50} value as in: 35 mg/kg; oral, rat. Such an entry would inform the reader that the chemical of interest has an LD_{50} of 35 mg/kg of body weight when administered orally to rats (see Table 2-1).

The format of the notation used to indicate the LD_{50} is used by toxicologists for other similar applications. The two-letter acronyms are descriptive of the endpoint (lethality, effectiveness, etc.) of the dosage or exposure – they include: ED, effective dose; EC, effective concentration; LD, lethal dose; and LC, lethal concentration. Similarly, the numerical expression is used to indicate what proportion of the exposed population was found to exhibit the endpoint of interest. For illustration, the expression **ED_{25}** would indicate the effective dose for 25 percent of the exposed population.

The LD_{50} is an indicator of relative toxicity that is useful for comparing toxicities of two or more substances. It is determined in a laboratory and is based on an **acute exposure**, which is an exposure to a high level or concentration for a relatively short duration of time. Usually it is administered in a single, large dose to adult male and female test animals. Aside from determining the LD_{50}, these tests may provide other information about the material; for example the symptoms associated with a toxic exposure, and guidance in determining dosages for other studies, such as those for determining subchronic and chronic effects.

The terms **acute** and **chronic** are used to describe exposures as well as effects. For example, an acute exposure occurs when a worker spends a short time in an area of high concentration; any resulting effects, such as irritation, would be termed acute effects. A brief exposure to ammonia is an example: a worker could experience watering eyes and coughing almost immediately following an acute exposure. Chronic exposures are typically long-term exposures to low levels of contaminant, often below levels allowed by regulation. The resulting health effects are chronic effects. Emphysema is a potential chronic effect resulting from years of cigarette smoking.

The **subacute effects** are those associated with exposures or doses that are below the LD_{50} but still high enough to cause a toxic response. The **chronic** toxic effects of a chemical are those associated with a longer time of exposure, usually several months or more, to a relatively low level or dose. Testing the chronic toxicity of a chemical is one way that scientists determine its carcinogenicity and such studies may go on for as much as two years (the life expectancy of most laboratory rats). Other tests for carcinogenicity involve the use of bacteria. The genetic material (DNA) of exposed bacteria is examined to determine if the substance causes changes in the DNA, called **mutations**. Some mutations are believed to be associated with cancer development. The alternative to a multi-year study is extrapolation of data from other studies, which use shorter

times and higher levels of exposure. The issues associated with extrapolation are discussed later in the chapter.

Some toxic materials can produce death in minute dosages, while others require massive exposure to produce death. For this reason, categories of toxicity have been devised, which allow us to compare different materials in terms of their relative toxicity. These qualitative comparisons are not precise but they do allow – at least to some degree – a comparison between "apples and oranges." An example of one toxicity classification can be seen in Table 2-2.

Much of the toxicity information we have for many chemicals is based on tests done in laboratories under relatively controlled conditions. Laboratories use specially bred animal populations with well-known characteristics, including lineage; these populations have a similar genetic makeup. In some cases, a certain population is used because of its susceptibility or sensitivity to a chemical or outcome. For example, if testing for a tumorous cancer is the endpoint, a population that has been bred to provide a fairly sensitive level of response to tumor-causing chemicals might be selected as the test population. The exposures or dosages, on the basis of dose per unit of body surface, are designed to be in a range that is comparable to that of humans.

Other tests of toxicity include studies to determine effects on the reproductive system; these can involve multiple generations of test animals and generally include studies of pregnant females, their offspring, and the next generation's offspring. Reduced reproductive capacity and birth defects are the usual endpoints of these studies. Another study that is often done is one for **mutagenicity**, or the ability of the substance to produce changes in the genetic material of the test animal. These tests require the examination of DNA, often obtained from cells in bone marrow or other tissues of exposed animals; some tests use bacteria, which allows easier access and observation of the DNA. The observed changes in the DNA help the toxicologist identify the mutagen that might be linked to malfunction of the cells, which may lead to various reactions, such as the incapability for energy conversion or for cell division, sometimes leading to cancer and even death.

Animal tests are not designed to prove that a chemical, drug, food additive, or other such material is "safe." They are used to determine the toxic effects that a material will produce. The specific tests that are used will vary depending on the intended end use of the material, its chemical structure, and the expected toxic effects (based on what is known

Toxicity Classification	LD_{50} Oral mg/kg	LD_{50} Inhalation mg/kg	LD_{50} Skin mg/kg
Supertoxic	<5	<250	<250
Extremely toxic	5-50	250-1,000	250-1,000
Very toxic	50-500	1,000-10,000	1,000-3,000
Moderately toxic	500-5,000	10,000-30,000	3,000-10,000
Slightly toxic	>5,000	>30,000	>10,000

Table 2-2: An example of one system of toxicity classification. Toxic materials may be classified according to their relative ability to cause harm, usually based on the dose-response relationship. This table describes several categories of toxicity; most toxic substances can be placed into one of these.

Route of Administration	Pentobarbital[1] LD$_{50}$ mg/kg	Procaine[1] LD$_{50}$ mg/kg	DFP[2] LD$_{50}$ mg/kg
Oral	280	500	4.0
Subcutaneous	130	800	1.0
Intramuscular	124	630	0.9
Intraperitoneal	130	230	1.0
Intravenous	80	45	0.3

[1]Mouse toxicity data
[2]Di-isopropylfluoro phosphate; rabbit toxicity data.

Table 2-3: Effect of route of administration on the toxicity of various compounds. The relationship between dose and response is affected by the route of exposure. This table illustrates how toxic effect varies depending on the route of exposure.

Species	Route of Administration	LD$_{50}$
Rat	Oral	10 mg/kg
Rat	Intravenous	6.6 mg/kg
Mouse	Oral	117 mg/kg
Mouse	Intravenous	87 mg/kg
Dog	Oral	1,200 mg/kg
Dog	Intravenous	51 mg/kg

Table 2-4: Toxic substances may be more toxic to some organisms than others, as this table illustrates for the pesticide chlorfenvinfos. It is important for the industrial hygienist to know what species was used for laboratory testing in order to properly evaluate the relative hazard posed by a toxic material.

about other similar materials). The use of animal testing to determine toxic responses has some obvious benefits; in fact, this extrapolation, or application of animal data to humans, is one of the basic tools used by toxicologists. There are many instances in which the use of animal tests to predict negative health effects on humans has proven to be an effective and useful mechanism – vinyl chloride, cadmium, and organophosphate pesticides are examples of these. However, there are problems related to applying animal data to a human population. For one thing, there may be differences between the route of exposure in humans and the manner in which the test chemical was administered to the animals. The animals may have been exposed to much higher levels or dosages compared to the probable or expected levels of human exposure that would be encountered in practical use of the material. While some test methods are intended to simulate lifelong human exposures, they are sometimes "compressed" in terms that are more workable for the laboratory and the species used in the study.

Another issue with extrapolation of animal data is the mechanism by which the toxin exerts its effect in the body of the test animal, which may be quite different from the effects on the human body. The ability of a toxin to exert a specific effect in test animals does not mean the same effect will manifest itself in all exposed species. This inter-species variability is illustrated by the fact that nearly all known human carcinogens are also carcinogenic in some

other species – but not in all laboratory animals that are tested. However, on a body weight basis, humans are usually more susceptible to the toxic effects of a material, sometimes by a factor of ten or more.

An important consideration is the application of a **safety factor** to the dosage; this provides a wide margin between the lowest anticipated human exposures and the lowest levels of exposures tested on animals. Because of the potential differences, animal data are generally extrapolated with caution, often using safety factors, to ensure the conservative application of test data between species. Also, the tests performed on new chemicals are conducted according to guidelines established by agencies such as the United States Food and Drug Administration and the Environmental Protection Agency.

For many materials we do have human epidemiological study data; it may be data from long-term exposures to old hazards (such as carbon monoxide), from a controlled study, or data that was obtained following an accidental exposure to a hazardous chemical. Epidemiological studies may be descriptive, that is, they may focus on a group of workers found to exhibit similar symptoms following a common exposure or experience with a hazardous material. Other studies may involve a group of individuals – called a cohort – with a common exposure and follow them through time to see if they develop disease; this is a **prospective epidemiological study**. A **retrospective epidemiological study** attempts to trace a cohort with a disease or condition back through time to determine if there was a common exposure that could be attributed with causing the condition. All of these studies present challenges to the researchers. It is difficult to keep track of individuals, document exposures, and take into account other confounding factors such as the health of cohort members, exposures to other possible toxins, and personal lifestyle factors such as smoking. Despite these difficulties, such studies are valuable for detecting long-term health effects that might occur as the result of exposure to a material. Some diseases may not develop for many years; an example is lung cancer, which may occur as much as 30 years after an exposure to asbestos fibers. This delay between exposure and disease is called the **latency period**.

Although there are volumes of toxicity data available to the industrial hygiene professional, one should keep in mind that for most chemicals in use there are limited toxicological data available. Only a fraction (less than 10 percent) of the thousands of chemicals presently in use have regulatory or other recommended standards for human exposure, and these are based mostly on airborne concentrations. The use of such standards in determining whether or not an exposure is "safe" is not a straightforward process; it requires careful consideration of toxicity data as well as the specifics of the exposure situation; also, individual susceptibility cannot be discounted. The comparison of a hazardous material or agent exposure to a regulatory or other standard as an indicator or yardstick of the acceptability of that level of exposure is discussed in Chapter 3.

Toxicity versus Risk

The assessment of a material in terms of its toxicity, as in its LD_{50} value, is one indication of the intrinsic toxic properties of a substance, but it does not provide the occupational health professional with a complete picture of the hazard posed by the chemical. There are two components to the equation that should be included in our discussion, namely **risk** and **safety**. Risk is the probability that harm will occur; safety is the probability that harm will not occur under a certain set of conditions. These factors allow us to consider the conditions under which a substance is used and take into account the benefits associated with its use. For instance, a potentially toxic substance may have great benefit when

used as a therapeutic drug but pose a threat to human health when used as a food additive. As another example, there may be significant risks associated with the use of a chemotherapeutic drug for cancer treatment; however, the threat of disease may be less desirable than the negative side effects of the drug.

As a society we must deal with these issues of safety and determine what level of risk is acceptable to us. The debate is often long and complex, and involves consideration of numerous factors, such as:

—What benefits might be gained from using the substance?

—Are there other materials available that pose a lesser risk and provide similar benefits?

—How many people will be exposed to the material and experience the benefits or the negative effects?

—Will manufacture of the material provide needed jobs? What occupational hazards are associated with manufacture of the material?

—What are the economic costs associated with using or not using the substance?

—Will the substance adversely affect the environment or consume natural resources to an unacceptable degree?

While these questions cannot be addressed here, the issues they raise are on some level the issues that must be considered by toxicologists, regulators, workers, employers, and occupational health professionals. Ostensibly the risk associated with the use of chemicals in the workplace is an acceptable one only when the benefits to society are worth the hazards posed by the materials to people and the environment. The ultimate value of that worth is dictated by social, economic, and political forces that are subject to change and are difficult, if not impossible, to control.

Checking Your Understanding

1. Define the following terms: dose, toxicity, LD_{50}, chronic, acute.

2. What term is used to refer to changes in the DNA?

3. Explain the difference between risk and safety.

2-3 Routes of Exposure

Toxic materials encountered in the workplace typically are from the hazard categories of chemical or biological agents. The chemical hazards present by far the largest number of potentially toxic exposures, as solvents, raw materials, hazardous wastes, and flammable or explosive mixtures. Chemical agents may be in the form of dusts, mists, fumes, vapors, fibers, and other particulate hazards. Biological toxins include such items as the familiar poison oak and ivy, infectious bacteria and viruses, as well as venoms and pathogenic agents carried by rodents, insects, and other animals.

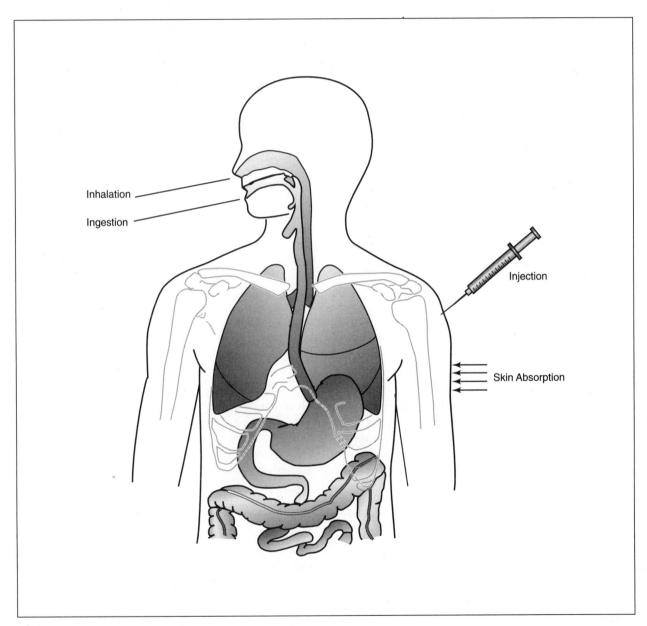

Figure 2-4: Toxic materials may enter the human body through one or more routes, including inhalation, ingestion, and absorption through the skin. Absorption through mucous membranes of the eyes and possibly the ear canals are less common, but can contribute to the overall exposure in some situations.

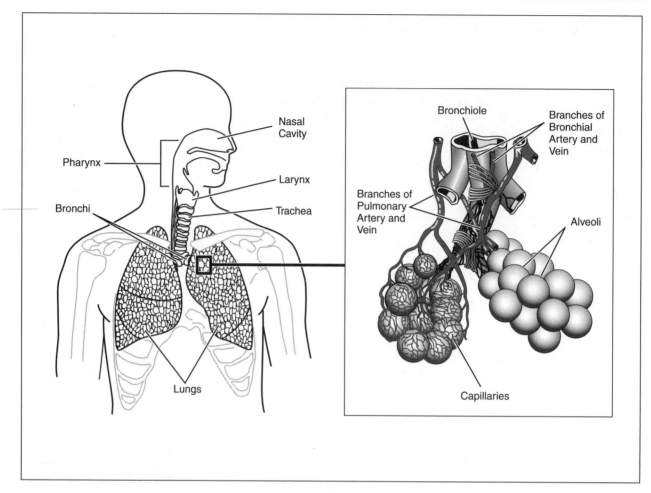

Figure 2-5: Basic structure of the respiratory system, including the alveolar region of gas exchange. The human respiratory tract provides an efficient gas-exchange mechanism, which may turn to a disadvantage when the air contains hazardous materials. Inhalation is considered the most significant route for occupational exposures to airborne contaminants.

Toxins enter the human body through one or more **routes of entry**: inhalation, ingestion, or absorption through the skin. Other less common routes include injection and absorption through moist surfaces surrounding the eyes and in the ear canal.

Of all the routes of entry, inhalation is the most common and, for the occupational health professional, the most important route of entry. The lungs are designed to provide an efficient gas exchange between the air and blood. They contain a very thin surface, with an area of roughly 300 to 1,000 square feet; a normal day's breathing equals roughly eight cubic feet of air. These facts illustrate the potential for a large amount of airborne toxins to come into close contact with the lungs – and the bloodstream – of the person breathing the contaminated air.

By contrast, the skin has a surface area of about 20 square feet and is several millimeters thick. Although the surface area is less than that of the lung, the skin still represents a fair amount of surface, which is potentially exposed to any toxic materials that might be present in the air. There are also countless examples of workers placing their hands directly into contact with a hazardous material; cleaning solvents, cutting oils, and process chemicals are but a few. Lipid-soluble materials are easily absorbed by the skin; this is due to the fact that a significant portion of the skin tissue is comprised of lipid (fat) molecules. For example, cleaning solvents used to

Box 2-1 ■ Case Study

An industrial hygienist was called upon to investigate a situation where there seemed to be an unexplained source of lead exposure to workers at a processing facility. The investigation was requested by management following medical surveillance test results (repeated twice), which indicated several workers had to be rotated out of the production area until lead levels in their blood dropped to acceptable levels. Management was concerned about the source of exposure since the facility had recently installed ventilation systems and implemented increased controls and training, and regular quarterly air sampling indicated lead levels were within acceptable limits.

The industrial hygienist spent several days in the shop watching work practices and obtaining personal monitoring samples. Results of the sampling showed that airborne levels of lead dust in the shop were well below allowable levels. Work practices seemed in order and consistent with good housekeeping standards for minimizing accumulation of lead dust in the work area. The personal hygiene practices of the employees were also consistent with accepted standards and regulatory requirements, and personnel were conscientious in their personal decontamination at the end of each work shift, changing out of their work clothes and showering each day. Break areas, shower, and locker rooms were also found to be well maintained.

After weeks of investigation and consideration, the industrial hygienist sought out the occupational health nurse who had been involved in the examinations. The schedule for blood testing was examined and it was noticed that the workers whose blood lead levels were high had come in for testing during late evening, which indicated they had been working an afternoon shift at the time. On a hunch, the industrial hygienist came in the next day on the afternoon shift and observed the work in the shop. Although there were no significant differences in housekeeping, work practices, or decontamination, the hygienist did notice a number of lunch containers sitting on top of one of the hot process vats. The workers had in fact been warming their lunches for the past few weeks in this manner, without realizing that they had been ingesting lead from the fumes released from the vat. The practice was stopped, and the workers were able to resume their duties in the shop after blood tests indicated lead levels had dropped.

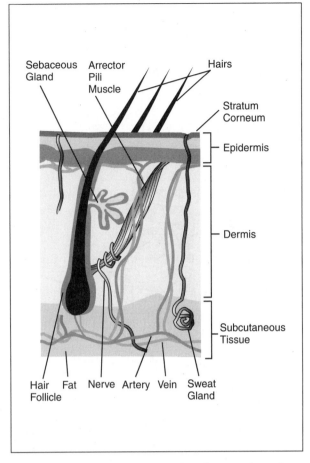

Figure 2-6: Cross-section of the skin showing the various layers. The skin provides an important first line of defense against many hazardous materials entering the body. Injuries and diseases of the skin are among the most commonly reported occupational disorders.

remove oils damage or remove the layer of fat cells. These types of materials have a special notation included with their exposure limit to indicate that the skin may present a significant route of exposure. On the other hand, water-soluble materials are not easily absorbed by the skin, because the lipid layer affords insulation and provides an effective barrier against them.

Ingestion is the third most important route of exposure. No worker will readily admit to eating the materials they work with, and few probably intentionally do such a thing. However, failure to wash hands and face before meals, eating or drinking in an area where airborne hazards exist, or lighting up a cigarette with dirty hands, are more subtle ways for ingestion to occur. Other methods for ingesting

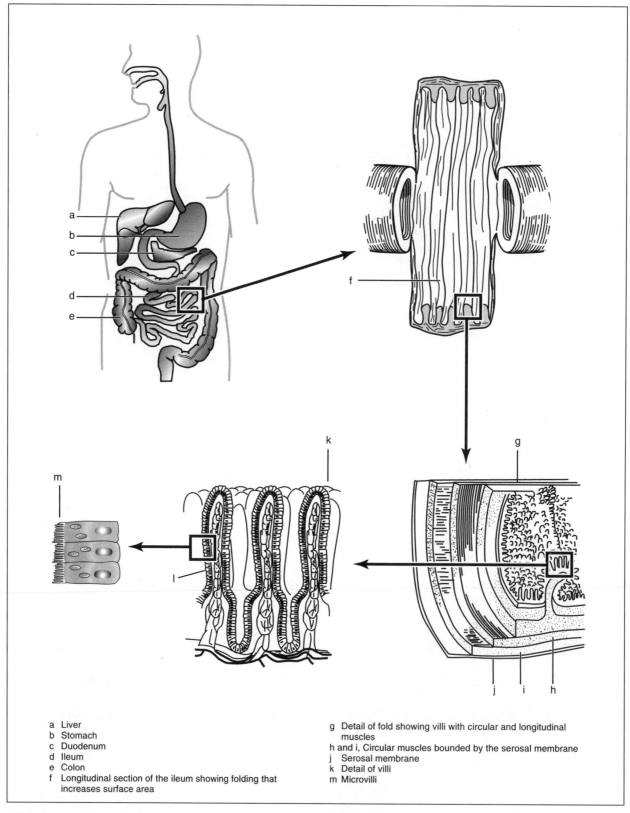

a Liver
b Stomach
c Duodenum
d Ileum
e Colon
f Longitudinal section of the ileum showing folding that increases surface area

g Detail of fold showing villi with circular and longitudinal muscles
h and i, Circular muscles bounded by the serosal membrane
j Serosal membrane
k Detail of villi
m Microvilli

Figure 2-7: Cross-section of digestive tract showing important features of the small intestine. Like the lungs, the intestines contain structures that increase surface area, resulting in an increased efficiency for absorption of materials through the membrane lining. The villi in the intestines have a function similar to that of the alveoli in the lungs.

substances include application of cosmetics or use of chewing tobacco in areas where contamination is present or suspected. The digestive tract, like the respiratory tract, is lined with a moist membrane designed for efficient absorption. The surface area in the intestinal regions is increased through small projections called villi; the absorptive surfaces of the villi are thin and highly vascularized to increase the efficiency of absorption of materials into the blood. As in the lung, contaminants that come into contact with and pass through this absorptive surface may produce local or systemic effects.

Injection is a less common way for materials to enter the body but it can be important for workers whose duties involve outdoor work, at construction sites, landfills, hazardous waste sites, or for individuals who work with plants, animals, or reptiles. Any breach of the skin – bites, stings, abrasions, puncture wounds, and cuts – can create a site for toxins to enter the body.

Although not considered a route of entry with as much importance as the others, absorption of materials through moist surfaces, such as the eyes and ear canals, still represent a possible route of exposure. Persons with damaged or perforated eardrums are sometimes prohibited from working in highly hazardous environments.

Checking Your Understanding

1. Name four routes by which hazardous materials may enter the body.

2. Which route of entry is most significant and why?

3. Why are lipid-soluble substances absorbed through the skin more easily than water-soluble ones?

2-4 Distribution of Toxins

We have seen how toxic materials may enter the body; once inside, there are several mechanisms that come into play for movement of materials from the site of initial entry to the site of action. In the industrial work setting, inhalation is the most significant route of exposure to toxins, followed by skin contact. The tissues of the lungs are delicate and provide direct contact with the blood. Inhaled substances may exert their toxic effect directly on the lungs; examples include irritation, scarring, or edematous reactions. Inhaled substances may also pass through the thin cells lining the lungs and enter the bloodstream; they are then transported via the blood to other parts of the body. Similarly, substances that are absorbed through the skin may affect the site of absorption, causing tissue damage ranging from redness and irritation to severe corrosion or chemical burns. The toxin may also pass through the skin with little effect, but reach and affect underlying tissues or enter the bloodstream.

Once absorbed, foreign substances can move from the site of absorption to other tissues and organs. In order to do this, the toxin must pass across cell membranes; this is done through one of several mechanisms: filtration, diffusion, active transport, or phagocytosis.

Cell membranes appear to be solid, but in fact have small openings, called pores, between the protein molecules that are part of the membrane. Small molecules can pass through these pores, usually along a concentration gradient (from higher to lower concentration). Filtration occurs in the kidneys, for example, where molecules pass from the blood into the urine.

Diffusion is the movement of a substance from a higher to a lower concentration. In order for a substance to diffuse across a cell membrane certain conditions must exist. First, since cell membranes are composed primarily of lipid molecules, if the material is fat-soluble, its movement through the membrane will be facilitated. Second, the compound must not be ionized. A positive or negative charge could result in formation of chemical bonds between the foreign molecule and cellular contents or structures and prevent movement through the membrane. Third, there must be a concentration gradient across the cell membrane; that is, there must be a difference in concentration of the substance from one side of the membrane to the other. The rate at which the material passes through the membrane is proportional to the difference in concentration on either side of the membrane; as the concentrations become more nearly equal, the rate of diffusion slows. Diffusion takes place in the lungs, for example, where oxygen and carbon dioxide are exchanged in the alveoli.

Facilitated diffusion is diffusion that occurs across a concentration gradient, but that cannot occur unless a specific carrier molecule is present on the cell membrane. The rate at which facilitated diffusion occurs is limited by the availability of the carrier molecule; if these molecules are in short supply, or if the carrier sites are occupied by other molecules, the rate at which facilitated diffusion progresses will be affected. An example of this is the transport of glucose from the intestinal cells into the blood.

Active transport is a term used to describe the movement of a molecule across a membrane that would otherwise be impermeable to the molecule. The transport mechanism may be a chemical reaction or a carrier molecule that attaches to and transports the molecule across the membrane. Energy is expended by the cell to make this happen. It is possible for active transport to move a substance against a concentration gradient. Again, this process can be affected if there are interferences with the chemical reaction, or if the carrier molecules are in short supply or bound to other molecules.

Phagocytosis (or **pinocytosis**, from Greek roots *phago*, meaning "to eat" and *pino*, meaning "hungry"), is the engulfing of a molecule by the cell membrane, or by another cell such as a white blood cell. This mechanism comes into play for absorption of solid materials such as silica and asbestos, and may also have a role in the development of occupational diseases, as in silicosis.

Checking Your Understanding

1. Name three passive (no energy expended) mechanisms that result in movement of chemical molecules within the body tissues.

2. Name a distribution mechanism that requires energy to be expended in order to move molecules.

3. Which distribution mechanisms can move molecules against a concentration gradient?

2-5 Biotransformation and Excretion of Toxins

Many foreign substances that enter the body are not water soluble, which makes it difficult for the body to eliminate them. The process by which these materials are chemically altered to make them easier to eliminate is called **biotransformation**.

Biotransformation accomplishes a number of things to make foreign substances easier to eliminate from the body. First, the molecule is made more water-soluble by causing it to become ionized. Another possible change is an increase in the size and weight of the molecule. These changes make the substance easier for the body to eliminate through excretion via the intestines, urine, or exhaled breath. Benzene is an aromatic hydrocarbon that may enter the body through inhalation. Benzene is very lipid-soluble, so to facilitate its elimination it is biotransformed into phenol, which is a benzene ring with an OH group attached. Phenol then reacts with a sulfate group to form phenyl sulfate, which is very water soluble and can be excreted in the urine.

Sometimes these changes actually increase the hazard posed by the substance by creating molecules that are less soluble in water and therefore more difficult to eliminate. Acetaminophen, a widely used aspirin substitute, is normally metabolized via two routes, one of which produces a toxic result. The toxic compound is usually produced at such low amounts that it is further detoxified by the body and eliminated. However, if a high enough dosage is taken, the toxic compound may be produced at higher than normal levels and cause damage to the liver. Many other therapeutic drugs can cause serious, irreversible organ damage or death, if taken at too high a dose.

Biotransformation and Excretion via the Liver

The liver is an important organ with an indispensable role in metabolism, energy storage, and protein synthesis. The liver receives blood from the

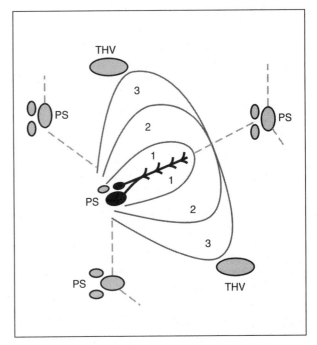

Figure 2-9: Liver acinus and zonal regions. The liver plays an important role in eliminating toxic materials from the body. It may concentrate, excrete, or biotransform materials to facilitate their removal; because of this role, it is susceptible to damage. In this illustration, PS is the portal space, consisting of a branch of the portal vein, a hepatic arteriole, and a bile duct; THV is the terminal hepatic venule (central vein); 1, 2, and 3 represent the various zones drifting off the terminal afferent vessel.

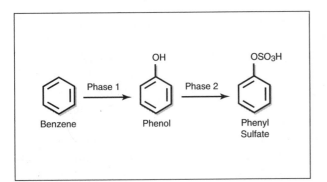

Figure 2-8: Biotransformation reaction of benzene. Benzene, a carcinogen, is lipid-soluble, which makes it difficult for the body to excrete as such. A series of chemical reactions inside the body converts the benzene to phenyl sulfate, a water-soluble substance that can be excreted in urine.

digestive tract and may concentrate, transform, or excrete substances – including toxins – that are in the blood. The bile that is produced by the liver contains compounds that have been removed from the blood by the liver; the bile is passed into the intestines and the compounds are excreted. This process of biliary excretion is an important one for some toxic materials: for example, mercury is eliminated from the body via this process. The functions of the liver make it vulnerable as a **target organ**. The term target organ refers to a specific organ where the toxic effect of a substance is manifested. For example, the liver is said to be the target organ of vinyl chloride, which has been shown to cause liver cancer. The liver receives blood from the digestive system via the portal vein and is potentially exposed to concentrated levels of ingested toxins, although inhaled toxins may also reach the liver via the bloodstream.

The functional unit of the liver is called an **acinus**. Each acinus consists of a mass of tissue surrounding a portal venule, a hepatic arteriole, a bile ductule, some lymph vessels, and some nerves. Many liver toxins, or **hepatotoxins** (from the Greek word

hepa, for liver) are classified according to the region of the acinus where the toxic effect occurs. These are referred to as Zone 1, the periportal zone; Zone 2, the midzone; and Zone 3, the centrilobular zone (shown in the diagram). Centrilobular necrosis, then, would be a condition characterized by damage to tissues in the centrilobular zone.

Damage to the liver is classified according to the **histological changes** that occur in the liver tissues; these are observable changes in the shape or appearance of the cells or the organ itself. Toxins may produce one characteristic change, or a combination of them. The histological changes that occur in liver disease include **necrosis**, or cell death; **steatosis**, which is intercellular fat accumulation, or fatty liver; **cholestasis**, the interference with the production of bile and biliary excretion; **immune cell infiltrate**, the presence of abnormally high numbers of immune cells; and **neoplasia**, or cancer (from neoplasm, "new growth"). Other common liver conditions include **fibrosis** or **cirrhosis**; these terms refer to a condition where **collagen** has been deposited in the liver to the point where it interferes with normal liver function and internal architecture. Collagen is a proteinaceous connective tissue that is not normally present in the liver; its normal place in the body is as a component of tendons, ligaments, and bones. These fibrotic changes usually follow long-term exposures; examples of toxins that cause such changes are ethanol and carbon tetrachloride.

Filtration and Excretion through the Kidneys

The kidneys receive about one-fourth of the cardiac output and therefore are exposed to a large proportion of the blood and any foreign molecules it may contain. Excretion from the bloodstream through the kidneys into the urine is the primary method for elimination of small, water-soluble molecules. Large molecules such as proteins, and lipid-soluble materials, are reabsorbed through the tubules of the **nephron**, which is the name given to the functional unit of the kidneys. The **glomerulus** is a bed of capillaries located near the proximal end (where the incoming blood arrives) of the tubules; other small blood vessels surround the tubules. Materials pass to and from the blood and urine as blood passes through the nephrons. This exchange of materials

Substance	Hepatotoxic Effect			
	Necrosis	Steatosis	Cirrhosis	Cancer
Acetaminophen	+			
Beryllium	+			
Carbon tetrachloride	+	+	+	+
Chloroform	+	+		+
Ethanol		+	+	
Phosphorus	+	+		
Tetrachloroethane	+	+		
Urethane	+			
Vinyl Chloride				+

Table 2-5: Many materials can severely damage the liver, resulting in tissue changes that prevent the organ from functioning properly. This table lists some toxic materials and their effects on the liver.

occurs via filtration, diffusion, facilitated diffusion, or active transport, depending on the material's molecular size, solubility, the concentration gradient, the pH of the fluid in the tubules, and other factors.

The rate of excretion or elimination of a substance through the kidneys depends upon the mechanism of transport as well as on the concentration of the substance in the blood. For example, if the mode of transport is via passive diffusion, the rate of elimination will increase as the concentration in the blood increases. Lipid-soluble materials will passively diffuse from the blood into the urine if a concentration gradient exists. Water-soluble materials may diffuse out of the blood, but will diffuse back into the blood if they are not ionized once in solution in the urine.

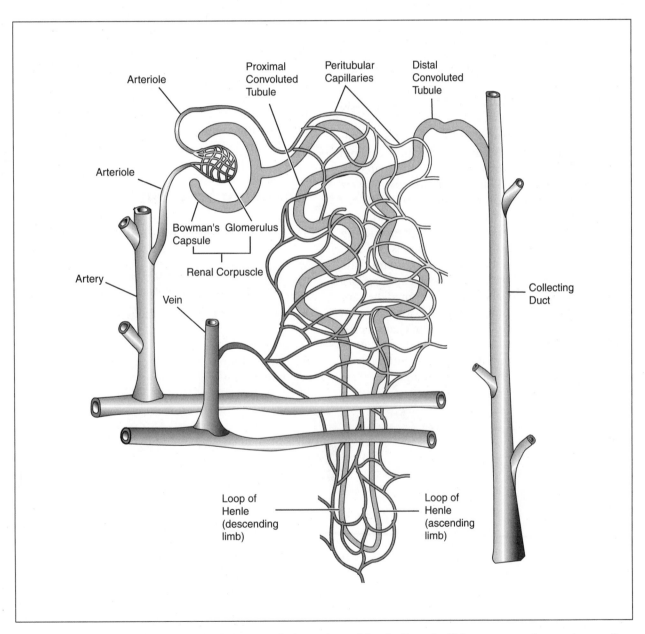

Figure 2-10: The functional unit of the kidney is called a nephron. Like the liver, the kidney processes large volumes of blood and plays an important role in elimination of toxic materials from the body. This role also makes kidney tissues susceptible to damage.

If the mode of transport into the tubules is via active transport, the rate of elimination will be limited by the number of carrier molecules available to perform the transport. An example of this is ethanol, which requires active transport for elimination. If the concentration of ethanol exceeds the available concentration of carrier molecules, the rate of elimination, once maximized, remains constant, and continued intake of ethanol results in toxic levels being reached in the plasma and tissues, accompanied by nervous system effects and other unpleasant symptoms familiar to many. Diet affects urinary pH, and this has been used to induce acidic or basic urine to facilitate excretion of certain compounds. Fluid intake affects urinary output, and high fluid intake tends to facilitate urinary excretion.

Because of the arrangement of capillaries and tubules, materials that pass into the urine have an opportunity to be reabsorbed into the blood. This is most likely to occur when the molecules are lipid soluble. Thus, toxins may fail to be eliminated in their pass through the kidneys. Kidney toxins exert their effects by altering the permeability of the tubules or glomeruli, or both, affecting blood flow, filtration, fluid pressures, and otherwise interfering with the kidney's filtration process. There is evidence that some toxins, such as heavy metals, accumulate in kidney tissues following an acute exposure. In many cases there appears to be no immediate, apparent damage to the nephrons. The formation of metal-protein compounds is thought to have a protective effect; cellular damage is evident, but nephron function is not initially affected. Subsequent exposures may result in nephron damage.

Excretion via the Lungs

Although designed for absorption, the lungs may be a site for elimination of gaseous toxins. These may be the toxins themselves, or metabolic products of the toxins. Toxins that are eliminated via exhaled breath typically include volatile compounds, such as benzene. Benzene molecules diffuse out of blood and into the alveolar space; this is passive diffusion down a concentration gradient. Lipid-soluble compounds, such as benzene, are suited for elimination through this mechanism since the thin membranes of the alveoli are composed of lipid molecules. The thinness of the membranes also allows gases dissolved in the blood to diffuse rapidly into the alveolar air. Other examples of gases that are eliminated in the breath are carbon monoxide and carbon dioxide.

Checking Your Understanding

1. Name three liver toxins and the type of damage or disease caused by each.

2. What are the names of the functional units of the liver and the kidneys?

3. How do degreasers damage the skin?

4. What is a glomerulus?

5. Name four ways in which the body can eliminate toxins.

2-6 Classes of Toxins and Toxic Responses

Toxins exert a range of responses, from reddening of the skin to liver damage, pulmonary edema, and in some cases, death. Although two materials may have different chemical or physical properties, the manner in which they affect the organism may be similar once inside the body. Materials with similar effects may therefore be conveniently classified or grouped together into one or more classes, according to their effect: irritants and sensitizers, systemic toxins, neurotoxins, reproductive toxins, and carcinogens.

Irritants and Sensitizers

This class of toxins is characterized by one of two responses: **irritation**, generally some physical damage to tissues that is localized (at the site of contact); or **sensitization**, which is an adverse reaction that occurs following more than one exposure event to a substance. Sensitization reactions may be localized or systemic, depending on the material and on the way the body responds.

Irritants

There are many substances that act as an irritant to the skin or to more sensitive tissues such as the eyes or the respiratory system. The severity of the irritant reaction depends on the strength or potency of the irritant, the circumstances of exposure, the site of contact, and the sensitivity of each individual. **Corrosion** is the most severe response of this type and is characterized by almost immediate changes in the exposed tissue: ulceration, tissue damage, and probable permanent damage to the affected area. A good example of a corrosive response is the severe chemical burn that results from exposure of skin to nitric acid or sodium hydroxide. Some strong irritants may produce a corrosive response after repeated or high exposures. An acute irritation, by contrast, is typified by redness and inflammation, and is usually a reversible condition. Repeated exposure to irritants may result in a cumulative response, where the individual was at one time not affected by the material, but has experienced repeated exposures and some redness and swelling is then produced on contact with the material. Such materials are often referred to as **marginal irritants**. This type of reaction does not involve the immune system and should not be confused with sensitization, which is addressed in the next section. Examples of marginal irritants are soaps, detergents, and other products; many of these are not detected in laboratory tests, which tend to look for corrosive or acute irritant responses.

Sensitizers

Sensitizers are substances that stimulate a response from the immune system. The immune system's role is to recognize and reject foreign objects in the body,

Type of Reaction	Antigen Type	Antibody Type	Reaction
TYPE I Anaphylactic Reactions	Free antigen	Reaginic antibody (IgE) fixed to membrane of mast cell	Degranulation of cell and release of mediators
TYPE II Cytolytic Reactions	Antigen associated with cell membrane	Free antibody (IgG, IgM, IgA)	Agglutination with complement fixation and lysis
TYPE III Toxic Precipitin Reactions	Free soluble antigen in excess of antibody	Free antibody	Precipitin complex deposited in vascular epithelium
TYPE IV Cell-mediated hypersensitivity reactions	Antigenic component of cell membrane	Activated T-cell killer lymphocyte	Death of cell followed by phagocytosis

Table 2-6: Allergic responses that follow exposure to a sensitizing agent can range from the relatively mild Type I reaction, which is usually a localized response such as a skin rash, to the more severe Type IV reaction, which can cause the individual to go into shock, a life-threatening condition.

which can include items such as defective cells, infectious agents, and hazardous materials. The immune response may be a simple reaction to the molecules of a toxin. More often, the toxin acts as a **hapten**, which means it combines with protein molecules that are normally present in the body, to form a new molecule. An **allergic response** occurs when the body recognizes the toxin or hapten + protein molecules as being foreign to the body. The immune system then releases **antibodies**, which are specialized proteins that react only with certain other (foreign) molecules. The foreign molecule must have a compatible reaction "site" – such as a specific chemical structure – for the antibody reaction to occur. This allows antibodies to attack only specific foreign molecules; if the individual does not have antibodies that are specific for the toxin that is present, then no allergic reaction will occur. The nature of the reaction depends upon the substance and type of antibody involved. There are four different types of allergic reactions (see Table 2-6).

The response or reaction to a sensitizer usually occurs after exposure to a substance that the subject has been exposed to previously; the triggering exposure may be of a very low level and could be undetectable using conventional industrial hygiene sampling methods. For many sensitizers there appears to be no threshold level, that is, any exposure seems enough to cause a reaction. However, the dose-response relationship does exist for sensitizers as evidenced by the fact that a higher level of exposure will generally cause a more severe reaction. In humans, allergic reactions run the gamut from mild irritation to anaphylactic shock, and can be fatal. Allergic reactions are characterized by symptoms of varying intensity, such as irritation of the skin, watery eyes, or severe difficulty in breathing.

Systemic Toxins

Systemic toxins are materials that affect target organs or organ systems; examples include vinyl chloride, which causes liver damage and liver cancer; cadmium, which causes kidney damage; and benzene, which affects the blood marrow and causes leukemia.

Systemic toxins exert their effect through a variety of mechanisms specific to the material-target organ interaction; often the toxicity mechanism is related to the normal function of the target organ.

For instance, the liver has an important role in biotransformation, while the kidneys filter the blood. Because of the amount of blood that passes through them, both of these organs are exposed to potentially damaging toxins that may be present in the blood, and normal processing of blood through these organs can result in injury.

Neurotoxins

Neurotoxins are compounds that have a negative effect on the nervous system. As with other toxins, the response may be mild or severe, and is dependent upon the substance as well as the dose. Neuro-

Portion of Nervous System	Primary Function/Control	Examples of Agents that Cause Damage or Other Adverse Effects
Central nervous system (brain and spinal cord)	Body movement, memory, emotions and behavior	Mercury Lead Carbon monoxide Organic solvents Pesticides
Autonomic nervous system (sympathetic and parasympathetic)	Heartbeat, breathing, reflexes	Organophosphorous compounds Vibration Metals (Lead, mercury, arsenic)
Peripheral nervous system* (peripheral nerves, brain stem, and spinal cord) – Sensory – Motor	 – Senses (heat, cold, touch, pain, pressure, proprioception) – Voluntary motor control	Metals (Lead, organic mercury, arsenic) Organic solvents (hexane, carbon disulfide, methanol) Organophosphorous compounds Vibration Traumatic injury Compressive injury (repetitive motion/use)

* Agents listed for peripheral nervous system often present symptoms involving both sensory and motor function impairment.

Table 2-7: This table summarizes the functions of the different parts of the nervous system. Some materials have damaging effects on only one part of the system; others affect more than one. Damage to the nervous system may be manifested in striking ways, such as behavioral changes or changes in motor coordination.

toxins are a concern to the occupational health professional since their effects can impact thinking ability, motor control, and regulation of breathing and heartbeat.

The **central nervous system** is composed of the brain and spinal cord, and controls several important body functions such as coordination, emotion, speech, and memory. The nerve tissues lying outside the brain and spinal cord make up the **peripheral nervous system**. The functions of the peripheral nervous system include transmittal of sensory information (touch, heat/cold, proprioception, and pain) and motor impulses for movement of the limbs; **proprioception** is the ability to recognize the relative position of one's body and limbs. Heart rate and breathing are regulated by peripheral nerve centers that control motor functions and also stimulate most of the major internal organs; collectively these centers comprise the **autonomic nervous system**, which is usually considered to be made up of two

Substances Associated with Central Nervous System Effects	
Organic solvents	Arsenic
Lead and lead compounds	Organic mercury
Asphyxiant gases	Organochloride compounds
Carbon monoxide	Organophosphate compounds
Cyanide compounds	Manganese
Substances Associated with Peripheral Nervous System Effects	
Arsenic	Lead
Organic mercury	n-Hexane
Methanol	Organophosphate compounds
Ethylene glycol	Organochloride compounds

Table 2-8: Many substances that workers are exposed to can affect the nervous system. These include metals, solvents, and pesticides. Effects of these materials include temporary dizziness from solvent exposure, loss of motor control from pesticides, and developmental defects from exposure to metals.

subdivisions: the sympathetic and parasympathetic.

Neurotoxins exert their effects through different mechanisms. One way is by interfering with the transmittal of nerve impulses. This can be through blocking, which prevents impulse transmission, or depolarizing, which eliminates the electrochemical gradient that is present in the cell. Other agents act to increase or decrease a neuron's sensitivity to nerve impulses; these are stimulants and depressants. The term **neuropathy** is used to refer to a toxic effect characterized by the progressive decline and death of nerves. Another important group of occupational neurotoxins is comprised by the acetylcholinesterase-inhibiting agents, which cause an increase in nerve stimulation by reducing production of or inhibiting the enzyme acetylcholinesterase. This enzyme acts to stop or slow stimulation of a neuron. Inhibition of acetylcholinesterase can result in uncontrolled stimulation of nerves that control muscle function, leading to shaking and tremors in exposed individuals. Organophosphate and carbamate pesticides are compounds that cause acetylcholinesterase inhibition.

Organic solvents are examples of neurotoxins that affect the central nervous system, causing dizziness, nausea, and disorientation, aside from other effects such as irritation of eyes and respiratory tract. Hexane, a toxin of the peripheral nervous system, causes a neuropathy resulting in loss of feeling in the hands and feet, a condition known as **glove and stocking syndrome**.

Lack of adequate oxygen can have a detrimental effect on nervous system tissues, as they rely heavily on oxygen supplied in the bloodstream. Anoxia is a term used to describe a condition of inadequate oxygen supply. Inadequate blood flow to the brain can result in severe damage to the brain or death within minutes. Asphyxiants such as carbon monoxide and methylene chloride can have a toxic effect on the nervous system.

Still another way to damage nerve tissues is through physical damage. This can be through physical trauma or via chemical agents that cause neuropathy. Many neurotoxins damage the insulative **myelin sheath** that surrounds nerve fibers; once the myelin sheath is damaged, the underlying nerve begins to die and eventually nerve impulses will be interrupted. Nerves in the central and peripheral nervous system may be affected. Examples of some demyelinating neurotoxins are lead and chronic carbon monoxide exposure.

Reproductive Toxins

Substances that affect the reproductive process are included in this class of toxins. The reproductive process includes germ cell or **gamete** (sperm and egg) production, fertilization, implantation of the developing embryo, and gestation. Reproductive toxins may affect males, females, or both – many are equal opportunity agents!

Probably one of the best known reproductive toxins is lead. Lead can affect males by causing decreased numbers of sperm or defective sperm. In pregnant females, lead can cause deformities in the developing fetus, and especially in the developing

nervous system. Toxins that cause abnormal development, or birth defects, are called **teratogens** (from the Greek word tero, meaning monster). Another well-publicized teratogen is thalidomide. This substance is a sedative that was advertised as being safe, nontoxic, and without an identifiable LD_{50}, based on laboratory tests on adult animals. It was actually advertised as being suitable for use by women suffering tension, nausea, and sleeplessness during pregnancy, and was sold over the counter for many months. Thousands of babies were born between 1959 and 1962 with severe deformities of the extremities, including no arms, no legs, deformed ears, and other malformations. The use of thalidomide was popular in Germany, Australia, Britain, and Ja-

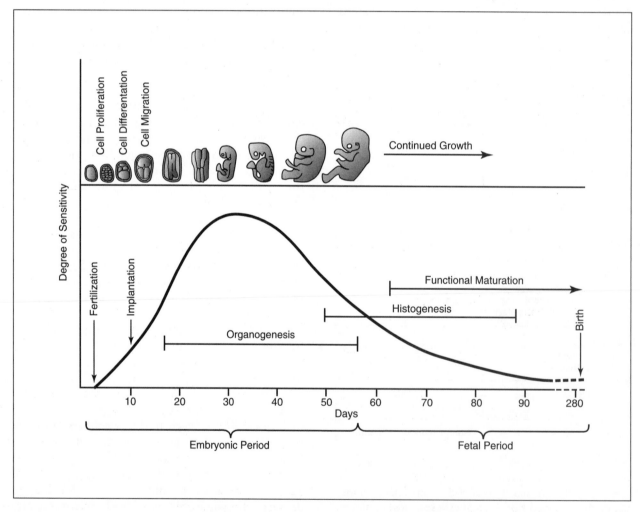

Figure 2-11: This diagram shows significant developmental stages of a human embryo, during which different organs and/or body parts are taking shape. Exposures to toxic materials that occur during critical development processes can have an adverse effect on the tissues and organs undergoing growth at the time of exposure.

pan; it was available in Canada as well. Its use was never approved in the United States due to actions by Frances Kelsey, of the U.S. Food and Drug Administration. Ms. Kelsey insisted that the manufacturer complete tests on pregnant animals before she would issue approval for marketing thalidomide. Before the studies were completed, the teratogenicity of thalidomide was proven and its administration to pregnant women was discontinued.

Another significant teratogen is diethyl stilbestrol, or DES. DES is a synthetic version of a compound belonging to a group of hormones called estrogens. It was approved for medical use in the United States in 1941 and was prescribed for women as an aid in the prevention of miscarriage as well as

for treatment of menopausal symptoms, prostate and breast cancer, and other medical uses. Its prescription for pregnant women was discontinued in 1972. It is linked to the development of cancer and genitourinary tract abnormalities of children (male and female) born to women who took DES during pregnancy. Among DES daughters, incidence of a rare form of vaginal cancer is higher than one would expect. The ability of DES to cause cancer and developmental deformities places it into the classes of teratogen and carcinogen. Its ability to cause cancer after **fetal exposure**, meaning the exposure occurred in the womb, raised many issues regarding transplacental exposure. The thalidomide and DES experiences demonstrated that substances posing a

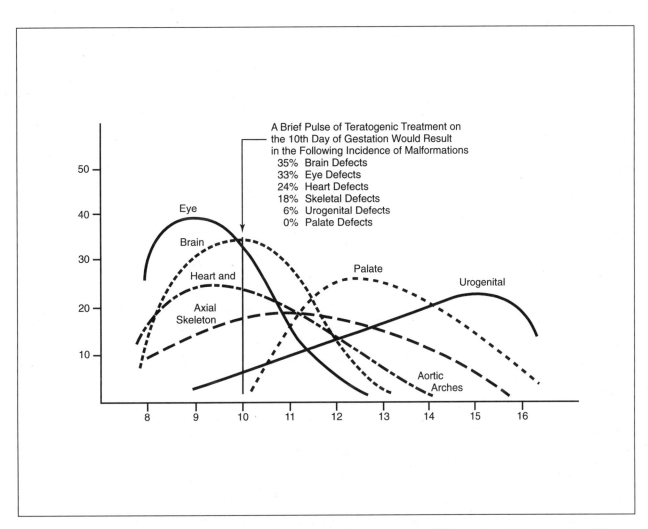

Figure 2-12: Incidence of malformations in rat embryos exposed to teratogens at different times during gestation. The effect of the toxic material is related to the particular growth or developmental stage during which the exposure occurs. For example, certain medications, such as tetracycline, are not recommended for pregnant women during the third trimester due to the yellowish discoloration of the teeth of the child that often results.

Agents Reported to Affect Male Reproductive Capacity*	
Steroids	
Natural and synthetic androgens (antiandrogens), estrogens (antiestrogens) and progestins	
Antineoplastic Agents	
Alkaloids	vinca alkaloids (vinblastine, vincristine)
Alkylating agents	esters of methanesulfonic acid (MMS, EMS, busulfan); ethylenimines (TEM, TEPA); hydrazines (procarbazine); nitrogen mustards (chlorambucil, cyclophosphamide); nitrosoureas (CCNU, BCNU, MNU)
Antimetabolites	amino acid analogs [azaserine (DON)]; folic acid antagonists (methotrexate); nucleic acid analogs (azauridine, 5-bromodeoxyuridine, cyctosine arabinoside, 5-fluorouracil, 6-mercaptopurine)
Antitumor antibotics	actinomycin D, adriamycin, bleomycin, daunomycin, mitomycin C
Drugs that Modify the Central Nervous System	
Alcohols	
Anesthetic gases and vapors	enflurane, halothane, methoxyflurane, nitrous oxide
Antiparkinsonism drugs	levodopa
Appetite suppressants	
Narcotic and non-narcotic analgesics	opioids
Neuroleptics (antidepressants, antimanic, and antipsychotic agents)	phenothiazines, imipramine, and amitriptyline
Tranquilizers	phenothiazines, reserpine, monoamine oxidase inhibitors
Drugs that Modify the Autonomic Nervous System	
Antiadrenergic drugs (for hypertensive and cardiac disorders)	α- and β-blocking agents, clonidine, methyldopa, guanethidine, bretylium, reserpine
Other Therapeutic Agents	
Alcoholism	tetraethylthiuram disulfide (antabuse)
Analgesic and antipyretics	phenacetin
Anticonvulsants	diphenylhydantoin (phenytoin)
Anti-infective agents	amphotericin B, hexachlorophene, hycanthone, nitrofuran derivatives (furacin, furadroxyl), sulfasalazine
Antischistosomal agents	niridazole, hycanthone
Antiparasitic drugs	quinine, quinacrine, chloroquine
Diuretics	aldactone, thiazides
Gout suppressants	colchicine
Histamines and histamine antagonists	chlorcyclizine, cimetidine
Oral hypoglycemic agents	chlorpropamide
Xanthines	caffeine, theobromine
Metals and Trace Elements	
Aluminum, arsenic, boranes, boron, cadmium, cobalt, lead, mercury, methylmercury, molybdenum, nickel, silver, uranium	
Insecticides	
Benzene hexachlorides	lindane

Table 2-9: Many materials have been linked to negative effects on the male reproductive system. This is one area of toxicology, for both males and females, that warrants more investigation.

Agents Reported to Affect Male Reproductive Capacity* (continued)	
Insecticides (continued)	
Carbamates	carbaryl
Chlorobenzene derivatives	chlorophenothane (DDT), methoxychlor
Indane derivatives	aldrin, chlordane, dieldrin
Phosphate esters (cholinesterase inhibitors)	dichlorvos (DDVP), hexamethylphosphoramide
Miscellaneous	chlordecone (kepone)
Herbicides	
Chlorinated phenoxyacetic acids	2,4-dichlorophenoxyacetic acid (2,4-D), 2,4,5-trichlorophenoxyacetic acid (2,4,5-T), yalane
Quaternary ammonium compounds	diquat, paraquat
Rodenticides	
Metabolic inhibitors	fluoroacetate (fluoracetamide)
Fungicides, Fumigants, and Sterilants	
Apholate, captan, carbon disulfide, dibromochloropropane (DBCP), ethylene dibromide, ethylene oxide, thiocarbamates (cineb, maneb), triphenyltin	
Food Additives and Contaminants	
Aflatoxins, cyclamate, diethylstilbestrol (DES), dimethylnitrosamine, gossypol, metanil yellow, monosodium glutamate, nitrofuran derivatives	
Industrial Chemicals	
Chlorinated hydrocarbons	hexafluoroacetone, polybrominated biphenyls (PBBs) polychlorinated biphenyls (PCBs), 2,3,7,8-tetrachlorodibenzo-p-dioxin (TCDD)
Hydrazines	dithiocarbamoylhydrazine
Monomers	vinyl chloride, chloroprene
Polycyclic aromatic hydrocarbons (PAHs)	dimethylbenzanthracene (DMBA), benzo(a)pyrene
Solvents	benzene, carbon disulfide, glycolethers, hexane, thiophene, toluene, xylene
Miscellaneous	diethyl adipate, chloroprene, ethylene oxide cyclic tetramer
Consumer Products	
Flame retardants	tris-(2,3-dibromopropyl) phosphate (TRIS)
Plasticizers	phthalate esters (DBP, DEHP)
Antispermatogenic Drugs (Investigational)	
Derivatives of 1-benzylindazole-3-carboxylic acid, 1-p-chlorobenzyl-1H indazol-3-carboxylic acid, chlorohydrins, chlorosugars (6-chloroglucose), dichloracetyldiamines derivatives (Win 13,099, 17,416, 18,446), dihydronaphthalenes (nafoxidine), dinitropyrroles (ORF-1616), gossypol, 5-thioglucose, α-chlorohydrin, monothioglycerol	
Miscellaneous	
Personal habits	alcohol consumption, tobacco smoking
Agents of abuse	marijuana and other centrally acting drugs
Physical factors	heat, light, hypoxia
Radiation	α, β, and γ radiation; x-rays
Stable isotopes	deuterium oxide

* Both laboratory and clinical reports are included (Target Organ Toxicity Center Reproductive Toxicity Information File).

Table 2-9: Continued.

Agents Reported to Affect Female Reproductive Capacity*	
Steroids	
Natural and synthetic androgens (antiandrogens), estrogens (antiestrogens), and progestins	
Antineoplastic Agents	
Alkylating agents	cyclophosphamide, busulfan
Antimetabolites	folic acid antagonists (methotrexate)
Other Therapeutic Agents	
Anesthetic gases and vapors	halothane, enflurane, methoxyflurane
Antiparkinsonism drugs	levodopa
Antiparasitic drugs	quinacrine
Appetite suppressants	
Narcotic and non-narcotic analgesics	opioids
Neuroleptics (antidepressants, antimanic, and antipsychotic agents)	phenothiazines, imipramine, and amitriptyline
Serotonin	
Sympathomimetic amines	epinephrine, norepinephrine, amphetamines
Tranquilizers	phenothiazines, reserpine, monoamine oxidase inhibitors
Metals and Trace Elements	
Arsenic, lead, lithium, mercury and methylmercury, molybdenum, nickel, selenium, thallium	
Insecticides	
Benzene hexachlorides	lindane
Carbamates	carbaryl
Chlorobenzene derivatives	chlorophenothane (DDT), methoxychlor
Indane derivatives	aldrin, chlordane, dieldrin
Phosphate esters (cholinesterase inhibitors)	parathion
Miscellaneous	chlordecone (kepone), mirex, hexachlorobenzene, ethylene oxide
Herbicides	
Chlorinated phenoxyacetic acids	2,4-dichlorophenoxyacetic acid (2,4-D), 2,4,5-trichlorophenoxyacetic acid (2,4,5-T)
Food Additives and Contaminants	
Cyclohexylamine, diethylstilbestrol (DES), dimethylnitrosamines, monosodium glutamate, nitrofuran derivatives (AF2), nitrosamines, sodium nitrite	
Industrial Chemicals and Processes	
Building materials	formaldehyde
Chlorinated hydrocarbons	polychlorinated biphenyls (PCBs), chloroform, trichloroethylene
Paints and dyes	aniline

Table 2-10: Effects on the female reproductive system include teratogens and other materials that affect fetal development. Most toxicity studies are limited in the number and type of reproductive effects they are designed to detect.

Agents Reported to Affect Female Reproductive Capacity* (continued)	
Industrial Chemicals and Processes (continued)	
Plastic monomers	caprolactam, styrene, vinyl chloride
Polycyclic aromatic hydrocarbons (PAHs)	benzo(a)pyrene
Rubber manufacturing	chloroprene
Solvents	benzene, carbon disulfide, chloroform, ethanol, glycol ethers, hexane, toluene, trichloroethylene, xylene
Miscellaneous	cyanoketone, hydrazines
Consumer products	
Flame retardants	TRIS, polybrominated biphenyls (PBBs)
Plasticizers	phthalic acid esters (DEHP)
Miscellaneous	
Personal habits	alcohol consumption, tobacco smoking
Agents of abuse	marijuana and other centrally acting drugs
* Both laboratory and clinical reports are included (Target Organ Toxicity Center Reproductive Toxicity Information File).	

Table 2-10: Continued.

hazard to the fetus were capable of crossing the placenta. Until these dramatic demonstrations of transplacental toxicity, it was popularly believed that the placenta acted as a protective barrier, preventing such substances from crossing from the mother's blood to the fetus.

The mechanism by which a teratogen acts on a developing fetus varies with the substance involved and is also linked to the time period – during gestation – in which the exposure occurs. Fetal development is marked by rapid cell growth, specialization of cells into organs, then growth and maturation of the tissues and organs of the fetus. Teratogens may act to interrupt or alter any of these processes, resulting in embryonic death, malformations (as in thalidomide), or functional deficits such as mental retardation (as in lead).

Carcinogens

Carcinogens is a name given to a class of toxins that cause cancer. This ability is defined by whether the substance is able to cause a **neoplasm**, which literally means "new form." Occupational carcinogens consist of a variety of chemical and physical agents, including organic and inorganic solvents, heavy metals, solid materials such as asbestos fibers, "natural" substances such as hormones and nitrosamine, and materials that suppress the immune system.

The mechanism by which a material causes growth of abnormal cells is dependent upon the substance as well as the age, sex, and overall health of the exposed individual. Recent efforts to map the human genome have revealed that there are specific genes that are associated with some cancers, and there may be other inherited factors that play a part in cancer development. Popular theory among cancer researchers is that there are at least two general classes of carcinogens. The **genotoxic carcinogens** are those that react with the **DNA**, which is the genetic material that contains information for cellular function, metabolism, and growth. For example, the use of sodium nitrite as a preservative in processed meats such as bacon may produce nitrosamines in the body; these are believed to cause cancer via a genotoxic mechanism. Another category or class of substances that cause cancer includes substances that do not appear to interact with the genetic material. In these cases the cause appears to be through some other interaction. This second class of carcinogens is called **epigenetic carcinogens**. DES and asbestos are examples of epigenetic carcinogens. Some carcinogens do not fit into either of these categories.

The mechanism of cancer causation is important to occupational health professionals because of the clear implications that there may be no threshold dose for many carcinogens, especially for those in the genotoxic category. Thus, a single exposure or DNA-damaging event could be enough to trigger cancer. This is sometimes called the **one-hit theory**. Genotoxins react with DNA to damage or change the genetic code; these changes, or mutations, cause changes in cell behavior such as uncontrolled growth. Mutations in germ cells, such as sperm, could result in the offspring inheriting damaged or altered genetic material and later in development of disease or dysfunction.

Another important concept in cancer causation is the theory of initiation and promotion. This is the theory that exposure to certain agents, called **initiators**, make the cell susceptible to the development of cancer. When the damaged tissue is later exposed to a **promoter**, which may be another chemical substance, or a set of specific conditions, the development of cancer is somehow activated. This theory supports the possibility of a threshold dose, since exposure to either the initiator or the promoter alone is not enough to cause cancer to develop, and in some cases it may take a specific dose or level of exposure to a promoter to trigger the development of cancer.

Checking Your Understanding

1. Explain the difference between irritation and sensitization.

2. What are antibodies?

3. Contrast systemic and localized toxic effects.

4. What is the name given to the part of the nervous system that controls the heart rate?

5. Explain the difference between genotoxic and epigenetic carcinogens.

6. Describe the one-hit and the initiation and promotion theories of cancer causation.

Summary

This chapter presented a review of basic toxicology terms and concepts as well as methods used to investigate and describe the toxicity of a substance. The dose-response relationship and the threshold dose are important descriptors of toxic effects and may be used as an indication of the relative toxicity of a substance. The use of standard toxicity tests and dosages, such as the LD_{50}, provides additional information about the range and type of toxic effect that a substance may have on an organism. Toxic effects may vary in severity from simple irritation to permanent injury to death. The response is dependent upon the substance, the dose, the conditions of exposure, and the individual's susceptibility to the effect. Toxic effects may occur immediately upon exposure, or may be delayed for several hours, weeks, or even years.

Toxins may enter the body through one of several routes; inhalation is the most significant in terms of occupational exposures, followed by absorption through the skin. Once inside the body, toxins may cause injury at the site of entry, or travel to other tissues or organs to exert their effects. The liver and kidneys are potential target organs to ingested and blood-borne toxins due to their important roles in metabolism and filtration. Some volatile compounds are eliminated through exhaled breath.

For ease of organization and reference, toxins are often classified by effect. These include: irritants and sensitizers; systemic toxins; neurotoxins; reproductive toxins; and carcinogens. Effects of toxins may be localized (as in irritants) or systemic; toxic effects may manifest themselves after a long latency period; and prenatal exposures may result in delayed effects even into adulthood.

Critical Thinking Questions

1. Explain the dose-response relationship and the threshold dose. Illustrate each with an example.

2. Explain what, if any, conclusions could be drawn from the following toxicity data:

   Compound A: LD_{50} 100 mg/kg; mouse, oral;
   Compound B: LD_{50} 25 mg/kg; rat, intravenous;
   Compound C: LD_{50} 120 mg/kg, rat, oral.

3. List the major classes of toxins and give an example of each.

4. Explain the threshold dose concept as it relates to the one-hit theory of cancer causation. How does the threshold dose relate to the initiation and promotion theory?

5. Explain why exposure to a toxic agent does not necessarily mean one will develop a disease.

3

Occupational Health Standards

Chapter Objectives

Upon completing this chapter, the student will be able to:

1. **Explain** the historical need for worker health protection regulations and the origin of OSHA regulations and describe the process used for establishing OSHA legislation.

2. **Explain** the role of NIOSH and other advisory committees and industry groups in establishing health and safety regulations.

3. **Define** the various terms and acronyms as they apply to worker exposure to hazardous agents.

4. **Explain** why exposure limits do not represent a "safe" level of exposure.

5. **Describe** the OSHA standards that have significant impact on industrial hygiene programs.

Chapter Sections

3-1 Introduction

Hazardous substances capable of causing illness were first recognized and recorded more than 2,000 years ago, but it was not until 1833, for example, that the Factory Acts were passed in Great Britain, restricting the employment of children in textile mills. Regulations to protect workers in the United States were not enacted until 1908, when the first worker compensation laws were passed. The Walsh-Healy Act, passed in 1936, did impose some limited requirements on government contractors, but broader regulatory measures aimed at preventing worker illness and injury did not come about until passage of the Federal Coal Mine Health and Safety Act of 1969, and the Occupational Safety and Health Act in 1970. The latter of these laws, often referred to as the OSHAct or simply as the Act, is perhaps the broadest in scope of any regulation ever passed. It applies to all employees whose employer engages in business affecting commerce, except for government employees. To quote directly from the preamble, the OSHAct was intended: "To assure safe and healthful working conditions for working men and women; by authorizing enforcement of the standards developed under the Act; by assisting and encouraging the States in their efforts to assure safe and healthful working conditions; by providing for research, information, education, and training in the field of occupational safety and health; and for other purposes."

In this chapter we will take a look at the origins of worker health and safety regulations in the United States, specifically the formation of the Occupational Safety and Health Administration (OSHA), and its role in the development and enforcement of these regulations. The organization and content of a typical OSHA standard will be reviewed; terms used to describe exposure will be defined; and the difference between enforceable standards and recommended levels will be explored.

3-2 The Occupational Safety and Health Act

The OSHAct describes how OSHA will function to develop and enforce regulations aimed at protecting the health and safety of workers in the United States. It also lists specific responsibilities of employers for ensuring worker protection; among them is a requirement "to provide a workplace free from the recognized hazards that are causing or likely to cause death or serious physical harm." This requirement, found in section 5(c) of the OSHAct, is better known as the **General Duty Clause**. The Act requires employers to comply with the standards promulgated by OSHA; it also requires employees to comply with the requirements of the specific regulations that apply to them, which includes following safety rules at their workplace and making proper use of assigned protective equipment.

In addition to the above requirements, the OSHAct allowed for the creation of the Occupational Safety and Health Administration (OSHA) to fulfill the actions authorized by the Act. The powers and authorities of the agency were also defined by the Act, including authorization to develop regulatory standards as a means for controlling or eliminating workplace hazards and protecting worker health. The Act also authorized the creation of the National Institute for Occupational Safety and Health, or NIOSH. NIOSH's role was defined as that of researcher and advisor to OSHA and continues to be an important source of data and information relative to worker health. NIOSH has specific responsibilities in the areas of methods development for exposure sampling and analysis, respiratory protection, and issuance of recommended levels for exposure to hazardous substances.

When the OSHAct went into effect on April 28, 1971, one of the first goals of the new agency was to propose legislation for worker protection as mandated by the Act. Rather than go through a lengthy development process for these initial regulations, the Department of Labor chose to adopt many of the existing standards that were being followed by industry at the time. The criteria that were in use described minimum requirements for materials and applications ranging from compressed gas cylinders and ladders to workplace air quality standards. The adoption of these **consensus standards** allowed OSHA to enact a large part of its regulations in a relatively short time. The sources of these standards included groups of industry experts from many areas; a look at the sources section of a regulation reveals these roots – the Compressed Gas Association, the American National Standards Institute (ANSI), and others.

Significant contribution to chemical exposure standards came from the American Conference of Governmental Industrial Hygienists, or ACGIH. Many of **OSHA's permissible exposure limits (PEL)** that were adopted in 1971 were in fact the **threshold limit values (TLV®)** published in the 1968 *Threshold Limit Values* booklet. Other limits for chemical exposures were from a list published by ANSI, which was then known as the American Standards Association.

Some states have their own occupational safety and health standards, sometimes called state plans. OSHA approves state-administered worker health and safety programs that are at least as stringent as the federal OSHA requirements. Some individual state programs may have requirements that exceed OSHA's. Examples of areas where state requirements might vary include hazard communication and exposure limits for hazardous materials. Due to the amount of variation that exists between so-called state plans, we will not address them in this text. An individual practicing industrial hygiene in a state with its own plan, must absolutely be familiar with the provisions contained in that plan.

How OSHA Standards are Promulgated

Since passing the first OSHAct, Congress has enacted other regulations for protecting worker health and safety. Before issuing a new standard, the Secretary of Labor must determine that it is "reasonably necessary and appropriate to remedy a significant risk of material health impairment." The Supreme Court of the United States has held that this means there must be a finding by the Secretary that a workplace is unsafe; that is, significant risk(s) are present that can be eliminated or reduced by a change in practices. Part of the regulatory process includes public review and comment as well as considerations for economic and technological feasibility. Because the regulatory process is a unique blend of science and politics, it is helpful for occupational safety and health professionals to understand how OSHA regulations move from a proposed standard to a final rule that is enforceable as law. This process involves a number of steps:

— Proposal of the new standard;

— Receipt and evaluation of recommendations from advisory committees and input from other interested parties by OSHA;

— A notice of intended rulemaking is published in the Federal Register;

— OSHA receives input through public comments and/or hearings;

— The final version of the standard is published in the Federal Register.

Proposing a New Standard

Although any interested person can request or petition OSHA to issue a standard for worker health and safety, it is more common for an industry group, a state or local government, or an employer or labor representative organization to propose a new standard. For example, a labor union may ask OSHA to take action to protect workers who are exposed to a hazardous chemical. OSHA may also initiate the process, as may the Secretary of Health and Human Services, or a nationally recognized standards-producing organization, such as ANSI, and also NIOSH.

Input from Advisory Committees

There are two standing advisory committees that OSHA may call upon to develop recommendations for the proposed standard. The National Advisory Committee on Occupational Safety and Health (NACOSH), which serves as advisor and makes recommendations to the Secretary of Health and Human Services, as well as the Secretary of Labor. The Advisory Committee on Construction Safety and Health advises the Secretary of Labor primarily on construction safety and health standards. If necessary, other ad hoc committees are formed to assist in development of recommendations.

NIOSH, established to serve as the research branch of OSHA, has many opportunities through its projects to make workplace investigations, gather data regarding exposures to potentially hazardous materials, and collect statistics on the incidence of occupational illness. This data forms the basis for NIOSH recommendations to OSHA regarding workplace hazards. NIOSH recommendations are forwarded to the Secretary of Labor and also published and made available to the public. They are commonly referred to as "criteria documents," and are usually published with the subtitle "Criteria for a Recommended Standard." The NIOSH criteria document for a particular hazardous agent contains other information that is useful to the industrial hygienist, such as descriptions of the processes that were observed as part of the study, as well as detailed recommendations for controls, exposure monitoring, medical surveillance, training, and use of personal protective equipment (PPE). NIOSH recommends levels for worker exposure to hazardous agents based on its own studies and research. The NIOSH exposure levels are called **recommended exposure limits (REL)**.

Industry groups with an interest in the standard may also provide recommendations to OSHA regarding the proposed standard. Such groups may have exposure data that support or refute a proposed exposure level, or they may have concerns about the feasibility of proposed controls.

NIOSH produces and disseminates large quantities of information relative to health and safety. Publications include the *NIOSH Pocket Guide to Chemical Hazards*, Criteria Documents, NIOSH Alerts, Current Intelligence Bulletins, Health and Safety Guides, as well as symposium and conference proceedings, scientific investigations, data compilations, and a variety of worker-related booklets and other items. Complimentary copies of some publications are available. Others may be ordered directly from NIOSH, the U.S. Government Printing Office, or the National Technical Information Service (NTIS). The best way to obtain these is to refer to the current NIOSH Publications Catalog, which contains a comprehensive listing of available materials along with ordering information. The NIOSH home page on the World Wide Web also contains information on publications; the phone number is 800 35NIOSH.

Notice of Proposed Rulemaking

After receiving recommendations, OSHA decides whether to proceed with the proposal of a new standard based on its evaluation of the data and whether it supports the conclusion that the standard is necessary and appropriate to remedy a significant risk of material health impairment. OSHA must then publish an "Advance Notice of Proposed Rulemaking" or a "Notice of Intended Rulemaking" in the Federal Register. The Advance Notice may be used to solicit information that can be used to draft the proposed standard. The Notice of Proposed Rulemaking typically contains the terms of the new rule, or a proposed text, and allows some time, usually 60-90 days, for the public to respond. Comments that OSHA receives as a result of these notices become part of the body of information used in formulating the final standard.

Public Comment

When the proposed standard is one that will affect a large group, or has major implications for a specific industry such as significant changes in work practices or a lowering of an exposure limit (with concomitant changes in required controls), OSHA may schedule a public hearing. The time and place of the hearing is published in advance in the Federal Register. OSHA may also accept written comments from interested parties who are unable to attend the hearing. The written comments and information presented during the hearing are then published in the Federal Register, along with OSHA's response and conclusions as to whether the standard should be amended based on the comments.

Revisions of Proposed Standard and Issuance of the Final Standard

After the close of the comment period, and following any public hearing that was scheduled, OSHA may make changes in the proposed standard to reflect issues that were raised. OSHA must then ensure that the full and final version of the standard is published in the Federal Register along with the date on which it becomes effective. The final version is usually accompanied by an explanation of the standard and the reasons for implementing it in its final form. This informational section typically precedes the standard; because of this it is called the **preamble**. The preamble to an OSHA regulation contains much that is often overlooked by industrial hygienists (as well as by employers and others), who may be focusing only on the regulation itself. For example, the preamble to the proposed rules for revising the PELs in the construction, maritime, and agriculture industries included historical background illustrating the need for revising the PELs. It also contained OSHA's rationale for deciding on the provisions of the proposed standards and included specific questions for which OSHA was soliciting comment. Reading the preamble to a regulation is useful because it provides an explanation and a background for a standard provision.

The process for changing (amending) or revoking a standard is similar to the process for promulgating a new one. OSHA is required to publish advance notice of such actions in the Federal Register; there is a comment period; and there may also be hearings on the proposed changes. The final version of the standard, along with its effective date, is published in the Federal Register. All changes are subject to the same appeals process as the original regulation.

It is easy to conclude that the rulemaking process can be long and complex. There is no average time for an OSHA regulation to move from initial proposal to final rule – the confined space entry regulation took more than twelve years to wind its way through the system!

Emergency Temporary Standards – If OSHA determines that workers are in grave danger due to exposure to a toxic substance or agent or a physical hazard, it may establish an emergency temporary standard. The text of the emergency temporary standard is published in the Federal Register where it also serves as a proposed rule. It is then subject to the same process of review and comment, except that the final ruling is supposed to be made within six months. The validity of an emergency temporary standard may be challenged in a U.S. Court of Appeals. These types of standards are rarely enacted by OSHA, but the inclusion of the ability to enact and enforce them was considered a necessary part of the OSHAct.

Appealing a Standard – Anyone who is adversely affected by a final or temporary emergency standard may file an appeal, which is a petition for judicial review of the standard. An adverse effect might be a major change in processes or methods, requiring many resources and taking time to complete. The appeal must be filed with the U. S. Court of Appeals for the circuit in which the petitioner lives or has the principal place of business, within 60 days of the standard's promulgation. Filing an appeal does not delay enforcement unless the Court of Appeals specifically orders it.

Variances

A variance is an alternative to an OSHA requirement that ensures that the employer's workplace is as safe as it would be if the employer did comply with the OSHA requirement. There are three main types of variances: temporary, permanent, and experimental.

A **temporary variance** is usually requested by an employer who cannot comply with a standard by its effective date, due to unavailability of technical/professional personnel, materials, or equipment, or because major construction or changes in facilities are necessary but will not be completed in time. When requesting a temporary variance, the employer must demonstrate to OSHA that it has a plan for coming into compliance with the standard, and that all available measures are being taken to protect employees in the meantime. The employer must also identify the particular standard or provision from which the variance is being requested, and explain why compliance cannot be achieved. Also required is a description of the steps that will be taken to achieve compliance and the dates by which they will be accomplished. Employees must be informed about the application for variance and a copy of the application must be provided to the bargaining unit or union, if one exists. Employees also have the right to request a hearing on the application for variance. A summary of the variance application must be posted in a location where other notices are normally placed.

A temporary variance is granted for the time needed to achieve compliance or one year, whichever is shorter. The employer may request an extension or renewal; these are granted in six-month intervals and are limited to two renewals. Generally speaking, OSHA will not grant a temporary variance to an employer who cannot afford to come into compliance.

A **permanent variance** is the use of an alternative which replaces the OSHA requirement. When requesting a permanent variance, the employer must prove to OSHA that the alternative method, work practice, and/or process operation provides a safe and healthful workplace as effectively as would full compliance with the standard. Again, employees must be informed about the application for variance and their right to request a hearing to address their concerns. OSHA will review the application and any supporting evidence, make a site inspection, and hold a hearing if appropriate. If OSHA approves the variance request, it will issue a written statement detailing the specific exemptions of the standard that have been granted, and describing the employer's responsibilities for maintaining a level of protection that is at least as effective as meeting the standard. Within six months following issuance of a permanent variance, the employer or the employees may request the variance be changed or revoked. OSHA may also initiate these actions.

An **experimental variance** may be granted by OSHA to an employer who is participating in an approved experiment to demonstrate or validate new safety and health techniques. Approvals of such experiments are issued by the Secretary of Labor or the Secretary of Health and Human Services. The variance is granted to permit the experiment to proceed but terminates with the study completion unless another type of variance is applied for and issued by OSHA. These types of variances are rare, but the agency does have the ability to use this tool to protect workers who are in an unusual or unique situation.

Other OSHA Orders

OSHA may issue an **interim order** to allow an employer to continue operations under existing conditions while an application for a variance is being considered. The employer must apply to OSHA for such an order and include in the application the reasons justifying the granting of such an order. The request may be included as part of an application for variance. If OSHA issues an interim order, its terms and conditions are published in the Federal Register. As with other requests for exceptions to a standard's requirements, employees must be informed by posting and by giving a copy of the order to the bargaining unit representative.

From time to time OSHA may also grant other variances, for example in the interest of national defense. However, granting of a variance does not give an employer *ex post facto* protection – it is not retroactive! Also, employers are not to use the variance process as a means for settling a citation for violation of standards requirements, although they may apply for a variance while they have outstanding citations.

OSHA regulations exist for construction and general industry and can be generally described in terms of the approach that is taken with regard to specifying means for compliance. One approach is to describe precise requirements and detailed directions for compliance; an example is the regulation

Box 3-1 ■ Code of Federal Regulations – Organizational Structure

OSHA regulations for general industry are found in Title 29 of the Code of Federal Regulations (CFR). Title 29 is the section of federal regulations dealing with labor; other Titles apply to other areas; for example, Title 40 contains the regulations enforced by the Environmental Protection Agency, while Title 49 contains the Department of Transportation regulations. Part 1910 of Title 29 contains the occupational safety and health standards for general industry; Part 1926 contains health and safety regulations for construction. Within each Part are sections or subparts that address specific topics; specific regulations pertinent to each subpart are included and assigned a number in sequence for ease of reference. For example, the regulation for occupational exposure to asbestos in general industry is located in 29 CFR 1910, Subpart Z; the regulation's numerical identifier is 1910.1001.

Each regulation is organized in a more or less similar fashion. Main subject or topic areas are denoted by small letters in parentheses, (a), (b), (c); further division into specific points or topics within these are assigned consecutive numbers, again in parentheses, (1), (2), (3); the next subdivision is designated using a lower-case roman numeral, (i), (ii), (iii); and further subdivisions are assigned capital letters, (A), (B), (C), always in parentheses.

Citing a specific regulatory requirement entails identifying its location within the regulation using the sequence of letters and numbers that corresponds to that location. For example, the requirement to include the name, social security number and the measured exposure of a worker that is exposed to asbestos during removal of asbestos-containing materials is found in 29 CFR 1926.1101 (n) (2) (ii) (F).

See Appendix 1 and 2 for a list of the contents of each Part (current as of the date of publication of this book).

---

for scaffolding found in Subpart L of 29 CFR 1926, which specifies, among other things, minimum widths for planks used for building scaffolding platforms. There are similarly detailed regulations for hazardous materials; an example is the regulation for coke oven emissions, 29 CFR 1910.1029.

Other regulations take a less detailed approach; these state clearly the required end-points of the regulation, but leave room for employers to determine the best way to reach a state of compliance; an example would be the hazard communication regulation, 29 CFR 1910.1200. This regulation tells the employer what is required for a hazard communication program, but does not tell the employer specifically how to do it; any program that meets the requirements would be acceptable. Another good example is the benzene standard, 29 CFR 1910.1028. In large measure, these different approaches for legislating worker protection are a reflection of the maturity of OSHA, its relationship with industry at the time the legislation was passed, and the process that is followed for producing the regulations. In the years immediately following passage of the OSHAct, there was a desire on the part of industry for OSHA to define requirements in a fairly detailed

manner, leaving little room for interpretation and therefore making it easy for employers to know what to do to comply with the regulations. Another factor is the subject matter being regulated. Fall protection devices, for example, are easier to describe in terms of use and performance specifications than are the specific actions that might be needed in a manufacturing setting for control of lead fumes. Recent OSHA regulations combine these two approaches, including specifics where necessary and appropriate, but specifying a desired level of performance in other areas, leaving open the means by which employers may comply.

Checking Your Understanding

1. Name and describe the three types of variances.

2. Under what conditions does OSHA schedule a public hearing when proposing a new standard?

3. What factors can make proposing and enacting an OSHA regulation a long and complex process?

3-3 Exposure Limits

In this section, we will take a closer look at the terms that are used by occupational health professionals to describe worker exposure to levels of hazardous materials and physical agents. Like other areas of industrial hygiene, many of these terms have an acronym that, in most cases, is a more convenient means of making reference. However, before applying these terms it is important that the industrial hygiene professional understands the definition and application of each, especially when the terms are used to describe worker exposures. We will describe each term individually. A table that summarizes the important types of exposure limits in use by occupational health professionals in the United States is included at the end of this section.

Exposure limits are usually thought of as air quality values that are applied to work environments. They represent an air concentration below which health hazards are unlikely to occur among most exposed workers. These limits are based on scientific studies involving animal, and sometimes human, exposures. Industrial hygienists are continually challenged by the introduction of new substances – many of which are without exposure limits – into the industrial environment. Exposure limits also exist for hazardous physical agents such as noise, electromagnetic fields, and ionizing radiation; these limits are expressed in terms appropriate to the agent involved and represent a level of exposure that is believed to be protective of most workers who are exposed. These limits are addressed in later chapters dealing with these specific hazardous agents. For purposes of this chapter, we will focus on the chemical exposure limits.

Sources of Exposure Limits

As mentioned earlier in the chapter, the original sources of the chemical exposure limits that were adopted by OSHA were many. Several of the groups continue to publish exposure limits. OSHA is probably the most familiar source of such limits in the United States; other limits come from NIOSH, the American National Standards Institute (ANSI), American Society for Testing and Materials (ASTM), the ACGIH, and the American Industrial Hygiene Association (AIHA). While these standards and limits often agree, they may be different for a specific chemical or hazardous agent. Some of the limits have been developed for application to a specific situation, such as an emergency response to a hazardous material release, while others represent recommendations for exposure to materials that are not currently regulated by OSHA. The OSHA limits are the only ones that are enforceable as law.

Exposure Limit Terms

Exposure limits are typically expressed as a **time-weighted average (TWA)**. This means that the measured levels are averaged over the time period during which the sampling took place. The term TWA may also be applied to an employee's measured exposure to a hazardous agent or chemical. In order to be representative of an 8-hour work day, a sampling time of seven hours is the minimum amount of time that is acceptable, and OSHA inspectors are instructed to sample for at least this amount of time when evaluating worker exposures against allowable regulatory levels. The lead standard, for example, requires employers to obtain seven-hour samples – see 29 CFR 1910.1025 (d)(1)(ii). Exposure limits may be based on 8-hour TWAs, 15-minute TWAs, and even five-minute or instantaneous limits. Computation of an 8-hour TWA is by the following equation:

$$8\text{-hr. TWA} = \frac{(C_x T_x) + \dots (C_n T_n)}{8}$$

Where

C_x = the concentration measured during time interval T_x;

n = the total number of intervals measured; and

8 = the total time of exposure, in hours.

Other TWAs are figured in a similar fashion, using the appropriate time interval and total time of exposure. For example, a ten-hour TWA would use the number 10 in the denominator of the above equation. A 15-minute **STEL, short-term expo-**

Box 3-2 ■ Calculating a TWA – An Example

Air samples for a worker's exposure to oil mist were found to indicate the following levels:

	Measured Amount	Time of Sample
Sample 1	3 mg/m^3	2 hours
Sample 2	8 mg/m^3	6 hours

Using the equation:

$$\text{8-hr. TWA} = \frac{(C_x T_x) + \dots (C_n T_n)}{8}$$

Substituting the sample values:

$$\text{8-hr. TWA} = \frac{(3 \text{ mg/m}^3 \times 2 \text{ hours}) + (8 \text{ mg/m}^3 \times 6 \text{ hours})}{8 \text{ hours}}$$

Computing the values:

$$\text{8-hr. TWA} = \frac{6 + 48}{8}$$

The 8-hour TWA = 6.75 mg/m^3

This value exceeds the OSHA PEL for oil mist, which is 5 mg/m^3.

sure limit, would likely have only one 15-minute sampling interval, and the denominator would be 15 minutes. It is essential that the same units be used to express the sampling interval and total time of exposure. Concentration units may be expressed as parts per million (ppm) for gases and vapors, or milligrams per cubic meter (mg/m^3) for solids such as fumes, dusts, and mists. Exposure limits for physical agents are not expressed as a concentration in air. For most physical agents, the exposure limit is a time value related to the level of exposure. For example, for noise, the exposure limit is given in hours or minutes, and varies with the sound level; the higher the level, the lower the amount of time workers may be exposed. Other physical agents have limits that are expressed in units appropriate to the agent; for instance heat stress TLV values are temperature-based. The ACGIH TLV booklet contains a section for physical agents of general interest to industrial hygienists, including radiation, vibration, laser energy, electromagnetic fields, and others. TLVs and PELs for physical agents are addressed in later chapters dealing with the specific agent(s).

The OSHA PELs are typically stated as 8-hour TWAs; this means that the concentration is averaged over the work shift, generally an 8-hour interval. In setting a PEL or exposure standard, the OSHAct requires that it be one that "most adequately assures, to the extent feasible, on the basis of the best available evidence, that no employee will suffer ma-terial impairment of health of functional capacity even if such employee has regular exposure to the hazard dealt with by such standard for the period of his working life." Exposure to any material regulated by OSHA may not exceed the PEL during any 8-hour work shift, over any 40-hour workweek.

For some regulated materials, a short-term exposure limit (STEL) and/or a ceiling (C) limit may be listed. OSHA defines a STEL as an employee's 15-minute TWA exposure, which shall not be exceeded at any time during a workday, unless another time limit is specified in a notation below the limit, in which case the TWA exposure over the specified time period shall not be exceeded at any time during the working day. In order to prove compliance with a STEL, the employer has to obtain a sample over a 15-minute interval, during which a level approaching the STEL might be reasonably expected.

A ceiling limit, according to OSHA, is one that shall not be exceeded during any part of the working day. If instantaneous monitoring is not feasible to determine compliance with the C value, OSHA allows use of a 15-minute TWA sample to assess whether or not the C limit has been exceeded.

There is another OSHA exposure limit that is referred to as a peak limit. This term is used to refer to a level that exceeds the ceiling limit, but that OSHA allows for a specific limited time during the work shift. These peak levels must never be exceeded and must be compensated for by periods of exposure during which the concentration is low enough so that the employee's cumulative 8-hour TWA is below the PEL. These peak limits are listed in Table Z-2 of Subpart Z of CFR 1910 under the column "Acceptable maximum peak above the acceptable ceiling concentration for the 8-hour shift."

OSHA also implements a **mixture rule**. The definition for this rule can be found in 29 CFR 1910.1000 (d)(2)(i). The purpose of the rule is to provide protection for workers who are exposed to two or more substances that affect the same organ or cause the same disease or impairment. This type of interaction is called **synergism**. For example, carbon tetrachloride and acrylonitrile both can cause damage to the liver. Under the mixture rule, workers exposed simultaneously to such chemicals cannot be exposed to the maximum allowable limit for each; and workplace levels must remain at a fraction of the PEL so that the sum of the exposure fractions does not exceed unity, or 1. The mixture rule can be illustrated mathematically by the following equation:

$$PEL_{mixture} = \frac{C_1}{PEL_1} + \frac{C_2}{PEL_2} + \dots \frac{C_n}{PEL_3}$$

Where

C = the measured concentration of a particular contaminant; and

PEL = the OSHA PEL for the contaminant

The value of $PEL_{mixture}$ must not exceed 1.

The OSHA PELs are found in Tables Z-1 and Z-2 of Subpart Z of CFR1910; others are included in substance-specific standards that address a single hazardous material, for example benzene, lead, and acrylonitrile. Besides specifying the allowable levels of worker exposure, substance-specific regulations typically contain a series of requirements for ensuring worker protection against the particular agent. These include:

—Employee training on hazards associated with exposure to the agent.

—Establishment of regulated areas to limit the number of employees who are exposed.

—Use of engineering or other control measures to reduce or control the level of hazardous agent present in the work environment.

OSHA attempted to update the PELs in January 1989 with passage of the Air Contaminants Final Rule. The Rule adopted 164 new PELs and established 212 PEL values that were more protective than the previous levels. For some 160 PELs, the updated values represented no change. A large portion of the input into the updated PELs was from the ACGIH, specifically information contained in the 1987-88 TLV booklet. (One of the criteria that OSHA considered in determining which PELs needed updating was whether or not they were significantly different from the 1987-88 TLVs.) The new OSHA regulation was soon challenged in the Eleventh District Court of Appeals. The cases, brought by the AFL-CIO, charged that some of the proposed PELs were too low, and that others were too high. There were also challenges on some of the more general aspects of the proposed updates to the PELs. The revised PELs were remanded by the Court; OSHA must reevaluate its position on each PEL and demonstrate that the proposed PELs are consistent with OSHA's role in identifying and eliminating significant risk(s) that are present in the workplace. Although they were remanded, there is a possibility that OSHA can enforce the proposed PELs under the General Duty Clause of the original OSHAct. Many employers have taken the approach that they will voluntarily comply with the proposed PELs.

—Use of PPE (personal protective equipment, for example, respirators or hearing protection).

—Ongoing medical surveillance to monitor the health of exposed workers.

—Periodic measurement of worker exposure using accepted industrial hygiene methods and equipment.

—Provisions for maintaining exposure and medical surveillance records.

—Written programs, policies, and procedures that are followed to ensure compliance.

A term that is often seen in substance-specific regulations is the **action level**. This term is used to describe airborne concentrations that trigger certain provisions of a regulation. Generally, but not always, the action level is one-half or 50 percent of the PEL value; and it is usually an 8-hour TWA. For example, the PEL for vinyl chloride is one ppm, and its action level is 0.5 ppm. By contrast, the PEL for lead

Box 3-3 ■ Applying the Mixture Rule – An Example

A process involves the use of two solvents that act synergistically on the body. Air sampling indicates that the individual PELs for each solvent are not exceeded. However, does the exposure exceed the allowable level under the OSHA mixture rule?

The following information is obtained by the industrial hygienist:

Solvent	OSHA PEL	8-hr. Measured Exposure
Ethyl acetate	400 ppm	250 ppm
Benzyl chloride	1 ppm	0.75 ppm

Using the equation:

$$PEL_{mixture} = \frac{C_1}{PEL_1} + \frac{C_2}{PEL_2} + \dots \frac{C_n}{PEL_n}$$

Substituting values for the solvents:

$$PEL_{mixture} = \frac{250}{400} + \frac{0.75}{1.0}$$

Solving:

$$PEL_{mixture} = 0.625 + 0.75$$

The value:

$$PEL_{mixture} = 1.375$$

This worker's exposure exceeds 1.0, and therefore the OSHA PEL for the mixture of these two materials is also exceeded.

---

is 50 $\mu g/m^3$, but the action level for lead is 30 $\mu g/m^3$, not 25 $\mu g/m^3$. According to OSHA, the 8-hour TWA must never exceed the PEL; if employee exposures reach the action level, there are specific steps that the employer must take. For example, in the lead standard OSHA mandates a medical surveillance program for all employees who may be exposed to lead at or above the action level for more than 30 days per year. Some regulations require additional measures such as the use of administrative controls to limit the number of exposed employees, training, hygiene programs, periodic monitoring, and use of PPE. Other OSHA regulations that must be followed may be cross-referenced in the standard; for example, if respirators are required, there will be a reference to the OSHA regulation for respiratory protection, 29 CFR 1910.134.

Another common term used by OSHA is the **excursion limit**. This is a limit that is analogous to the peak limit. It is a time-weighted average, the length of which is specified by OSHA, that cannot be exceeded during the workday. As with the peak and ceiling limits, the employee's total 8-hour time-weighted average exposure may not exceed the PEL. For example, in the case of asbestos, OSHA has specified an excursion limit of 1.0 fiber per cubic centimeter (f/cc), averaged over a sampling period of 30 minutes. To determine whether exposures are in compliance with this limit, the employer has to obtain samples over 30-minute intervals that are

OSHA Term	Use/Application
PEL	Eight-hour TWA exposure limit
STEL	15-minute TWA exposure limit
Excursion limit	30-minute TWA exposure limit
Action level	Some fraction (usually 50 percent) of the PEL for a substance. Exposures at this level trigger some additional action on the part of the employer, such as worker training, use of PPE, and medical surveillance.
Ceiling limit	Level of exposure that cannot be exceeded at any time during the work shift.
Peak limit	Level of exposure above the ceiling level that is allowed to occur one time for a short interval. Usually applied to situations where a necessary process fluctuation occurs, for example where a ventilation system must be shut down for a moment to accommodate a specific task or action.

Table 3-1: This table summarizes the many terms used by OSHA to describe allowable levels of exposure to hazardous chemicals. There are terms for instantaneous, short-term, and full-shift exposures.

representative of the employee's highest exposures of the workday, and the employee's 8-hour TWA has to be at or below the PEL of 0.1 f/cc. Table 3-1 summarizes the terms that OSHA uses to describe exposure limits in various regulatory standards.

ACGIH TLVs

OSHA has established PELs for many chemical substances (e.g., asbestos, lead, and vinyl chloride) and hazardous agents (noise is an example); as stated earlier, a large number of these are in fact former ACGIH TLV values. The ACGIH TLVs are not, however, equivalent to the OSHA PELs. Besides the PELs being regulatory levels that are enforceable as law, the PELs and TLVs are defined very differently. According to the 1996 ACGIH publication "Threshold Limit Values for Chemical Substances and Physical Agents" (often referred to as "the TLV booklet" or simply "TLVs"), the TLVs refer to airborne concentrations of substances, and represent conditions under which it is believed that nearly all workers may be repeatedly exposed, day after day, without adverse health effects. A caution is included: due to the variation of individual susceptibility, there may be a small percentage or workers who experience discomfort when exposed at levels below the TLV. Others may suffer due to aggravation of a pre-existing condition, or by development of an occupational illness, even if exposed at or below the threshold value. And still others may be hyper-susceptible, or extremely sensitive, due to genetic factors, personal habits (such as smoking), age, physical condition, or as a result of previous exposures. Clearly, the TLVs are not intended to describe safe levels of exposure for all workers!

The ACGIH publishes a companion to the TLVs called *Documentation of the Threshold Limit Values and Biological Exposure Indices (BEIs)*. The *Documentation* contains the scientific studies and data that were used in deciding each TLV and BEI; this is a valuable source of information to the practicing industrial hygienist, since it contains a summary of the considerations that were used in determining each limit, and provides references to the literature and other sources used in the determination. It should be noted that each TLV is based on the information that is available at the time; this information comes from industrial experience, experimental studies, both animal and human, and sometimes from a combination of these. Because of this, the amount and nature of the information that is used varies from substance to substance, over time. This means that there is a certain amount of uncertainty inherent in each TLV.

Another caveat when applying TLVs is that the user be aware of the underlying reason for setting each value. Some TLVs are set to prevent disease or health impairment; others are set to minimize the effects of irritation, narcosis, or nuisance effects. The toxic effects of the substance generally influence the setting of the TLV. However, this process can become complex when a substance produces multiple effects, synergistic effects, or if there are other concerns, such as carcinogenesis or reproductive effects, for which little toxicological data are available.

The TLVs are recommendations and should be used as guidelines for good practice, not relied upon as fine lines between safe and unsafe concentrations. They are not intended to be used for evaluation of community air pollution; evaluation of long-term, uninterrupted exposures; as proof (or disproof) of an existing disease or physical condition; or for use in situations where working conditions, substances, and processes are different from those in the United States. In short, the TLVs are intended solely for use by trained industrial hygienists.

> The definition of a TWA and the procedure for time-weighting an exposure that is compared to an ACGIH value are the same as that used for determining compliance with an OSHA PEL.

It is now appropriate to take a look at the categories of TLVs that are specified by the ACGIH. The first is the **threshold limit value–time weighted average**, or **TLV®–TWA**. This is the concentration for a normal 8-hour workday, 40-hour week, to which it is believed nearly all workers may be repeatedly exposed, day after day, with no adverse effect.

The second is the short-term exposure limit, or STEL. This represents the concentration to which workers can be exposed continuously for a short period of time without suffering from adverse effects such as irritation; chronic or irreversible tissue damage; or narcosis increasing the likelihood of accidental injury, impairing a worker's self-rescue capability, or materially reducing work efficiency. It is assumed that the 8-hour TWA exposure of workers remains at or below the levels of the TLV-TWA.

Agency	Eight-hour TWA Term	Short-term Exposures	Instantaneous Exposures	Enforceable as Regulation
OSHA	PEL	excursion limit or STEL	C (ceiling), peak ceiling	Yes
NIOSH	REL*	STEL	C (ceiling)	No
ACGIH	TLV	STEL	C (ceiling)	No
* The NIOSH RELs are usually for a 10-hour TWA.				

Table 3-2: OSHA, NIOSH, and the ACGIH issue exposure limits, although only those issued by OSHA are enforceable by regulation. This table compares the terms used by each to describe and refer to the various limits. When arranged in this manner, the similarities and differences between each group's terms are more obvious.

The STEL is not a separate exposure limit that is independent of the TLV-TWA; rather, the STEL is a 15-minute TWA exposure that should not be exceeded at any time during a workday, even if workplace concentrations are at or below the TLV-TWA. This means that exposures above the TLV-TWA are limited to the level defined by the STEL. These exposures should not be longer than 15 minutes and should not occur more than four times each day; there should also be at least 60 minutes between each successive exposure in the STEL range. In some instances, an averaging period other than the 15-minute one is used; if this is the case, it should be specified. All substances do not necessarily have a STEL, since it may not be appropriate for some. In general, STELs are recommended only for substances where toxic effects have been reported from high-level, short-term exposures.

The third threshold level is the **TLV-ceiling**, or **TLV-C**. This concentration should not be exceeded during any part of the workday. Evaluation of exposure relative to this TLV requires instantaneous sampling; however, in many instances this is not possible and a longer sampling period is required. For some substances, for example highly irritating ones, the TLV-C is the most relevant of the three values.

Special Terms Used in Exposure Limits

Skin Notation

NIOSH, OSHA, and the ACGIH utilize what is called a **skin notation** in their exposure limits; this is when the word "skin" literally is included as part of the exposure limit entry. For these substances, absorption through the skin is considered to be a significant route for the material to enter the body. The exposure limits that are listed for these substances are airborne concentrations; it would be difficult to sample the material as it passes through the skin! Substances that have a skin notation require the industrial hygienist to consider not only the measured airborne concentration, but the entire process, including work habits, use of PPE, effectiveness of engineering controls, and other factors, as part of evaluating the hazards associated with exposure. For substances with a skin notation, contact with the skin should be prevented through the use of protective gloves, aprons and impermeable clothing; goggles; good work habits; and other appropriate measures.

Biological Exposure Indices (BEI)

Included in the ACGIH TLV booklet is a section called **biological exposure indices (BEI®)**. These can be useful in assessing the overall amount of a chemical that is absorbed into the body of an exposed individual. Biological specimens are collected from exposed individuals and then analyzed for a specific chemical that is an indicator of the amount of the substance of interest that has entered the body, either through inhalation, skin absorption, or some other route of exposure. The indicator chemical, or **determinant**, can be the chemical itself, a metabolic product of the chemical, or a change in the body's chemistry that is induced by the chemical. The determinant is measured in biological samples –

exhaled air, blood, urine, or other specimens, all collected from exposed workers. Measurement of the determinant can indicate the intensity of a recent exposure, an average daily exposure, or a chronic, cumulative exposure, depending on the specimen, time of sampling, and the determinant measured. Like the TLVs, BEIs are reference values intended as guidelines, not definite indicators of safety. They are not intended for use in measurement of adverse effects or to diagnose an occupational illness.

Biological monitoring data is inherently variable and subject to influence by many factors. The health status of the individual plays a role, as does age, gender, and pregnancy status, as well as any chronic conditions or medications. Work conditions, including intensity of work load, the temperature and the humidity of the work area, and fluctuations in levels of airborne contaminants will affect the measurements. Other possible sources of variation include exposures outside of work, community air pollutants, and lifestyle habits. The process of obtaining the samples also may contribute to variations through sample contamination, improper handling and storage, and the analytical method used.

The existence of a BEI is not an indication that one must do biological monitoring. However, BEIs can be useful in determining the significance of skin absorption, determining whether PPE is effective, or detecting a nonoccupational exposure. Air monitoring should precede and accompany biological monitoring. Like the TLVs, the basis for setting a BEI should always be evaluated as part of the monitoring process; BEIs are included in the ACGIH publication *Documentation of TLVs and BEIs*. The current ACGIH TLV booklet should be consulted for additional information.

Immediately Dangerous to Life and Health

There are some materials for which an **IDLH (immediately dangerous to life and health)** level has been established. According to the NIOSH Respirator Decision Logic, an IDLH condition is one that poses a threat of exposure to airborne contaminants that is likely to cause death, immediate or delayed permanent adverse health effects, or prevent escape. The logic used in setting these levels was that the IDLH was the maximum concentration from which the worker could escape without loss of life or irreversible health effects. For some substances, severe

effects such as eye and respiratory tract irritation, incoordination, and disorientation were also considered as these could affect a person's ability to escape. The IDLH values were originally based on the effects of a 30-minute exposure to provide a safety margin, not to indicate that a person should – or could – stay in the area for that amount of time. In an IDLH environment, immediate evacuation is recommended.

IDLH does not apply only to airborne levels of contamination. Atmospheres containing less than 19.5 percent oxygen are considered by OSHA to be immediately dangerous to life and health. Firefighting, emergency response at a hazardous material release, confined spaces, explosive atmospheres, and entry into uncharacterized or unknown atmospheres, may also be considered to be IDLH. Only a self-contained breathing apparatus operated in pressure-demand mode should be worn in such situations. As with other exposure limits, not all substances have an assigned IDLH value.

Carcinogens

Carcinogens are substances known to cause cancer or suspected of causing it in exposed humans (see Chapter 2 for more information about carcinogens). The notation **Ca** is used by NIOSH to indicate that a substance is considered a potential occupational carcinogen. While NIOSH may list recommended exposure limits (RELs) for some carcinogens, it does not do it for all of them. For the ones without RELs, NIOSH recommends that any exposure be avoided. OSHA generally addresses carcinogens through substance-specific regulations that may or may not include a PEL. An example is 4-nitrobiphenyl, which is a confirmed human carcinogen. The OSHA standard that regulates its use in manufacturing, processing, and handling is found at 29 CFR 1910.1003. No PEL or action level is given; however, the standard does require the provision of a clean change room for employees, defined as an environment that is "free from 4-nitrobiphenyl". This implies that any detectable level is not acceptable.

The ACGIH uses a classification system that involves five categories of carcinogenicity. This system has been in effect since 1992 and replaces an earlier, similar classification system. The categories for carcinogenicity as listed in the 1996 TLV booklet are as follows:

A1   This designation indicates an agent that is a confirmed human carcinogen. The classification is assigned to substances for which there is evidence from epidemiological studies or convincing clinical evidence in humans.

A2   This designation is for suspected human carcinogens. These agents are carcinogenic in experimental animals at dose levels; by routes of administration; sites; histological types; or by mechanisms that are considered relevant to worker exposure. The epidemiological data available are often conflicting or insufficient to confirm an increased risk of cancer in exposed humans.

A3   These are animal carcinogens. These agents cause cancer in experimental animals at relatively high doses, or by routes of administration, at sites, of histological types, or by mechanisms that are not considered relevant to worker exposure. Available epidemiological data do not confirm an increased risk of cancer in exposed humans, and evidence suggests that the agent is not likely to cause cancer in humans, except under uncommon or unlikely routes, or levels, of exposure.

A4   This designation indicates the agent is not classifiable as a human carcinogen, based on inadequate data for humans and animals.

A5   This designation is for agents that are not suspected as human carcinogens. The determination is based on properly conducted epidemiological studies in humans. This means that 1) the studies have gone on long enough to allow for any latency period associated with the cancer, 2) data have been collected in a reliable and scientific manner, and 3) the studies have adequate statistical power to conclude that exposure does not convey a significant risk of cancer. Animal studies may be used if they are supported by other relevant data.

The above categories are subject to change as more data become available relative to an agent's carcinogenicity. The designations are not assigned to substances for which there are no reported carcinogenicity data; therefore the lack of designation indicates lack of data, not necessarily a lack of carcinogenic potential. The ACGIH recommends that for A1 agents without a TLV, workers should be equipped to eliminate any potential for exposure to the fullest possible extent. For A1 agents with a TLV, as well as A2 and A3 agents, worker exposure by all routes should be carefully controlled and kept as low as possible below the TLV. The ACGIH publication, *Documentation of Threshold Limit Values*, contains additional information in the section "Guidelines for the classification of occupational carcinogens."

Checking Your Understanding

1.   Explain the main differences between OSHA PELs and ACGIH TLVs.

2.   What is the mixture rule?

3.   Differentiate among the concepts of ceiling limit, peak limit, and excursion limit?

4.   How are carcinogens indicated by NIOSH?

5.   What does a skin notation in an exposure limit indicate?

6.   What is an action level?

7.   Explain the use of a BEI in evaluation of a worker exposure.

3-4 OSHA Standards as Part of the Industrial Hygiene Program

Exposure limits represent a small part of the regulatory picture. There are entire OSHA regulations that need to be incorporated into almost every industrial hygiene program. These include regulations for recording and reporting occupational injuries and illnesses; employee exposure and monitoring records; and written compliance plans that address specific OSHA regulatory requirements. Some standards that are almost universally included as part of an organization's overall safety or industrial hygiene program are hazard communication, respiratory protection, hazardous waste operations and emergency response, confined space entry, and noise, along with any substance-specific regulations that apply to operations within the plant or location. A brief look at the requirements of some of the more generally applicable regulations follows.

Hazard Communication

The hazard communication standard, 29 CFR 1910.1200, has also been called the "right-to-know" standard because it mandates that employees be informed about the hazardous materials they are exposed to in their place of employment. The standard requires that employers take a number of actions, including:

— Identify and maintain a list of all hazardous materials that are present in the workplace.

— Maintain a material safety data sheet, or MSDS, (furnished by the supplier, manufacturer, or distributor of each hazardous material) in a location that is known and accessible to employees during their work shift.

— Train employees on the safe use, handling, and storage of the hazardous materials they work with. Employees must be informed about the health hazards associated with exposure, and the steps they can take to protect themselves. Additional training must be provided prior to introduction of new hazards.

— Ensure the ready identification of hazardous materials inside containers, process piping, and other vessels. This can be through the use of labels, process sheets, or some other mechanism that is understandable to employees.

— Inform employees about the regulation and its contents.

— Prepare a written program that describes how the requirements of the standard will be achieved.

Respiratory Protection

The respiratory protection standard is found at 29 CFR 1910.134. This standard outlines what OSHA considers to be a minimally acceptable program. Such a program addresses:

— Assignment of responsibility for the program's overall implementation, review, and maintenance. Typically the person who is chosen to act as the program administrator is a health and safety professional or a person with adequate training in the subject matter.

— Written standard operating procedures that address selection, use, and care of respirators.

— A medical surveillance program to evaluate employee's physical ability to work while wearing a respirator. Periodic re-evaluations are required.

— Employee training on the use, care, and limitations of the respirator.

— Fit testing according to an accepted protocol; fit tests may be qualitative or quantitative, depending on contaminant and other applicable requirements.

— Procedures to be used for cleaning, storing, maintaining, and inspecting respirators.

— Provisions for periodic monitoring of contaminant levels to ensure that the respirators provide an adequate level of protection.

— Periodic review of the program to evaluate its effectiveness. This includes regular inspections of respirators under actual use conditions to assure that procedures are being followed, as well as updating the program as conditions or regulations change.

Hazardous Waste Operations and Emergency Response (HAZWOPER)

Hazardous waste operations and emergency response activities are covered by the OSHA standard found at 29 CFR 1910.120 (1926.69 for construction activities). This standard, often referred to by its acronym, HAZWOPER, specifies minimum requirements deemed necessary by OSHA to ensure the health and safety of personnel working to characterize or clean up sites where hazardous materials have been either accidentally released or dumped (legally or illegally), or where they are treated, stored, or disposed of. Paragraph (p) contains provisions that apply to workers at treatment, storage, and disposal (TSD) facilities regulated by EPA under the Resource Conservation and Recovery Act (RCRA). Hazardous materials responders, or hazmat teams, are covered by paragraph (q). While there are some differences between requirements, there are several required program elements that are common to all three:

— employee training commensurate with assigned responsibilities;

— a medical surveillance program;

— selection and use of PPE;

— emergency and spill response; hazard analysis of work sites;

— air monitoring;

— decontamination; and

— an organizational structure, identifying the chain of command and specifying responsibilities of supervision and employees.

For waste site characterization and cleanup, the written safety and health program must name a site safety and health supervisor, who has responsibility for development and implementation of, and the authority to ensure compliance with, the safety and health program. A written PPE program is stipulated, which must cover proper use, fitting, and limitations of the protective gear. Also included is a requirement to establish and maintain control over access to the site. Employers are required to perform air monitoring for toxic materials to assure that engineering controls and PPE in use are adequate to prevent employees from being exposed in excess of PELs, or other published exposure levels, such as NIOSH RELs or ACGIH TLVs, for substances that have no PEL. Safe procedures for drum and container handling are also required to be addressed in employee training programs. Employees are required to receive 40 hours of classroom training followed by 24 hours of on-the-job training under the direction of a trained supervisor. Employees whose potential exposure to site hazards is limited must receive 24 hours of training with eight hours of on-the-job training under the direction of a trained supervisor. Annual 8-hour refresher training is required. Supervisors must receive a one-time supervisor course in addition to the regular 40-hour curriculum required for workers.

For TSD workers, the standard requirements are essentially the same as those for waste site workers, except that the total hours required for training are less (24 hours) and there is more emphasis on emergency response plans.

For hazmat workers covered under paragraph (q) the standard requires training specific to the duties and function of personnel, and recognizes five levels of response personnel. They are:

1. First responder awareness level; these are individuals who are likely to witness or discover a hazardous substance release; because they initiate the emergency response sequence, they are trained to recognize a hazardous material release but do not respond beyond notification of the proper authorities.

2. First responder operations level; these individuals are trained to respond defensively, but do not attempt to stop the release. They may take action to contain the material in the area where it is being released, keep it from spreading, and maintain control over the site so that others are not exposed.

3. Hazardous materials technicians take action to stop the release. They are the ones to install a plug, patch, or take other actions to stop release of the material.

4. Hazardous materials specialists have training and a role onsite similar to that of the hazardous materials technicians; however, they also are expected to be able to use instruments and measurement techniques to evaluate known and unknown materials at the site. They may interact with state or other local authorities to coordinate site response activities with others.

5. The on-scene incident commander is the senior official on the site who is responsible for controlling site operations. Typically, the response team will be comprised of operations level responders, and hazardous materials technicians and specialists, who take direction from the incident commander. These persons will work together with local hazmat teams and others who have the skills needed for an effective containment as well as for control and cleanup.

For all personnel, HAZWOPER specifies minimum qualifications for the instructors who provide the required training. Informational appendices to the standard address PPE selection and testing, guidelines for complying with the standard, and a list of references.

Confined Spaces

A **confined space** is defined by OSHA as any space that has these characteristics: 1) large enough so that an employee can enter and perform work; 2) limited or restricted means for entry or exit; and 3) not designed for human occupancy. Most of us think of tanks, bins, silos, and similar vessels as being confined spaces. It should be noted that trenches, parking garages, and other spaces may present hazards even though they do not look like a confined space. The configuration of enclosing walls or equipment, as well as potential processes or activities that might generate a hazardous atmosphere, need to be considered when assessing an area for potential confined space hazards.

OSHA's standard for entry into confined spaces is 29 CFR 1910.146, Permit-required confined spaces. The promulgation of this standard illustrates the legislative process; OSHA's original Advance Notice of Proposed Rulemaking was published in the Federal Register in 1975, but the final version of the confined space regulation did not become

effective until 1993. For a brief summary of the rulemaking process that produced this standard, see the preamble and final rule in the January 14, 1993 Federal Register.

The standard requires that employers identify all confined spaces in their facility and evaluate these spaces to determine whether any of the spaces are permit-required confined spaces, or permit spaces. A **permit space** is defined by OSHA as 1) a confined space containing a hazardous atmosphere, 2) a confined space containing a material that could engulf an entrant, 3) a space with a configuration that could trap an entrant, or 4) a space that contains any other recognized safety or health hazard. It should be noted that the potential for an unsafe condition to develop in a space should also be considered when determining whether or not a space is a permit space. The hazard may already be present in the spaces, or it may be created by activities performed in the space by entrants. Activities such as welding, painting, or use of solvents could lead to the development of a hazardous atmosphere inside a confined space. Examples of hazardous atmospheres include lack of oxygen, a flammable or combustible mixture, or levels of toxic materials that exceed a PEL. Examples of physical hazards include unexpected startup or operation of equipment located in the space as well as release of chemicals or other solid materials into the space from supply lines or piping. OSHA considers an **entry** to be the breaking of the plane at the opening of the space by any part of the body.

The standard requires that each space be assigned a unique identifier, such as a number or specific name. Danger signs must be installed at the space entrances, or some other means of communicating the presence and potential danger of the spaces must be implemented. Permit spaces cannot be entered until certain measures have been taken and documented on an entry permit. While OSHA does not dictate the format of the permit, it does specify its content. An entry permit must contain:

—The identity (name or number) of the space to be entered.

—The purpose for the entry, i.e., the work to be performed in the space.

—The date and duration of the permit.

—The names of personnel who are authorized to enter the space.

CONFINED SPACE ENTRY PERMIT

SPACE ENTERED	PURPOSE OF ENTRY	ESTIMATED TIME FOR COMPLETION OF TASK

HAZARDS PRESENT IN SPACE	AUTHORIZED PERSONS & ROLE (ENTRANT, ATTENDANT, ENTRY SUPVSR.)

ACCEPTABLE ENTRY CONDITIONS:

Atmospheric conditions
Oxygen 19.5-23.5%*
Combustible gas/vapor <10% LEL*
Toxics (list contaminant and limit)
Physical conditions/requirements
(lockouts, isolation, ventilation, etc.)

TESTING OF ATMOSPHERIC CONDITIONS

TIME	VALUE	INIT.	TIME	VALUE	INIT.	TIME	VALUE	INIT.
____	____	____	____	____	____	____	____	____
____	____	____	____	____	____	____	____	____
____	____	____	____	____	____	____	____	____
____	____	____	____	____	____	____	____	____
____	____	____	____	____	____	____	____	____
____	____	____	____	____	____	____	____	____

Name and signature of person performing testing

RESCUE AND EMERGENCY SERVICES TO BE SUMMONED (Location and contact name)	EMERGENCY Telephone # / radio frequency:	VERIFICATION OF AVAILABILITY of emergency services BY _____ TIME:

COMMUNICATION PROCEDURE TO BE USED DURING ENTRY (list specific equipment required)	COMMUNICATIONS EQUIPMENT IN PLACE ☐ INIT.

PERSONAL PROTECTIVE EQUIPMENT REQUIRED (Specify type where appropriate)

Gloves _____
Respirator _____
Eye/face protection _____
Clothing _____
Shoes/boots _____
Hearing Protection _____
Other _____

TRAINING AND OTHER QUALIFICATION OF PERSONNEL VERIFIED BY: (name, date and time, and method of verification)

ATMOSPHERIC TESTING EQUIPMENT TO BE USED

Type   Manufacturer & Model Number   Date of Calibration

FREQUENCY OF MONITORING

☐ PRE-ENTRY
☐ PERIODIC (SPECIFY INTERVAL)
☐ CONTINUOUS

RESCUE EQUIPMENT REQUIRED (harness/retrieval system, lighting, barriers, PPE, ladders)

Harness and retrieval system
Barricades around entrance
Other (list)

CONFIRMATION THAT EQUIPMENT IS IN PLACE   INIT.
☐   _____
☐   _____
☐   _____
☐   _____

RECORD OF ENTRANT ENTRY AND EXIT TIME

Entrant Name   Initial Entry time   Exit Time / Re-Entry   Exit Time / Re-Entry   Exit Time / Re-Entry   Final Exit Time _____

COMMENTS / SPECIAL NOTES / PROBLEMS WITH ENTRY

Pre-entry briefing completed: SIGNATURES of entrants	Entry authorized by Entry Supervisor Signature Date Time	Entry CANCELED by Entry Supervisor Signature Date Time

Figure 3-1: Prior to any worker entering a permit-required confined space, OSHA requires that hazards in the space be evaluated and either controlled or eliminated. Among the requirements are that workers must be properly trained, that plans must be in place for an emergency rescue, and many others. Completion of all spaces on a confined space entry permit provides assurance that worker protection and regulatory requirements are met.

—The names of personnel serving as attendants.

—The name and signature or initials of the person serving as the supervisor. This person authorizes entry after the permit has been filled out and cancels the permit after the task is completed.

—The hazards that are in the space.

—The measures used to eliminate or control the hazards in the space.

—A description of the acceptable entry conditions; this means adequate oxygen, along with the absence of a flammable and/or toxic atmosphere. Lockout/tagout procedures might also be applicable.

—The results of atmospheric testing performed prior to entry, and periodically during work in the space. The person performing the test must be identified on the permit, as well as the time at which the testing was performed.

—Instructions for summoning rescue or emergency service personnel.

—A description of the communication methods or procedures that will be used to maintain contact between attendants and entrants.

—A list of PPE, communications, rescue, and other equipment to be used.

—Other information, as needed, to ensure the safety of employees for the particular space.

—Reference to other permits, such as a hot work permit, that have been issued for the task.

—A space for describing any problems encountered during the entry.

Permits can be in the form of a standard procedure for spaces where entries are made often and regularly, as long as the space is checked according to the standard requirements before anyone enters. Depending on the nature of the hazard in the space, supplemental forced-air ventilation, continuous atmospheric monitoring, and standby emergency rescue personnel and equipment may be required.

Persons who participate in, or supervise entry into, permit spaces are required by OSHA to be trained in specific areas of responsibility, depending on their assigned role. Supervisors, entrants, and attendants are required to be trained to perform specific tasks, as are rescue personnel. OSHA also specifies minimum qualifications for personnel conducting atmospheric testing of confined spaces.

Employers whose facilities contain permit spaces are required to prepare a written permit space program that contains the overall procedures and approach used to regulate employee entry into permit spaces. The program must address all applicable requirements of the standard outlined above, as well as provisions for coordination between multiple employers, periodic evaluation of the program, and changes or updates that become necessary due to process or facility changes. Canceled permits must be retained for a year to facilitate program review. Problems with entries noted on the permits must be addressed through revision of the program, if appropriate.

The OSHA standard includes appendices containing sample entry permit forms, a decision flow chart for assisting in evaluation of confined spaces, a discussion of atmospheric testing, and two examples of written programs that meet the regulatory requirements.

Noise

Occupational exposure to noise is addressed in 29 CFR 1910.95. This standard requires protection against the effects of noise when levels exceed those shown in Table G-16 of the regulation.

The required protection consists of feasible administrative and engineering controls; if these do not reduce noise levels below the table values, then PPE must be provided and used. Noise exposures that equal or exceed an 8-hour TWA of 85 decibels (a noise dose of 50 percent) are referred to as the action level. If employees are exposed to noise at or above the action level, the employer must take a number of actions. These include instituting a monitoring program to measure the noise exposure of employees. Employers must include the monitoring as part of a continuing, effective hearing conservation program. The monitoring must be performed so that it identifies those employees who need to be included in the hearing conservation program. The monitoring must be repeated whenever there are changes in equipment, processes or production levels that result in more employees being exposed above the action level, or if the changes cause such an increase in the noise level that the hearing protection might no longer be adequate to reduce exposures to acceptable levels. Employees who work in areas where noise levels meet or exceed the action level must be informed about the results of the monitoring.

Permissible Noise Exposures[1]	
Duration per day, hours	Sound level dB(A) slow response
8	90
6	92
4	95
3	97
2	100
1 1/2	102
1	105
1/2	110
1/4 or less	115

[1] When the daily noise exposure is composed of two or more periods of noise exposure of different levels, their combined effect should be considered, rather than the individual effect of each. If the sum of the following fractions; $C_1/T_1 + C_2/T_2 + \cdots = C_n/T_n$ exceeds unity, then, the mixed exposure should be considered to exceed the limit value. C_n indicates the total time of exposure at a specified noise level, and T_n indicates the total time of exposure permitted at that level. Exposure to impulsive or impact noise should not exceed 140 dB peak sound pressure level.

Table 3-3: Table G-16 of 29 CFR 1910.95. The OSHA PEL for occupational noise exposure is 90 decibels (dB) as an 8-hour TWA. This table shows the allowable noise exposures for periods of exposure at different noise levels.

The hearing conservation program must also include initial audiometric tests to evaluate employee hearing; this is called the baseline audiogram, and must be performed within six months of an employee's first exposure at or above the action level. Audiograms must then be repeated annually, and employees are required to be informed about the results. Employees who participate in the hearing conservation program must receive training that addresses 1) how noise exposure causes hearing loss; 2) the different hearing protectors available for use, their noise attenuation, their advantages and disadvantages as well as their proper fitting, use, and care; and 3) an explanation of the audiometric testing process.

Other requirements of the standard include calibration of the audiometric testing equipment; procedures for evaluating audiometric tests; procedures for evaluating the attenuation provided by hearing protectors; and records retention and transfer. The standard has five mandatory appendices that address

noise exposure computation, estimation of hearing protector attenuation, and audiometric testing procedures and equipment. There are also four informational appendices that provide information on age-correcting audiograms, monitoring noise levels, and references and definitions.

Checking Your Understanding

1. Name three OSHA standards that are included in most industrial hygiene programs.

2. The OSHA respiratory protection standard mandates ten subject areas for a minimally acceptable program; name five.

3. What is OSHA's definition of a confined space?

Summary

In this chapter, we introduced OSHA and described how many of the first worker health and safety standards originated as consensus standards. This is no longer the method for acquiring OSHA standards. The current process by which OSHA establishes standards for worker health and safety protection is more complicated and takes more time. A new standard must be determined to be reasonably necessary; it is usually proposed by an industry group with an interest in some specific aspect of worker safety. OSHA then seeks technical input from NIOSH and standing advisory committees on worker safety and health. If the decision is made to pursue a new standard, a notice of intended rulemaking is published in the Federal Register, sometimes accompanied by proposed text for the new standard. Next, there is an opportunity for public comment; hearings may also be held as part of this step. Finally, OSHA makes revisions to the proposed standard and issues the final version in the Federal Register. OSHA standards can generally be classified into two groups: 1) performance-based standards, where the required/desired end-points are described by the rule, and 2) more detailed standards that describe specifically how compliance is to be achieved, often in great detail.

This chapter also discussed the basic terms and concepts used to evaluate and describe worker exposures to hazardous physical and chemical agents. OSHA limits, also called PELs, are enforceable as law and include limits for full-shift and shorter-term exposures. The OSHA exposure limits were compared to those issued by NIOSH and the ACGIH. Unlike OSHA, these agencies are not subject to the lengthy process of rulemaking and are able to review and revise their exposure limits on a regular basis. Many of the terms are used by all three groups; they include the STEL for 15-minute exposures and the ceiling limit for instantaneous exposures. Other terms are used to indicate special situations. For example, the skin notation is used for substances that are easily absorbed through the skin. The use of biological samples, such as urine, blood, and exhaled breath, for use in evaluation of exposure and absorbed dose, was discussed. The term IDLH was defined and illustrated. Finally, several OSHA standards that are an important part of most industrial hygiene programs were highlighted. These included hazard communication, respiratory protection, hazardous waste operations and emergency response, confined space entry, and occupational noise exposure.

Critical Thinking Questions

1. A worker is exposed to trichloroethylene during the workday as follows:

   3 hours . . . . 50 ppm
   2 hours . . . . 125 ppm
   1 hour . . . . 110 ppm
   1 hour . . . . 85 ppm

   What is the worker's 8-hour TWA exposure for the day?

   How does this level compare to current OSHA, NIOSH, and ACGIH exposure levels?

2. How do the OSHA, ACGIH, and NIOSH exposure limits for carbon tetrachloride compare? Can you provide some explanation for these differences?

3. What are the steps that OSHA must take in order to issue a new standard?

4. What is the significance of the skin notation? How do we measure exposures that occur through skin absorption?

5. Describe how OSHA, NIOSH, and the ACGIH handle exposure limits for carcinogens.

6. Where in the OSHA regulation on respiratory protection are the requirements for sanitizing of respirators? (Provide the citation that gives the location within the standard.)

7. What sources would you consult in evaluating worker exposure to a substance not regulated by OSHA?

$$50 \times 3 + 2 \times 125 + 1 \times 110 + 1 \times 85$$
$$8$$

$$\frac{150 + 250 + 110 + 85}{8}$$

$$\frac{595}{8} = 74.375$$

4

Airborne Hazards

Chapter Objectives

Upon completing this chapter, the student will be able to:

1. **Recognize** the basic structures of the human respiratory tract and know their functions.

2. **Describe** the mechanisms for clearing the lungs of airborne contaminants.

3. **List** the major classifications or groups of airborne hazards.

4. **Explain** the link between size, solubility, and site of action of airborne contaminants.

5. **Name** some materials that pose airborne hazards and give examples of processes during which airborne hazards might be encountered.

6. **Describe** some occupational diseases associated with inhalation of airborne contaminants.

7. **Understand** the meaning of specialized terms used in discussion of airborne materials and the occupational diseases they cause or are associated with.

8. **Understand** how spirometry is used to detect patterns of lung damage and evaluate lung function.

9. **Recognize** situations where oxygen-deficient atmospheres might exist.

10. **Explain** the difference between simple and chemical asphyxiants.

11. **Explain** the mechanisms that can lead to hypoxia.

4-1 Introduction

In previous chapters we explored and defined the role of the industrial hygiene professional in identification, evaluation, and control of workplace hazards; we looked at the various ways materials enter the body and interact with tissues and organs to cause harmful effects. We also addressed methods used to sample the air for use in evaluating the potential hazards posed by airborne materials.

In this chapter we will take a look at some of the airborne substances that can pose health hazards. We will begin with a review of the anatomy of the human respiratory tract and some of the protective mechanisms that exist to prevent damage from the materials present in the air we breathe. Next, we will look at some specific airborne contaminants that may be encountered in the occupational setting, including dusts, fumes, vapors, gases, metals, fibers, minerals, and organic particles. The occupational diseases and conditions associated with inhalation of these contaminants will be discussed, and the relationship between particle size and the site of effect on the respiratory tract will be explored. Methods for measuring lung function and evaluating lung impairment will also be presented. Finally, the potential health effects of oxygen-deficient atmospheres will be discussed.

4-2 Anatomy and Function of the Lungs

In Chapter 2 we discussed how airborne hazards present a significant threat to workers because so many materials become airborne in an industrial setting. Welding, grinding, spraying, hot processes – all of these are examples of ways in which raw materials, process chemicals, products, and by-products may become airborne. A worker without appropriate respiratory protection is potentially at risk of inhaling whatever is in the air in the factory, shop, workroom, or office. While the lungs do possess a number of protective features, they are designed to provide an efficient method for the transfer of gases into and out of the body. That same design also provides an efficient route of entry for many airborne substances. Figure 4-1 shows the basic structure of the human respiratory system.

It is convenient to separate the respiratory tract into regions when comparing the physical structures that comprise the respiratory system. The upper or **nasopharyngeal region** is comprised of the head, nose and nasal passages, sinuses and mouth, and all associated features such as the tonsils and epiglottis, including the back of the throat. This region, and in fact nearly the entire respiratory tract, is lined with specialized skin tissue called a **mucous membrane**. The mucous membrane produces a layer of **mucus**, a moist, sticky substance that captures many of the materials we inhale. The many small hairs in the nasal passages also help to trap particles that enter along with inhaled air.

The middle or **tracheobronchial region** of the respiratory system is generally considered to be the

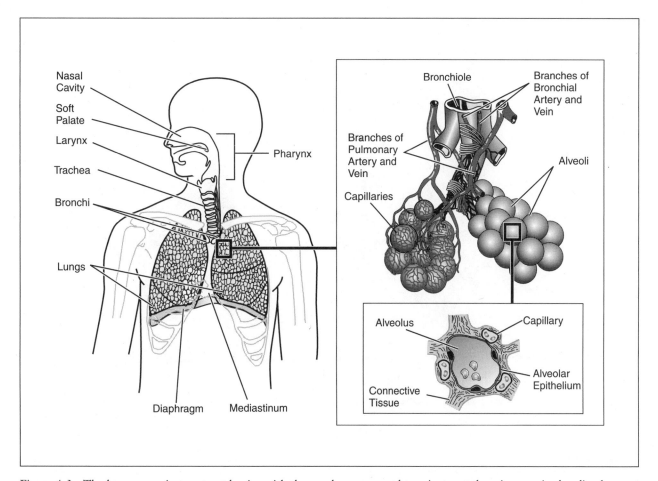

Figure 4-1: The human respiratory tract begins with the nasal passages and terminates at the microscopic alveoli, where gas exchange takes place. Airborne materials differ in their ability to penetrate and be deposited in different regions of the respiratory tract, based on their particle size and solubility.

trachea, or windpipe, and the larger air passages, called **bronchi** (singular, bronchus) of the lungs. The trachea and bronchi are constructed of rings of cartilage and muscle. The cartilage supports the windpipe and helps maintain its tubular shape. The muscles contract and can help force air, and with it the contaminants, out of the lungs through the cough reflex. The trachea and bronchi are lined with a mucous membrane and fine hairs, or **cilia**, which help capture and remove foreign particles. The cilia move in a wavelike manner to push the mucus, and any particles in it, upward and toward the larger air passages. Cigarette smoking can paralyze the cilia, impairing their ability to remove inhaled substances. Particle-laden mucus is removed from the body by coughing, expectorating, or swallowing.

In the lower, **distal region** of the lungs, the bronchi split, or **bifurcate**, into two smaller branches, which split repeatedly, for a total of about seventeen times, creating passages with increasingly

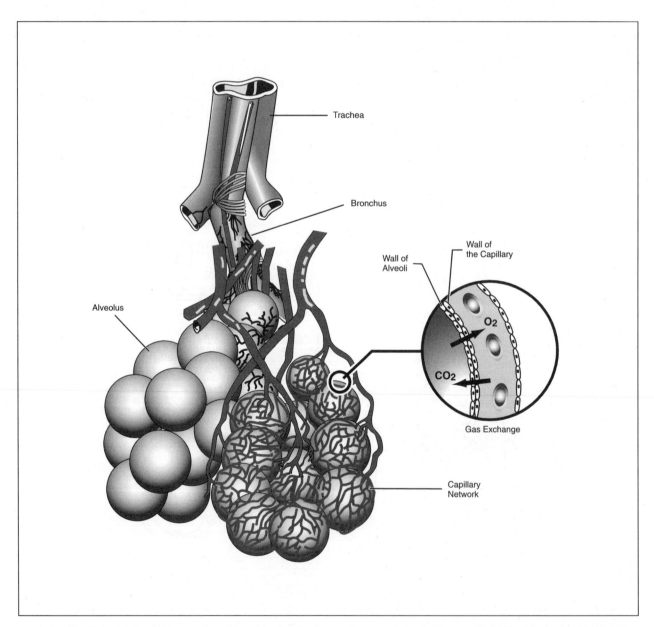

Figure 4-2: The many branches, twists, and turns of the lung passages provide surfaces on which airborne materials may impact and stick. This protective mechanism prevents most materials from reaching the gas-exchange region.

smaller diameters. Eventually, the rings of cartilage disappear altogether and the tiny air passages, now called **bronchioles**, are quite small. The airways end in microscopic sacs called **alveoli**, the shape of which resembles clusters of grapes. The alveoli are where gas exchange occurs. To allow for efficient diffusion of gases, the alveolar membranes – composed of specialized skin cells called **pneumocytes** – are only one cell thick, and are surrounded by a bed of tiny blood vessels called **capillaries**. As we saw in Chapter 2, passive diffusion is the mechanism by which most gases pass into and out of the blood (see Figure 4-2).

Our discussion of inhaled substances requires an understanding of the units of measure that describe the necessarily small dimensions of such substances. A **micron**, or **micrometer**, is one one-thousandth of a millimeter (0.001 mm) and is represented by the Greek letter μ (pronounced mew) accompanied by the letter m: μm. The μ is also used to represent other measures such as micrograms (μg) or microliters (μl).

Protective Mechanisms of the Respiratory Tract

Airborne pollutants come from many sources; manufacturing plants, power generation plants, and vehicle exhausts represent the majority of contributors. The burning of coal and wood, natural disasters such as volcanoes and forest fires, and something as seemingly innocent as the wind blowing across the earth can also generate airborne substances. Plants and animals also release particles, such as pollen, spores, dander, as well as gases such as carbon dioxide (CO_2) and methane (CH_4). Because we must breathe the air, we are potentially exposed to the particulate matter suspended in it, whether at work or outdoors enjoying a hike.

There are several ways in which the airways and lungs capture and/or remove these airborne contaminants. The larger particles, in the range of 10 μm and larger, are removed in the nose and upper airways. Particles ranging from 5 to 10 μm are captured in the tracheal region. Only particles 0.5 μm to about 3 μm are small enough to make it to the alveolar region, but only a fraction of these actually penetrate into the deep region of the lungs, due to

the body's efficient mechanisms for removing inhaled particles. Most particles in the size range of 3-5 μm eventually contact the walls of the airways and stick in the mucus lining of the tracheal region. This capturing system is enhanced by the many branches and splits of the smaller airways, forcing the air onto a tortuous path and providing many surfaces on which the particles can impact. Once the particles are stuck in the mucus, their removal is performed through the ciliary action described earlier; this removal process is referred to as the **muco-ciliary elevator** (also called the muco-ciliary ladder

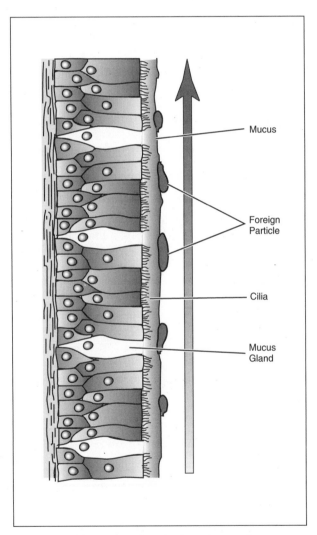

Figure 4-3: The mucus sheet in the upper respiratory tract. The cells lining the bronchioles, bronchi, and trachea are ciliated. The wavelike motion of the cilia moves a thin sheet of mucus (produced by the mucous glands) upward at a rate of about one inch per minute. Foreign particles in the air are trapped in the mucus and removed by the lungs.

or mechanism). The cough reflex is another possible way for irritating gases and particulates to be removed. These particulates cause the muscles surrounding the bronchi to contract, which forces the air – along with the contaminant – out of the lungs.

In the alveolar region, the body's primary defense is provided by specialized white blood cells, called **macrophages**, which engulf foreign objects and attempt to dissolve them. This mechanism does not always work, as we shall see. It is also possible for some foreign substances to pass through the cell membranes and lodge between cells, in what is called the **interstitial space**, and even exit the lungs in this way. Only the smallest particles have the potential to penetrate this far into the lungs.

Checking Your Understanding

1. Name three regions of the respiratory tract. What physical features or structures are associated with each region?

2. What is the role of the muco-ciliary ladder?

3. Where in the respiratory tract are macrophages found, and what is their function?

4. Where in the lungs does gas exchange take place?

4-3 Airborne Hazardous Materials

While there are many non-occupational sources of airborne contaminants, there are also many materials that become airborne in an occupational setting. Because of this, inhalation is generally viewed as the most significant route of entry for toxic materials in most workplaces. The specific airborne hazards that workers are exposed to will vary depending on their occupation; there may be a single contaminant, or a combination of them, in the air. For example, welding may produce exposure to metal fumes as well as to the shielding gases being used. Solvent cleaning operations may expose the workers to a single compound.

Aerodynamic Diameter

The terms particle and aerosol are generic and include a large variety of materials with differences in size, solubility, and other characteristics (see Figure 4-4). The **aerodynamic diameter** is especially useful for comparing particles with irregular shapes – such as dusts and fibers – to particles with regular shapes – such as droplets and mists. The aerodynamic diameter of a particle is considered to be equal to the diameter of a reference spherical particle with a unit density of one (1) that has the same settling velocity as the contaminant particle. For example, consider a chipping process. Let's assume the chips that are produced have uneven shapes, but they tend to fall out of the air, or settle, at roughly the same rate. Rather than try to measure these irregularly shaped particles, we relate them to a spherical particle that falls at the same rate as the irregularly shaped particles. Although the particles produced by the chipping process are not spherical, they are said to have an aerodynamic diameter that is equal to the diameter of the reference spherical particle.

The use of the aerodynamic diameter to describe particles allows us to compare masses that have very different sizes, shapes, and densities. In our example, if during our chipping process we were producing lead particles, they would be relatively dense and tend to fall out of the air quickly. One might reasonably expect the aerodynamic diameter of these to be relatively large. In contrast, chalk dust would be expected to be less dense than lead chips, which means that the chalk particles would have a smaller

aerodynamic diameter as compared to lead. Many aerosols are described using their aerodynamic diameter, which is also a commonly used term when discussing substances that are occupational health concerns.

Classes of Airborne Hazardous Materials

Airborne materials that pose health and safety hazards consist of particulates, gases, vapors, or some combination of these. We will also include oxygen-deficient atmospheres in our discussions of airborne hazards. Some gases, vapors, and particulates are capable of forming flammable or explosive mixtures with air. Others may displace the air, reducing oxygen content to unsafe levels. Some airborne hazards are considered to be nuisance hazards, which means they do not impair health, but may cause problems with visibility or operation of equipment. Some materials present serious health problems due to their irritating properties, or because they are causally linked to diseases like cancer. For the purposes of our discussions of airborne hazardous materials, we will use the following terms and classifications:

—Particulates/aerosols (solid particles, dusts, fibers, mists, droplets, fumes);

—Gases and/or vapors (gaseous contaminants, vapors);

—Oxygen-deficient atmospheres (containing less than 19.5 percent oxygen); and

—Combination (any combination of particulates, gases, and/or vapors, including oxygen deficient atmospheres).

In most situations an atmosphere will contain some combination of contaminants rather than a single one. Combinations of particulate and gaseous contaminants can pose different hazards from those that would be present with only one contaminant. For example, an airborne solvent could condense on the surface of a small (less than 1 μm in diameter) particle; the particle could then be inhaled,

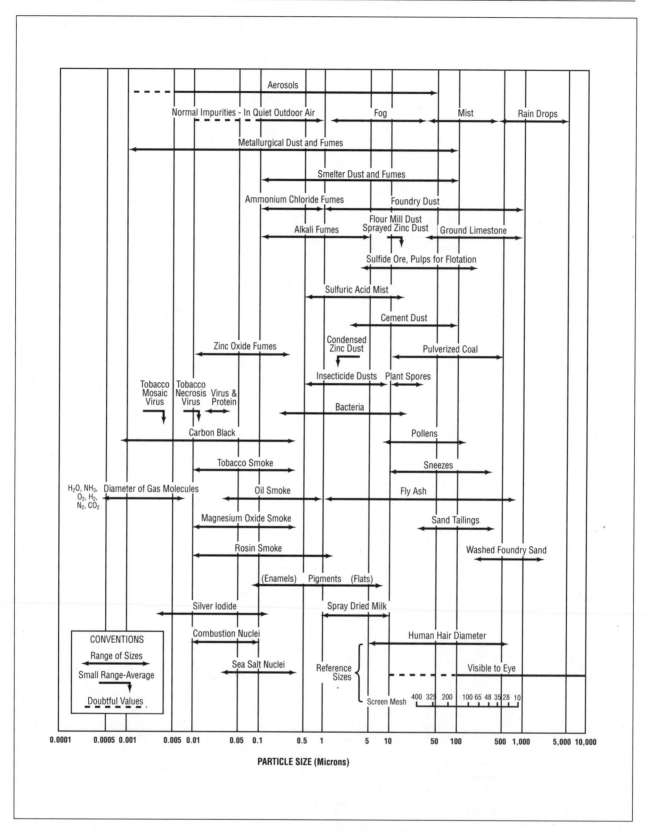

Figure 4-4: Sizes of various airborne particles. The size of a particle has a direct bearing on its ability to enter the respiratory tract. Very small particles (those with diameters of 10 μm or less) may reach the alveolar region. Some of these may be deposited in the tissues, while others may be exhaled from the lungs.

carrying the solvent deep into the lung. In the absence of these small particles, the solvent might only reach the middle region of the respiratory tract where it could dissolve in the mucus, which would effectively remove it from the inhaled air. Although the solvent may cause irritation to the mucous membrane in the middle region of the lung, it could pose a different – possibly a greater – hazard in the alveoli, where the protective mechanisms are not the same.

Because exposure limits can be based on such diverse properties as protection from irritation or preventing development of cancer years later, it is important to understand the basis for the exposure limits of the particular airborne contaminants that are a concern. (This is of course true of all exposures, not just those involving airborne materials!) Levels of airborne materials that are within regulatory limits may present other hazards, such as explosions or displacement of air (and oxygen). It is good practice, therefore, to acquaint yourself with all the hazardous properties of the airborne materials that are present.

Particulate contaminants represent a large group of airborne materials. The terms **particulate** and **aerosol** will be used to distinguish materials that occur as individual or discrete masses from gases and vapors. Particulates are often grouped according to their chemical and physical characteristics, which are related to their method of formation. The term aerosol usually refers to liquid droplets that are suspended in air, but the term is also used to refer to solids of any type that are small enough to remain airborne for an extended period of time.

Dusts are aerosols composed of (usually) dry particles. The particles may occur naturally, or be produced by some physical process, such as blasting, mining, grinding, polishing, or crushing. Dusts of concern to occupational health professionals vary in size; those with diameters of 10 μm or less are of special concern due to their ability to move beyond the upper airways. The term dust is used to refer to a variety of airborne particles, some of which exist naturally as fibers or powders, while others are generated during manufacturing, processing, or refining of raw materials. Depending on their size, inhaled dusts may be filtered by the body or be deposited somewhere in the respiratory system. Adverse health effects of dust inhalation range from irritation and allergic responses to debilitating diseases and cancers. Some airborne dusts present explosion hazards. Examples of dusts include airborne particles of silica, asbestos, coal, grains, and sawdust.

Mists are airborne droplets. Mists may be generated by any operation or process involving liquids: sprays, coolants, cutting fluids, paints, solvents, pesticides, and many others. Mists are created when air or another gas, usually under pressure, is introduced into a liquid, causing the liquid to break apart or **aerosolize**. The surface tension between the liquid molecules causes them to stick together and form droplets. It is possible for accidents or unusual events involving liquids to result in aerosolization; however, any process involving liquids is a potential generator of these mists. Aerosolization is a common method for biological agents, such as *Tuberculinum mycobacterium*, to become airborne.

The hazards posed by mist inhalation vary depending on the material. The size of the aerosol, its chemical properties – such as reactivity and solubility – and its toxicity, determine the resulting health effects. Because any process involving liquids can produce mists, it is impossible to address each of them specifically. Responses of respiratory tract tissue to inhaled mists include irritation, inflammation, stimulation of mucus production, **pulmonary edema** (accumulation of fluids in the lungs), and allergic or sensitization responses.

Fumes are produced when materials, usually metals, are heated to the point where they become a vapor or gas; the hot gas cools once it is airborne, and then condenses to form small particles (fume diameters range from 0.1 to 100 μm). Fume formation requires high temperatures, generally 2,000°C or more. Typical sources of fumes include welding, torch cutting, and sometimes buffing or grinding, depending on surface temperatures. (Carbon monoxide and gasoline are not fumes: their airborne phases are gas and vapor, respectively.)

Gases are among the most familiar of airborne contaminants. Air contains a mixture of gases: nitrogen, oxygen, a little argon, and some water vapor. Oxygen and carbon dioxide are gases that move into and out of our bodies through the lungs without much notice on our part. Carbon monoxide and acetylene are examples of gases that produce noticeable – and sometimes lethal – effects. Many airborne contaminants – too many to be addressed individually – are gases. Some of the more significant occupational hazards within this group include ammonia, chlorine, coke oven emissions, hydrogen fluoride, and sulfur dioxide. Table 4-1 lists some gases and vapors that are damaging to the respiratory tract.

Vapors are the gaseous phase of liquids; the tendency of a liquid to change from a liquid to a gaseous form depends on its **vapor pressure**, which is directly related to how quickly it will evaporate. Vapor pressures are expressed in millimeters of mercury (mm Hg), like barometric pressure. A higher vapor pressure indicates a faster rate of evaporation and can be useful as an indication of the likelihood that a liquid exposed to air will become an airborne vapor and thus a potential inhalation hazard. Substances that have high vapor pressure present potential airborne hazards such as toxicity, asphyxiation, and the formation of flammable or explosive mixtures in air. As is the case with mists and gases, there are too many vapors with potential adverse effects to be addressed specifically; they are also listed in Table 4-1.

Effects of Inhaled Materials

Airborne toxins may exert their toxic effect locally, that is, directly on the tissues with which contact is

Toxicant	Pulmonary Damage
Ammonia, NH$_3$	Irritation
Chlorine, Cl$_2$	Irritation
Coke oven emissions	Lung cancer
Hydrogen fluoride, HF	Irritation, edema
Nickel carbonyl, NiCO	Nasal and lung cancer; acute edema
Nitrogen oxides: NO, NO$_2$, HNO$_3$	Emphysema
Ozone, O$_3$	Emphysema
Phosgene, COCl$_2$	Edema
Perchloroethylene, C$_2$Cl$_4$	Edema
Sulfur dioxide, SO$_2$	Irritation
Toluene 2,4-diisocyanate	Sensitizer; edema
Xylene	Edema

Table 4-1: Once inhaled, gases and vapors may exhibit local effects – such as irritation and inflammation – on the lung tissues they contact. Other materials do not affect the lungs, but are transported by the blood to other locations in the body where they exert systemic or target organ effects.

made, as is the case with ammonia and other strong irritants of the respiratory tract. Systemic toxins that are inhaled reach their site of action through blood transport; for example carbon tetrachloride, a liver toxin, is inhaled and dissolves in the blood, which then carries it to the liver. Some materials are deposited in a particular region in the respiratory tract, where they cause damage to tissues, often resulting in disease. The location in the respiratory tract in which a toxic material exerts its effect is a function of the particular physical and chemical properties of the toxin, as well as of the dose, or the amount of exposure.

Solubility affects the absorption of gases and vapors along the respiratory tract and within the lungs. Soluble and reactive gases tend to affect the upper region of the respiratory tract as well as the moist tissues surrounding the eyes. Ammonia, alcohols, and many solvents produce strong irritating effects in this region. Gases that are less soluble tend to affect the middle and deep regions of the respiratory tract; this is due to their increased contact time with the surrounding tissues, allowing for slow dissolution. Phosgene gas is a classic example of a relatively insoluble material that affects the deep lung, while the very soluble ammonia gas affects the upper respiratory tract.

Size-selective Sampling

Because of the link between particle size and site of deposition in the respiratory tract in the development of occupational diseases, methods have been developed for sampling particulates that allow the industrial hygienist to capture particles with a particular size distribution, or with aerodynamic diameters that fall within a certain range. Sampling for a particular size range of particles is called **size-selective sampling**. The ACGIH has defined three particle size ranges, or fractions, that are used for this sampling; they are the **inhalable**, **thoracic**, and **respirable fractions**. Each fraction is defined in terms of a specific size range of particles, with aerodynamic diameters that relate to the region of the respiratory tract in which the particles are deposited when inhaled. As the aerodynamic diameter decreases, the potential for the particle to reach the deeper region of the lung increases. This is important when considering the possible negative health effects of particles. It is also useful when a specific region of the lung is the site of deposition for a specific material associated with an occupational disease.

ACGIH Size Fraction	Particle Aerodynamic Diameter, μm	Site/Region of Deposition
Inhalable	>10	Nasopharyngeal region
Thoracic	5-10	Thoracic region
Respirable	0.5-4	Alveolar region

Table 4-2: Occupational health professionals have long recognized a link between the site of deposition in the lung and the occurrence of occupational disease. This observation has led to the development of sampling methods that allow industrial hygienists to selectively sample particles within a specified range of aerodynamic diameters.

The inhalable fraction refers to substances that are hazardous wherever they are deposited in the respiratory tract: the nose and throat, the airways, or the deep lung. The thoracic fraction is used to indicate materials that are hazardous when deposited in the bronchi (airways) or in the gas-exchange (alveolar) region. The respirable fraction is used for materials that are hazardous when deposited in the gas-exchange region of the lungs, the alveoli. Samplers have been designed to capture only those particles that fall within the size range of each specific fraction, allowing industrial hygienists to sample for the sizes of interest. For example, for crystalline silica, the OSHA PELs have been established for the respirable fraction of silica-containing dust. This requires the industrial hygienist to perform exposure monitoring using samplers that will capture the respirable fraction. Specific air sampling techniques will be addressed in detail in Chapter 5.

Checking Your Understanding

1. Name four general categories or classes of airborne contaminants.

2. Explain how combinations of gases and particulate contaminants can pose hazards different from those posed by a single gas or particle.

3. What chemical or physical property is most important as it affects the depositing of inhaled particulates in the respiratory tract?

4. Name the three fractions or size ranges for airborne particles that pose inhalation hazards and define them in terms of penetration into the respiratory tract.

5. Why are some insoluble gases and vapors a serious hazard to the respiratory tract?

4-4 Occupational Diseases Associated with Airborne Particulates

The term **pneumoconiosis** literally means dust in the lungs; however, not all dusts that are inhaled cause an identifiable disease. Therefore, a more useful definition of pneumoconiosis includes the reaction of the lung tissue to the presence of dust. Diseases associated with the inhalation of dusts are often referred to as pneumoconioses (singular, pneumoconiosis). Inhaled dusts may simply remain as deposits in the lung tissue; they may be detectable using x-rays but without any resulting illness or adverse health effects. Other dusts can cause scarring and damage leading to serious impairment of respiratory function or to the development of a condition or response that is unique to the type of inhaled dust. Some well-characterized occupational diseases will be presented to illustrate the possible reactions to inhaled dusts.

> The word pneumoconiosis comes from *pneumo*, a Greek word for the lungs, and *-osis*, a Greek suffix that means disease. Pneumoconioses are better described by specific names that indicate the type of dust that was inhaled with the -osis ending; thus, silicosis is pneumoconiosis caused by inhaling silica dust. Other names that contain information about the type of disease-causing dust are: asbestosis (from inhaling asbestos dust), berylliosis (from inhaling beryllium), and aluminosis (from inhaling aluminum).

Physiological Responses Associated with Inhaled Dusts

As stated above, dusts that are inhaled and deposited in the lungs can induce one or more responses from the body, including no response at all. Inhaled dusts may stimulate an increase in the production or secretion of mucus, sometimes accompanied by an enlargement of the cells that produce the mucus. Another possible response is the engulfment of particles by macrophages. Irritating substances may cause an inflammation of tissues, sometimes accompanied by edema. Some dusts stimulate the formation of fibrous tissues, such as reticulin or collagen; the resulting growths or lesions can be **benign** (usually not associated with an adverse health effect). **Siderosis**, which is reddish discoloration from deposits of iron oxide in the lungs, is an example of a benign condition. Many dusts, when inhaled, can cause changes in lung tissue that affect the lung's breathing or gas exchange functions. The most severe response would be changes in the tissues that lead to severe lung damage or to the development of cancer.

Silicosis is a scarring of the lungs caused by inhalation of dust containing crystalline silica. The condition is characterized by the development of **nodules** that form in response to the dust. The nodules are made of strands of a thready, fibrous tissue called **reticulin**. Reticulin – as well as other connective tissue – is produced by a specialized cell called a **fibroblast**. In silicosis nodules, the reticulin is arranged in layers much like an onion, around an inner core, which can appear empty. The nodules form when macrophages – containing silica particles – die and release their digestive enzymes into the delicate lung tissue. The silica particle, released from the now-dead macrophage, is phagocytized by another macrophage, and the cycle begins again. The nodules form throughout the lungs, but may leave large areas of lung tissue unaffected. As more nodules are formed, they may become larger and clump together, causing localized areas of damage, which impair lung function. People who have silicosis are more susceptible to opportunistic infections such as tuberculosis and pneumonia. In some cases the development of tuberculosis may go unnoticed by the doctor since the physical symptoms associated with the disease are masked by the silicosis. Other complications associated with silicosis include **hypertension**, or high blood pressure, and an enlarged heart (the medical term is *cor pulmonale*).

The extent to which silicosis develops is proportional to the amount of crystalline silica present in the dust. These crystalline formations can occur as different minerals: quartz, tridymite; cristobalite; coesite, and stishovite. All of them contain silicon dioxide, SiO_2. The form stishovite is not associated with silicosis, while cristobalite and tridymite are more potent than quartz in causing a fibrotic response. The OSHA PEL for crystalline silica reflects these differences in toxic potential and varies depending on the type and percentage of crystalline silica present in the dust. To calculate the PEL, one must first determine the percent of the respirable fraction of the sampled dust that is crystalline silica. This is done through laboratory analysis of the sample, which also reveals which of the crystalline mineral forms, such as cristobalite or tridymite are present. Although crystalline forms of silica are capable of causing a fibrotic response in the lungs, noncrystalline or **amorphous** forms of these minerals – such as diatomaceous earth – are also of occupational concern.

The proposed 1989 revisions to the OSHA PELs included new values for amorphous and crystalline silica. As explained in Chapter 3, the proposed PELs were vacated as the result of a suit filed in the 10th Circuit Court of Appeals. A comparison of the vacated OSHA PELs and the current (1998) ACGIH TLVs:

	Vacated PEL	ACGIH TLV
Amorphous silica (total dust)	6 mg/m³	10.0 mg/m³
Crystalline silica (respirable quartz)	0.1 mg/m³	0.1 mg/m³
Cristobalite (respirable)	0.05 mg/m³	0.05 mg/m³
Tridymite (respirable)	0.05 mg/m³	0.05 mg/m³

If adopted, the PELs listed above would have done away with an old and somewhat unwieldy method for determining these PELs. The current PELs can be found in 29 CFR 1910.1000, Table Z-3. There are in fact several equations that are used by OSHA for calculating the PELs for this group of mineral dusts. All of the formulas are designed to weight the resulting PEL, based on the relative amounts of each mineral present. A unitless factor of 2 is added to the denominator in all cases (except for non-crystalline forms). The 2 in the denominator is there to limit the concentration of respirable dusts with less than 1 percent SiO_2 to 5 mg/m³.

The equations are listed below as they appear in the regulation:

Form	Equation to Calculate PEL
Amorphous (non-crystalline) silica	$\dfrac{80 \text{ mg/m}^3}{\% \text{ SiO}_2 + 2}$
Crystalline silica (respirable quartz)	$\dfrac{10 \text{ mg/m}^3}{\% \text{ SiO}_2 + 2}$
Crystalline silica (total quartz)	$\dfrac{30 \text{ mg/m}^3}{\% \text{ SiO}_2 + 2}$
Cristobalite (respirable)	$\dfrac{1/2(10 \text{ mg/m}^3)}{\% \text{ SiO}_2 + 2}$
Tridymite (respirable)	$\dfrac{1/2 \, (10 \text{ mg/m}^3)}{\% \text{ SiO}_2 + 2}$

If a sample is found to contain more than one type of crystalline silica, the dusts are considered to have an additive effect. In this case, the PEL is computed using the following formula, which equates the results based on the relative amounts and toxic potential of each mineral present:

$$\text{PEL} = \frac{10 \text{ mg/m}^3}{\% \text{ Quartz} + 2 \, (\% \text{ Cristobalite}) + 2 \, (\% \text{ Tridymite}) + 2}$$

The procedure for determining the PEL for crystalline silica might seem complicated. Fortunately, the current ACGIH TLVs for each type of silica (e.g., amorphous, cristobalite, and tridymite) are very close to the numbers that one would obtain using the equations mentioned above, when a value of 100 percent quartz is used in the denominator. For example, the ACGIH TLV for cristobalite is 0.05 mg/m³. If we were to calculate the OSHA PEL for a sample that had a respirable fraction of 100 percent cristobalite, using the appropriate equation from the OSHA standard for mineral dusts, we would obtain the following:

PEL for cristobalite (respirable)

$$= \frac{1/2\ (10\ mg/m^3)}{\%\ SiO_2 + 2}$$

$$= \frac{1/2\ (10\ mg/m^3)}{100\%\ SiO_2 + 2}$$

$$= \frac{1/2\ (10\ mg/m^3)}{102}$$

$$= 1/2\ (0.098\ mg/m^3)$$

PEL for cristobalite (respirable)

$$= 0.049\ mg/m^3$$

The PEL for respirable cristobalite dust that is calculated using the OSHA method is 0.049 mg/m³ , a value that is very close to the ACGIH TLV of 0.05 mg/m³. The other ACGIH values end up being very close to what would be OSHA's calculated exposure limits for a sample composed of 100 percent of each type of crystalline silica. However, since OSHA standards still reflect the calculation-based method for determining the PEL, it is wise to be familiar with it.

Many mineral dusts contain non-crystalline or **amorphous silica.** Examples include diatomaceous earth, clay, and talc. Dusts containing more than two percent quartz tend to produce lesions similar to the nodules associated with silicosis. Dusts that have small amounts (less than two percent) or no crystalline quartz, such as clay (kaolin), talc, and iron oxide, are associated with the formation of lesions that are more like coal macules. These lesions seem to occur in proportion to the amount of dust present; they contain small amounts of reticulin but do not necessarily impair lung function. Such dusts are called **nonfibrogenic**, since they do not cause a fibrotic response when deposited in the lungs.

Coal worker's pneumoconiosis is a well-documented condition, which – like silicosis – is characterized by a lesion that forms around inhaled particles – in this case, coal. The lesions, called macules, are composed of dust-filled macrophages that accumulate in the respiratory bronchioles and in the alveolar spaces. Some connective tissue also deposits in these macules. The macules have been observed to form a sleeve or cuff-like structure in the bronchioles. After a while, the smooth muscles **atrophy** or

deteriorate and they lose the ability to contract. As a result, the airways are permanently open, or dilated. The loss of elasticity in these localized areas of the bronchioles is called **focal emphysema**.

Another possible reaction of the lungs to the accumulation of coal dust is the formation of large deposits, usually in the upper regions of the lungs. These lesions are made of coal dust and macrophages, interspersed with connective tissue. Unlike the nodules that form around silica particles, these lesions are one large mass that continues to grow in size, until a large mass is formed. This condition is called **progressive massive fibrosis**.

Some dusts stimulate the development of a widespread fibrosis in the lungs. Unlike the nodules and macules that form around silica and coal particles, this reaction is characterized by fibrotic lesions that start in isolated areas, growing and expanding to spread among lung tissue in a diffuse rather than localized pattern. This condition is called **diffuse interstitial fibrosis**. Aluminum, beryllium, and asbestos are examples of dusts that are often associated with this type of response.

Mineral Fibers and Other Fibers

Inhaled fibers, though they are particles, demonstrate a few differences when compared to semispherical aerosols. For one thing, they are thinner or smaller in diameter along one axis, and they tend to orient themselves along that long axis in the air. This means that less-than-straight fibers will be removed from inhaled air more efficiently, since they are more likely to impact on the walls of the airways. It also means that straight fibers are more likely to penetrate deeper into the lungs. However, the length of these straight, rigid fibers also affects their ability to pass into the alveolar region; the majority of fibers in the deep regions are less than 50 µm long.

Asbestos is the best known of the mineral fibers that pose an inhalation hazard. Other fibers that might be encountered in the occupational setting include glass fibers and ceramic fibers. OSHA considers fibrous glass to be a suspected human carcinogen. This classification includes mineral wool or glass wool, a common insulation and filler material. This listing is based on animal studies involving injection of glass fibers into the peritoneal space, a route of exposure that is unlikely to occur in hu-

Box 4-1 ■ Determining a PEL for Crystalline Silica

Two air samples were taken over an 8-hour time period and submitted to the laboratory for analysis. The following data is known for each sample:

Sample	Volume Sampled	Respirable Dust	Laboratory Analysis
1	425 L	0.855 mg	5.2% quartz, 2.3% cristobalite
2	275 L	0.619 mg	4.8% quartz, 1.7% cristobalite
Totals	700 L (0.700 m³)	1.474 mg	

What is the PEL for this dust, and what was the employee's 8-hour TWA exposure?

First, we must calculate the percentage of quartz, cristobalite, and tridymite present in the respirable fraction of each of the two samples. (Since no tridymite was detected in analysis, we do not need to include it in our calculation.).

To determine the percentage of quartz in the respirable fraction:

% Quartz in respirable fraction:

$$5.2\% \times \frac{0.855 \text{ mg}}{1.474 \text{ mg}} + 4.8\% \times \frac{0.619 \text{ mg}}{1.474 \text{ mg}}$$

% Quartz in respirable fraction:
3% + 2% = 5%

To determine the percentage of cristobalite:

% Cristobalite in respirable fraction:

$$2.3\% \times \frac{0.855 \text{ mg}}{1.474 \text{ mg}} + 1.7\% \times \frac{0.619 \text{ mg}}{1.474 \text{ mg}}$$

% Cristobalite in respirable fraction:
1.3% + 0.7% = 2%

Second, we calculate the PEL for this mixture using the equation:

$$PEL = \frac{10 \text{ mg/m}^3}{\% \text{ Quartz} + 2\,(\% \text{ Cristobalite}) + 2\,(\% \text{ Tridymite}) + 2}$$

Substituting our calculated values in the equation:

$$PEL = \frac{10 \text{ mg/m}^3}{5\% \text{ Quartz} + 2\,(2\% \text{ Cristobalite}) + 2\,(0\% \text{ Tridymite}) + 2}$$

$$PEL = \frac{10 \text{ mg/m}^3}{5\% + 2\,(2\%) + 2\,(0\%) + 2}$$

$$PEL = \frac{10 \text{ mg/m}^3}{11}$$

$$PEL = 0.91 \text{ mg/m}^3$$

Now, we must determine the employee's 8-hour TWA exposure:

$$TWA = \frac{(\text{mg Respirable Dust Sample 1}) + (\text{mg Respirable Dust Sample 2})}{\text{Total Volume Sampled}}$$

$$TWA = \frac{(0.855 \text{ mg}) + (0.619 \text{ mg})}{0.700 \text{ m}^3}$$

$$TWA = \frac{1.474 \text{ mg}}{0.700 \text{ m}^3}$$

$$TWA = 2.10 \text{ mg/m}^3$$

The employee was exposed above the PEL of 0.91 mg/m³ for this mineral dust containing crystalline silica.

mans. OSHA is required to identify substances that are proven animal carcinogens as suspect human carcinogens. Some ceramic fibers and materials that are being manufactured to be used as asbestos substitutes appear to possess some of the physical characteristics that are suspected to play a role in the toxicity of asbestos fibers: they are very thin and relatively insoluble in the lungs. Additional research is necessary to prove the safety of these materials.

As mentioned earlier, asbestosis is a fibrotic scarring of lung tissue that occurs in response to inhaled asbestos fibers. Asbestos is a general term that is used to refer to some naturally occurring crystalline minerals with a physical form that most closely resembles long, thin structures or fibers. The two common families of minerals for these asbestiform minerals are the amphibole and the serpentine groups. Amphiboles have a straight, rigid structure and are represented by the asbestos forms amosite and crocidolite; they are also called brown and blue asbestos, respectively. The serpentine family is represented by chrysotile, or white asbestos; chrysotile fibers have a wavy appearance in contrast to the stick-like formations of the amphiboles (see Figure 4-5). Chrysotile asbestos is by far the most common type of asbestos in use. Other asbestos minerals – anthophyllite and tremolite – are encountered less frequently.

When removed from the ground, asbestos minerals bear only a slight resemblance to the fine, thin bundles of fiber that are used as fillers, insulators, and reinforcements in products as diverse as insulation, concrete pipes, and fabrics. The minerals are milled or otherwise processed to separate the fibrous structures, which are then used in the manufacture of various building materials and consumer products. During the milling and manufacturing, the fibers are separated and some are short and thin enough to be of respirable size. Once inhaled, the fibers that are short enough (10 µm) to penetrate

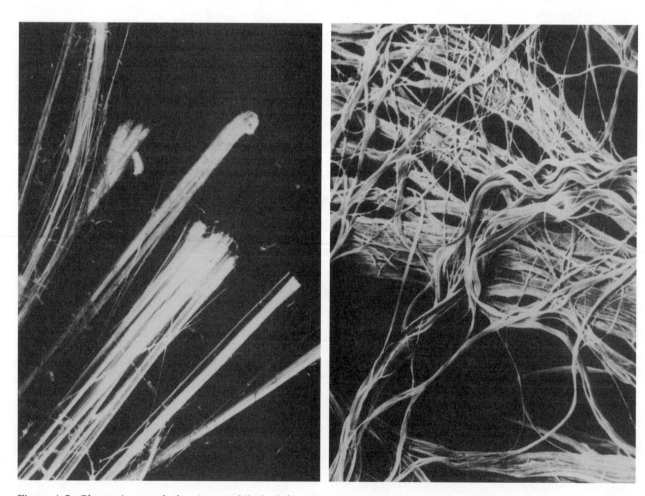

Figure 4-5: Photomicrograph showing amphibole (left) and serpentine (right) asbestos.

deep into the lung are engulfed by macrophages. The longer fibers, which travel through the air on their long axis to reach the smaller airways, cannot be engulfed by macrophages. These longer fibers may become lodged in the alveolar walls or between the cells, in the interstitial space. These sites are eventually invaded by fibroblasts, and the production of connective tissue in the alveolar region that follows can be severely disabling to the individual. The fibrotic pattern of asbestosis seems to involve the lower lung, and progresses along a diffuse pattern.

It is possible for some asbestos fibers to become coated with an iron-containing substance; these coated fibers are called **ferruginous bodies**. Typically, these are the fibers short enough to be engulfed by the macrophages. Shorter fibers – those less than about 5 μm – tend to behave more like particles than fibers and are less likely to cause fibrosis.

There are two cancers associated with exposure to asbestos: **bronchogenic carcinoma** – or lung cancer – and **mesothelioma**. Lung cancer occurs with a higher frequency among persons who also smoke cigarettes; that is, exposure to both results in a combined effect that is worse than exposure to a single agent. This exacerbating type of association is called synergism (see Chapter 2). Mesothelioma is a relatively rare cancer in the general population, but it occurs with a higher frequency among persons exposed to asbestos. This cancer can develop in the thin membranes that line the body cavity, or in those surrounding the lungs. The amphiboles appear to have a strong causal relationship with mesothelioma.

There is relative agreement among researchers and occupational health professionals that there is a dose-response relationship for asbestos-related diseases; that is, the severity of the response seems to be proportional to the amount of dust deposited in the tissues. Of the different forms of asbestos, chrysotile is apparently the form associated with the lowest risk. However, the threshold dose necessary to cause a response – in this case a disease or fibrosis – is not well defined, and many would argue that there is no threshold dose. Because of the difficulties that would arise in attempting to differentiate between the different types of asbestos using current field monitoring methods, a single exposure limit has been established by OSHA. Not only is this easier for industrial hygienists and for laboratories, but the single PEL is also more protective of workers. The ACGIH currently has established separate TLVs for each of the three most common types of asbestos.

Metals

Inhalation of metal fumes is one of the oldest known causes of occupational disease and was recognized and documented by Agricola and other early pioneers in occupational health (see Chapter 1). Metals may be inhaled either as fumes or in the form of dusts produced by grinding, machining, sanding, or sawing. Metal oxides, which are metal atoms combined with oxygen atoms, are another form of potentially inhalable metals; iron oxide – or rust, as it is commonly known – is an example.

Inhaled metals differ in their site of effect. Many are respiratory irritants; others exert their toxic effects on target organs other than the lungs. Some metals are deposited in one tissue, with their toxic effects manifested in another; for example, lead accumulates in bone but exerts toxic effects on the nervous system. Table 4-3 summarizes the sites of effect for some metals.

Metal	Organs/Systems Affected
Aluminum	Nervous system, respiratory tract
Beryllium	Respiratory tract, skin
Cadmium	Kidneys, nervous system, gastrointestinal tract, respiratory tract, bones, heart
Chromium	Kidneys, nervous system, liver, respiratory tract, skin, teeth
Copper	Gastrointestinal tract, hematopoietic system
Iron	Nervous system, liver, gastrointestinal tract, respiratory tract, hematopoietic system
Lead	Kidneys, nervous system, gastrointestinal tract, hematopoietic system, skin, reproductive system
Mercury	Kidneys, nervous system, gastrointestinal tract, respiratory tract
Nickel	Nervous system, respiratory tract, skin
Thallium	Kidneys, nervous system, liver, gastrointestinal tract, respiratory tract
Tin	Nervous system, gastrointestinal tract

Table 4-3: Metals are among the oldest recognized occupational hazards. The effects of metals on the body vary from local skin irritation to systemic effects including cancer or nervous system damage.

Inhalation of metal fumes during welding is associated with a set of symptoms that may be mistaken for other illnesses, such as the flu. The ensuing condition – characterized by fever, nausea, coughing and wheezing, and muscle aches – is called **metal fume fever**. Although generally attributed to exposure to a mixture of metal fumes, exposure to a single metal may produce similar symptoms. Some metals are associated with other occupational diseases. A discussion of selected metals with potential for occupational exposure follows.

Aluminum

Inhalation of aluminum fume can occur during smelting and refining of the metal. Aluminum dust can become airborne during metalworking operations such as cutting and grinding. Aluminum and its oxides are used in making paints and coatings, ammunition and explosives, abrasives, ceramics, as well as manufacturing of aircraft, vehicles, and building materials.

Aluminum salts are irritants of the respiratory tract. Particles of aluminum that are deposited in the eye can cause damage to the cornea. Inhalation of aluminum oxides is capable of producing alveolar edema following acute exposures. Chronic exposures can cause an interstitial fibrosis and emphysema, which is called **Shaver's disease** or **bauxite lung** (aluminum is refined or extracted from bauxite ore). Aluminum is not currently regulated by OSHA except as a nuisance dust.

Lead

Lead has many uses in industry and is therefore frequently encountered by industrial hygiene professionals. Processes that potentially expose workers to lead, either as fume or as an oxide (lead dust) include lead mining and smelting; cutting and welding on surfaces covered with lead-containing paints or coatings; manufacture/recycling of lead-containing batteries; and production of lead-containing paints and other coatings. Recent legislation in many states requiring identification of lead-based paint in residential buildings has spawned an industry of lead removal. The workers who remove lead-based paint as well as the people who live in these buildings are at risk for exposure to lead, especially if precautions were not taken during removal to contain the lead dust. Children appear to be particularly susceptible to the damaging effects of lead on the nervous system, as lead poisoning in children can lead to mental retardation.

Once inside the body, most lead – 90 percent of it – accumulates in bones, while the remainder ends up in the liver and kidneys. In bone, lead has a **biological half-life** of 10-20 years; this means that it takes about that long for half of the accumulated lead to be removed by biological processes from the bony tissue in which it has accumulated. In soft tissue, the biological half-life of lead is measured in months. Symptoms of lead toxicity include muscle weakness, insomnia, lassitude, weight loss, colic, constipation, headache, memory loss, anemia, irritability, paralysis of extensor muscles in the wrist, and the appearance of a dark line of discoloration, called a lead line, on the gums. Some of the physical symptoms of lead poisoning can be confused with other causes. In men, sterility may occur due to a decrease in the number of sperm produced. Lead is also teratogenic, making pregnant women a special concern for protection against lead exposure.

Because of the potential for so many serious negative health effects on exposed workers, there are several requirements imposed by OSHA on industrial and construction activities that involve lead. For example, if levels of airborne lead exceed the action level of 35 $\mu g/m^3$, employees must be trained about lead hazards and methods to protect themselves. Showers and change rooms must be available for worker use, so that the lead is not carried home on street clothes. The employer is also required to monitor the levels of lead present in workplace air and to provide engineering controls or personal protective equipment, if necessary, to protect workers from exposure above allowable levels. A medical surveillance program is also required, according to which levels of lead in the workers' blood are tested on a regular basis. Workers whose body burden of lead, as indicated by blood lead levels, exceeds the allowable levels must be removed from exposure until the levels drop. The specific regulatory standards can be found in 29 CFR 1910.1025 (general industry) and 1926.69 (for construction).

Cadmium

Cadmium is present in lead and zinc ores. It has many uses in alloys, electroplating, pigments, corrosion-resistant coatings, batteries, and fungicides.

Persons at risk for inhalation of cadmium include smelter workers, workers in production processes involving cadmium such as electroplating, as well as workers in welding or soldering. Some foods also contain trace amounts of cadmium.

Once inside the body, cadmium is transported via the blood to the liver, then to the kidneys, where it accumulates. The biological half-life of cadmium is about 20 years. Since we accumulate cadmium in our bodies over a lifetime, it is not unusual to find concentrations of 10-15 µg per gram of organ weight in the kidneys of a 50-year-old. However, occupational or other sources of exposure may lead to accumulations of much higher amounts of cadmium. The renal tubules, which have a role in filtration of salts, are damaged by concentrations of about 200 µg/g or more.

An acute exposure to cadmium can result in death. Exposure generally produces coughing, chest pain, and a metallic taste in the mouth. Delayed reactions can occur within 24 hours or up to a week following the exposure, with the exposed individual experiencing difficulty breathing, nausea and abdominal pain accompanied by diarrhea, wheezing, coughing, **pneumonitis** (irritation and inflammation of the lung tissues), and possibly pulmonary edema. Long-term effects are serious and may include anemia, liver and kidney damage, emphysema, and heart damage. The reproductive systems of men and women are potentially affected; in men, testicular atrophy can occur. Cadmium is also a teratogen – causing developmental defects in the fetus – and a human carcinogen. Increased amounts of zinc in the diet may reduce some effects of cadmium toxicity, since it is the replacement or displacement of zinc by cadmium in the body's enzymatic reactions that is responsible for some of its deleterious effects. Cadmium exposures in general industry are regulated by OSHA under 29 CFR 1910.1027. Air monitoring, employee training, medical surveillance, and protective measures such as engineering controls and the use of protective gear are required by regulation.

Chromium

Exposures to chromium and its various compounds can occur during smelting; roasting and extraction of chromate from ore; chemical and refractory processing; and in the manufacture of alloys containing chromium. Chromium salts are used in pigments, wood preservatives, photographic chemicals, and as anti-corrosive additives in boilers and cooling systems. The oxides of chromium exist in positively charged oxidation states ranging from Cr^{2+} to Cr^{6+}, but the forms existing as Cr^{3+} (trivalent) and Cr^{6+} (hexavalent) are the most important ones biologically.

Trace amounts of Cr^{3+} are essential for normal metabolic processes. However, hexavalent chromium (Cr^{6+}) is irritating and corrosive. Exposure to hexavalent chromium compounds can occur through inhalation of dust and metal fumes, inhalation of mists produced during electroplating and treating of metals with chromium-containing solutions, or through skin contact. Acute exposures can cause coughing, wheezing, headaches, difficulty and pain in breathing, fever, and weight loss. Respiratory tract irritation may persist after the other symptoms disappear. Exposure to chromium-containing mists can result in eye irritation; nasal congestion, nose ulcers, and perforation of the nasal septum; chronic bronchitis; severe skin irritation and skin discoloration; and erosion of the teeth. Exposure to chromium and its compounds has been associated with an increased risk of lung cancer. Liver and kidney damage may also occur. Medical surveillance programs and pre-employment screening tests for chromium exposure will include evaluation of respiratory function, examination of skin and nasal mucosa, and liver and kidney function tests. Exposure to chromium and its compounds is regulated by OSHA in 29 CFR 1910.1000.

Organic Particles

Inhalation of some organic particles is associated with lung impairments sometimes called **reactive airway diseases**. These diseases are characterized by a sensation of tightening of the chest, wheezing, and shortness of breath. The symptoms occur a few hours after exposure begins, but disappear after work, and the individual experiences few, if any, of the respiratory symptoms for the remainder of the work week. A weekend away from the exposure allows enough recovery so that subsequent exposure upon returning to work causes the symptoms to reappear. The reactive airway disease associated with exposure to natural fibers such as cotton, linen, hemp, and flax, is called **byssinosis** (from the Greek word *byss*, meaning fine threads of linen). In the early stages, the symptoms are intermittent and disappear, but if exposure continues the symptoms may become persistent and a permanent narrowing of airways occurs. As with asbestos, byssinosis seems to be made worse by cigarette smoking. Byssinosis can be prevented by identifying reactive individuals through medical screening evaluations and the use of effective engineering controls. Respiratory protective devices may also be used.

It is possible to inhale spores from fungi and molds as well as protein molecules produced by animals. These foreign particles elicit an allergic response in the lung called **allergic alveolitis**, which generally involves the small terminal branches of the bronchioles, just outside the alveolar sacs. The symptoms are coughing, increased production of mucus, fever, fatigue, and muscle aches. Because the symptoms resemble those of other illnesses, such as pneumonia, the diagnosis may be missed allowing the disease to progress and cause permanent, sometimes severe, lung damage. Some examples of organic agents and their associated diseases are summarized in Table 4-4.

Causative Agent	Associated Disease
Actinomyces spores	Farmer's lung; may develop into non-febrile form, characterized by gradual lung impairment with potential for severe damage if untreated
Proteins in bird droppings	Histoplasmosis; bird fancier's lung/pigeon handler's lung
Redwood sawdust	Sequoiosis; allergic response to proteins in redwood sawdust – a similar response has been seen among persons exposed to cedar sawdust
Moldy sprouted barley	Malt worker's lung
Wheat flour proteins	Wheat weevil disease
Cork dust	Suberosis
Maple bark	Maple bark disease

Table 4-4: Some inhalable organic particles – such as pollen, spores, plant and animal proteins – and their associated occupational diseases, which may also occur outside of the workplace.

Evaluating Lung Impairment

Scarring and damage of lung tissue causes two distinct patterns of impairment of lung function. The

Term	Description	Usefulness
Vital capacity (VC)	The volume of air that is exhaled using normal force after taking deepest possible breath.	May appear normal in persons with severe impairment. Reductions indicate restrictive pattern of impairment.
Forced vital capacity (FVC)	Same as VC, only exhaling as forcefully as possible.	Used with FEV_1 to evaluate obstructive impairment
Forced expiratory volume (FEV_1)	Volume of air exhaled during first second of FVC.	Used with FVC to evaluate obstructive impairment.
$\dfrac{FEV_1 \times 100\%}{FVC}$	Comparison of FEV_1 to FVC as a percentage	Should exceed 70 percent; if not, indicates obstructive impairment.

Table 4-5: It is possible to evaluate lung function indirectly by measuring the volumes of air that can be inhaled or exhaled and comparing these to predicted volumes based on the person's gender, stature, and age.

Agent	Pathology	Pattern of Respiratory Impairment
Silica/silicates	Nodular fibrosis	Restrictive
Cement dust	Nonspecific bronchitis	Obstructive
Coal	Macules; focal emphysema	Obstructive
Coal	Massive fibrosis	Obstructive, restrictive
Aluminum	Interstitial fibrosis	Restrictive
Asbestos	Interstitial fibrosis	Restrictive
Beryllium	Interstitial fibrosis	Restrictive
Iron	Accumulation of dust	None known
Tin	Accumulation of dust	None known
Organic textile fibers	Reactive airway disease	Restrictive

Table 4-6: Inhalation of foreign substances may result in changes in lung tissue, with or without impairment of lung function. Some occupational inhalation hazards are associated with both types of effects.

obstructive pattern of impairment occurs as a result of damage to the small airways or bronchioles, resulting in a decreased ability to exhale air. For example, coal deposits can cause permanent dilation of the small air passages, reducing the volume of air that can be forced out of the lungs. The volume of air that can be exhaled is called the **expiratory volume**; it can be measured to help evaluate the extent of lung damage. The focal emphysema associated with coal miner's pneumoconiosis is an example of an obstructive pattern of impairment.

The **restrictive** pattern of impairment describes a condition in which there is a reduction in the volume of air that can be taken in and then pushed out of the lungs (a breath). This volume of air is called the **vital capacity (VC)**. This type of damage may go unnoticed until the condition progresses to the point where physically demanding tasks become difficult due to the decreased efficiency in gas exchange that accompanies the restrictive pattern. Changes in the lungs that can cause this condition are due to the formation of the fibrotic lesions, which reduce the surface area in the alveoli where the gas exchange takes place. As the fibrosis progresses, ventilatory capacity will continue to decrease. Reactive airway diseases may also produce a restrictive pattern of airway impairment. (See Tables 4-6 and 4-7.)

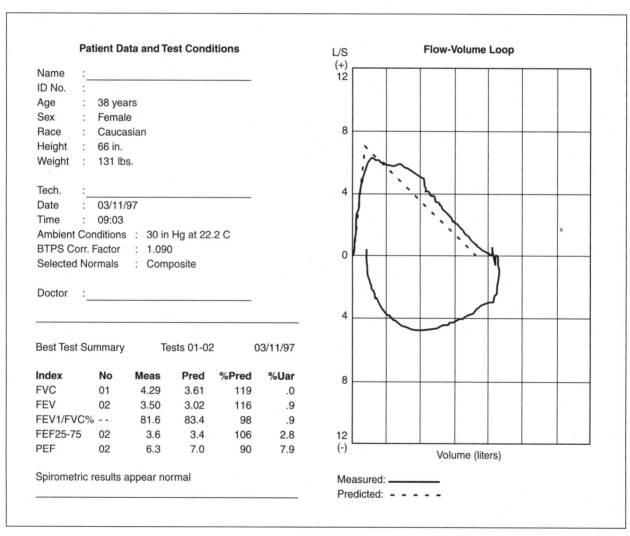

Figure 4-6: A printed output showing the results of spirometry for a female worker. It shows patient (worker) data and test conditions, a summary of the best outcome (usually two tests are done), and a graph that compares the measured and predicted volumes of air exhaled during a forced expiration.

It is possible for both patterns of impairment to occur at the same time. Pre-existing conditions, such as asthma, chronic bronchitis, and allergies, can aggravate the situation. The added damage from cigarette smoking may contribute to the overall pattern.

The testing process commonly used to evaluate lung function is called **spirometry**. It involves exhaling into a tube attached to a machine that measures the volume of air passing through the tube. (See Figure 4-6.) Test results are compared to expected values based on the age, gender, and weight of the individual. Several comparisons are made between the volumes of air that a person is able to inhale and/or exhale; the tests involve normal breathing and maximum efforts on the part of the test subject to determine the maximum volumes that can be achieved. The volumes are a relative indication of how well a person's lungs are functioning. Although compared against standard reference values, the test results are also useful for tracking changes that occur over time. Lung function tests by themselves are not used for diagnosing occupational disease; some association with a toxic material as the causative agent has to be established or suspected, and the physician will include consideration of any other physiological factors. Lung function tests are also used to evaluate a person's ability to wear a respirator.

Checking Your Understanding

1. Define the following terms: pneumoconiosis; fibrogenic; emphysema; ferruginous body; alveolitis; pneumonitis.

2. Name three metals that may be inhaled in dust or fume form. What processes may present exposure hazards to each? Describe the potential negative health effects associated with exposure to each, including acute and chronic effects.

3. Name and describe two diseases associated with exposure to organic particles. During which occupations or activities might these exposures occur?

4. Describe how lung function may be used for evaluating damage to a worker's respiratory system. What practical applications does this type of testing have for occupational health and safety?

4-5 Oxygen-deficient Atmospheres

Normal atmospheric oxygen averages about 21 percent. OSHA has defined an atmosphere containing less than 19.5 percent oxygen as **oxygen-deficient**. Without adequate oxygen, workers will become dizzy, uncoordinated, and eventually will pass out. Unless removed from such an environment quickly, permanent brain damage, or death, can occur. An atmosphere containing more than 23 percent oxygen is termed **oxygen-enriched**; these atmospheres can pose fire or explosion hazards. Oxygen-deficient atmospheres present occupational hazards for a large number of workers in the United States. OSHA's confined space regulation requires testing of the atmosphere inside a confined space to verify oxygen content prior to entry. The OSHA regulation for confined spaces is found at 29 CFR 1910.146, Permit-required confined spaces. Some examples of situations or locations where oxygen-deficient atmospheres can exist or develop include:

— Confined spaces: tanks, tunnels, vats, bins, trenches;

— Interiors of metal tanks which have oxidized (rusted) inside, using up available oxygen;

— Decomposition of vegetation or other organic matter;

— Enclosed areas containing inert gases, chemical asphyxiants, or gases displacing air;

— The use of evaporative cleaning solvents in tanks or other poorly-ventilated areas;

— Welding or cutting inside of tanks or other confined spaces;

— Processes that use or generate chemical or simple asphyxiants.

The condition within the body in which an inadequate amount of oxygen is available to the tissues is called **hypoxia**; severe hypoxia is called anoxia, which may result in permanent damage. Hypoxia may be caused by chemical asphyxiation but also by a decrease in pressure or flow of blood.

Differences in oxygen due to elevation are not due to variations in concentration, but pressure. Standard barometric pressure at sea level is 760 millimeters of mercury, or 760 mm Hg. In an atmosphere containing 21 percent oxygen, about 159 mm Hg is due to oxygen – this is called the **partial pressure** of oxygen, symbolized as P_{O_2} (mm Hg). (Other components of air, such as nitrogen and water vapor, also exert partial pressures.) A P_{O_2} of at least 60 mm Hg is necessary to assure adequate uptake of oxygen by the blood. Normally the P_{O_2} is much higher than this minimum value.

Oxygen-deficient atmospheres can exist or develop in a wide variety of locations and situations. They pose a particular hazard when they occur in a small, enclosed area, or in a space that has poor natural ventilation. Oxygen-deficient atmospheres may exist as part of the normal operation of a process: for example, an enclosed process that requires flooding with liquid nitrogen to maintain low temperatures. They can also develop following the introduction of a solvent or gas acting as an **asphyxiant**. Asphyxiants interfere with the body's absorption of oxygen from the surrounding air. Simple asphyxiants displace air, and therefore the oxygen it contains, resulting in an environment without enough oxygen to support life. Chemical asphyxiants interfere with the absorption or utilization of oxygen through reactions on a molecular level, and hypoxia resulting from chemical asphyxiation may occur in the presence of adequate oxygen.

Chemical asphyxiants can prevent oxygen from binding with the blood's oxygen-carrying protein, **hemoglobin**, as is the case with carbon monoxide. This occurs because carbon monoxide binds about 200 times more strongly to the hemoglobin than does oxygen. Thus, sites that would normally carry oxygen carry carbon monoxide instead. It takes a concentration of only 0.1 percent CO to inactivate half of the body's oxygen-carrying capacity in this way. The hemoglobin-CO molecule that is formed is called **carboxyhemoglobin**.

Miners used to take canaries into mines to use as "living oxygen meters." The basic premise was that a small bird, with metabolic and respiration rates

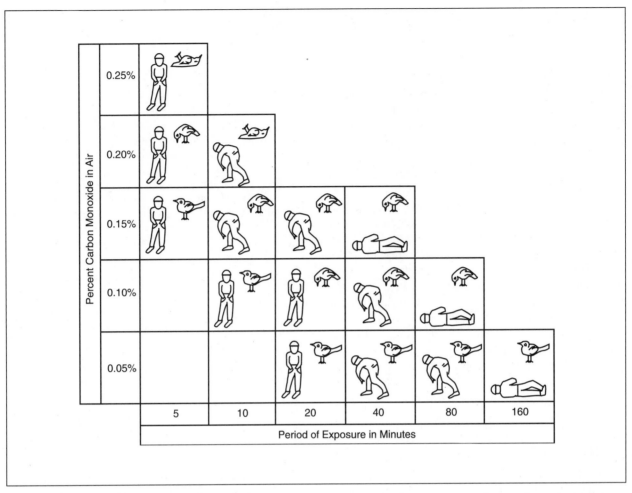

Figure 4-7: Effects of CO concentrations on miners and birds. Canaries, used as living gas meters for years by underground miners, are not necessarily good predictors of the effects of CO on humans. A monkey would have been a better choice!

much higher than that of the miners, would be susceptible to potentially critical changes in the atmosphere before the men would. The collapse of the bird was a warning for the miners to leave. However, this is not a fail-safe method for testing the safety of an atmosphere for human occupation. In canaries, the hemoglobin's affinity for carbon monoxide is about half as strong as in humans (110 vs. 200 times). This means that the amount of carboxyhemoglobin that forms in the bird's blood is always going to be lower than that for humans. Because of the difference in binding affinities, birds can exist in atmospheres containing concentrations of CO that are dangerous for humans. Figure 4-7 illustrates this point.

Another mechanism of chemical asphyxiation causes the iron present in hemoglobin to change from a ferrous state (Fe^{2+}) to a ferric state (Fe^{3+}). The resulting molecule is called **methemoglobin**, which cannot bind either oxygen or carbon monoxide. Asphyxiation resulting from such a reaction is called **methemoglobinemia**. Nitrates, amines, chlorates, and some nitrogen compounds can cause this type of hypoxia. The best known of these is hydrogen cyanide, HCN.

Checking Your Understanding

1. Explain the difference between simple and chemical asphyxiants.

2. What is hemoglobin?

3. What are carboxyhemoglobin and methemoglobin?

4. Give two names for the condition in which the tissues are lacking adequate oxygen.

Summary

The respiratory tract is a significant and efficient route for airborne dusts, mists, fumes, gases, and vapors to enter the body. Since all air contains suspended particles and gases, the respiratory tract needs protective mechanisms: these are coughing, the muco-ciliary ladder, and engulfment. Penetration into the lung by particles depends largely on size; for gases and vapors, solubility is important. The toxic effect of an inhaled material depends on where in the respiratory tract it is deposited, as well as on the dose and the chemical and physical properties of the substance.

Because of the link between site of deposition and development of disease, methods for size-selective sampling have been developed. These techniques allow industrial hygiene professionals to measure the amounts of specific sizes of airborne particulates that are present. The response of the lungs to the deposition of inhaled dusts varies; some dusts with occupational exposure significance include silica, asbestos, and cotton dust. Inhaled metals can cause a range of responses and disease; some affect the respiratory tract while others have target organ effects. Metals with occupational exposure significance include lead, beryllium, cadmium, and chromium. Lung function is evaluated using spirometry. Spirometers measure specific volumes of inhaled or exhaled air, which are compared against standard values. Spirometry may be used as a screening tool, as part of medical surveillance to detect changes in lung function over time, and as part of a medical evaluation for respirator use.

Oxygen-deficient atmospheres may develop in a variety of occupational settings. Simple asphyxiants displace air, creating an oxygen-deficient atmosphere. Chemical asphyxiants produce hypoxia due to interference with the absorption or utilization of oxygen on a molecular level. Chemical asphyxiation can occur even in the presence of adequate atmospheric oxygen. Examples of chemical asphyxiants include carbon monoxide and hydrogen cyanide.

Table 4-7 summarizes the site of action and associated disease or condition for some inhaled substances.

Toxicant	Common Name of Disease	Site of Action	Acute Effect	Chronic Effect
Asbestos	Asbestosis	Parenchyma		Pulmonary fibrosis, pleural calcification, lung cancer, pleural mesothelioma
Aluminum	Aluminosis	Upper airways, alveolar interstitium	Cough, shortness of breath	Interstitial fibrosis
Aluminum abrasives	Shaver's disease, corundum smelter's lung, bauxite lung	Alveoli	Alveolar edema	Fibrotic thickening of alveolar walls, interstitial fibrosis and emphysema
Ammonia		Upper airway	Immediate upper and lower respiratory tract irritation, edema	
Arsenic		Upper airways	Bronchitis	Lung cancer, bronchitis, laryngitis
Beryllium	Berylliosis	Alveoli	Severe pulmonary edema, pneumonia	Pulmonary fibrosis, progressive dyspnea, interstitial granulomatosis, *cor pulmonale*
Boron		Alveoli	Edema and hemorrhage	

Table 4-7: Inhalation is the most significant route of entry to occupational airborne hazards. The effects associated with inhalation of the material depend on where in the respiratory tract the inhaled particles are deposited.

Toxicant	Common Name of Disease	Site of Action	Acute Effect	Chronic Effect
Cadmium oxide		Alveolus	Cough, pneumonia	Emphysema, *cor pulmonale*
Carbides of tungsten, titanium, tantalium	Hard metal disease	Upper airway and lower airway	Hyperplasia and metaplasia of bronchial epithelium	Fibrosis, peribronchial and perivascular fibrosis
Chlorine		Upper airways	Cough, hemoptysis, dyspnea, tracheobronchitis, bronchopneumonia	
Chromium (VI)		Nasopharynx, upper airways	Nasal irritation, bronchitis	Lung tumors and cancers
Coal dust	Pneumoconiosis	Lung parenchyma, lymph nodes, hilus		Pulmonary fibrosis
Coke oven emissions		Upper airways		Tracheobronchial cancers
Cotton dust	Byssinosis	Upper airways	Tightness in chest, wheezing, dyspnea	Reduced pulmonary function, chronic bronchitis
Hydrogen fluoride		Upper airways	Respiratory irritation, hemorrhagic pulmonary edema	
Iron oxides	Siderotic lung disease: silver finisher' s lung, hematite miner' s lung, arc welder' s lung	– Silver finisher' s pulmonary vessels and alveolar walls; – Hematite miner' s upper lobes, bronchi and alveoli; – Arc welder' s bronchi		– Silver finisher' s subpleural and perivascular aggregations of macrophages; – Hematite miner' s diffuse fibrosis-like pneumoconiosis – Arc welder' s bronchitis
Kaolin	Kaolinosis	Lung parenchyma, lymph nodes, hilus		Pulmonary fibrosis
Manganese	Manganese pneumonia	Lower airways and alveoli	Acute pneumonia, often fatal	Recurrent pneumonia
Nickel		Parenchyma (NiCO), nasal mucosa (Ni2S3), bronchi (NiO)	Pulmonary edema, delayed by two days (NiCO)	Squamous cell carcinoma of nasal cavity and lung
Osmium tetraoxide		Upper airways	Bronchitis, bronchopneumonia	
Oxides of nitrogen		Terminal respiratory bronchi and alveoli	Pulmonary congestion and edema	Emphysema
Ozone		Terminal respiratory bronchi and alveoli	Pulmonary edema	Emphysema

Table 4-7: Continued.

Toxicant	Common Name of Disease	Site of Action	Acute Effect	Chronic Effect
Phosgene		Aveoli	Edema	Bronchitis
Perchloroethylene			Pulmonary edema	
Silica	Silicosis, pneumonconiosis	Lung parenchyma, lymph nodes, hilus		Pulmonary fibrosis
Sulfur dioxide		Upper airways	Bronchoconstriction, cough, tightness in chest	
Talc	Talcosis	Lung parenchyma, lymph nodes		Pulmonary fibrosis
Tin	Stannosis	Bronchioles and pleura		Widespread mottling of x-ray without clinical signs
Toluene		Upper airways	Acute bronchitis, bronchospasm, pulmonary edema	
Vanadium		Upper and lower airways	Upper airway irritation and mucus production	Chronic bronchitis
Xylene		Lower airways	Pulmonary edema	

Table 4-7: Continued.

Critical Thinking Questions

1. Describe the responses and associated diseases of the lung to deposits of inhaled dusts containing silica, coal, and asbestos.

2. What are the major groups or classes of airborne hazardous materials? Give examples of each.

3. Compare airborne fibers with airborne particles from an inhalation hazard viewpoint.

4. Explain how particle size is related to occupational diseases of the respiratory tract.

5. Describe the pattern of illness that follows an acute exposure to zinc fumes. *Page 206 of Toxicology*

6. Explain how hydrogen sulfide, a chemical asphyxiant, can exert its effect even in the presence of adequate atmospheric oxygen.
Page 211 of Toxicology and 83

5

Sampling for Airborne Contaminants

Chapter Objectives

Upon completing this chapter, the student will be able to:

1. **List** the main reasons for performing air sampling in an occupational setting.

2. **Describe** the difference between area samples and personal samples as well as integrated, direct-reading, and passive sampling methods within each.

3. **Describe** at least one method used for sampling the following types of air contaminants: particulates such as dusts, mists, and fumes; vapors and gases; and oxygen-deficient and flammable atmospheres.

4. **Explain** the importance of documenting sampling events and list some of the major items that should be recorded.

5. **Interpret** and evaluate air sampling results.

6. **Explain** the basic principles of analytical instruments and methods used in laboratories to evaluate air samples.

7. **List** five potential sources of error in sampling.

8. **Discuss** three of the problems or limitations associated with the use of air samples to represent worker exposures.

5-1 Introduction

Previous chapters presented background on airborne contaminants and explained some of the effects of toxic materials in the human body. The reader will recall that inhalation is the most significant route by which airborne materials enter the human body. Measurement of airborne materials is therefore of great interest to industrial hygienists, whose role is to evaluate the potential health hazards to workers breathing contaminant-containing air. Our discussion will begin with some basic terminology and concepts; it will look at some of the devices and methods used for sampling, and will then move on to sampling methods. We will also briefly discuss some of the analytical methods used for evaluating air samples once they are collected. Interpretation of results and comparison with regulatory limits will be explored and, finally, sources of error will be reviewed and some of the limitations of air sampling results will be discussed.

Although there are several reasons for sampling air – and other environmental substances, such as groundwater and soils – our focus will be on methods that are used by occupational health professionals.

5-2 Why Sample the Air?

The presence of airborne toxins in the workplace has long been recognized. Primitive methods of respiratory protection included the use of goat bladders strung across the workers' faces; more recent examples include firefighters' use of their neckerchief or bandana. Neither of these methods are effective for removing airborne contaminants that are microscopic in size. OSHA regulations and good industrial hygiene practice rely on and/or require three basic approaches (listed in the preferred order of implementation) to protect workers from hazardous exposures: 1) an engineered control; 2) an administrative or behavior-based approach; and 3) personal protective gear – usually as a last resort. We will take a closer look at these three approaches in Chapters 11 and 12, which deal with control of airborne contaminants and selection and use of PPE.

The preferred approach to dealing with airborne hazards is to engineer them out of the process, usually through process design changes, such as containing or enclosing the material. One example of an engineered method for controlling a contaminant is capturing it with a local exhaust ventilation system that removes it from the work area. Another alternative is to substitute less hazardous materials for those that present real or potential hazards to the workers. The ideal substitute or replacement would be one that eliminates or greatly reduces the health and physical hazards to the workers; often these materials pose fewer environmental hazards as well. Regardless of the method used for worker protection, air sampling is used 1) to determine the concentration of contaminants present before design changes; 2) to monitor levels while waiting for implementation of controls; and 3) after controls have been established, to monitor their effectiveness.

In manufacturing settings, where the processes tend to be stationary and well defined, engineering out the hazard or installing controls are often feasible and can be very cost-effective when compared to the costs of worker's compensation, health care, and loss of productivity, not to mention the personal cost to employees whose health is affected as the result of inhalation of a hazardous or toxic material. There are, however, several settings where engineered controls are either not feasible or not effective in reducing or controlling levels of airborne contaminants. Examples include cleanups at hazardous waste sites, where the airborne hazards may vary daily and exist only temporarily. In these situations, air sampling is used to characterize the site, that is, to provide information about which toxic materials are present and at which levels. These data are used to plan subsequent air sampling efforts and to select the protective gear (respirators, gloves, special clothing) that will protect the workers from the materials and concentrations that have been identified.

Sampling is also performed at sites where a response to a hazardous materials release or spill has occurred. The sampling information may be used to establish safe distances, formulate evacuation plans, and select protective gear and equipment to

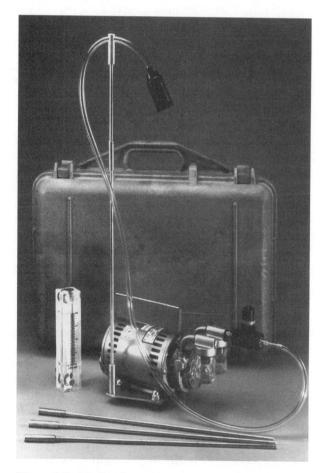

Figure 5-1: High-volume (2-15 lpm) air sampling pumps such as this are used for area sampling because they allow a relatively large volume of air to be sampled in a short time. They are also used for clearance sampling following removal of asbestos materials or lead-based paint.

be worn by responders entering the spill area. Air sampling is often done at locations where the existence of an airborne hazard is the result of the activities being performed, as in lead or asbestos abatement. In fact, sampling outside of an enclosed asbestos abatement work area is required by OSHA whenever these outside areas are occupied. These perimeter air samples provide important information about the effectiveness of the enclosure in containing the asbestos fibers. Sampling inside the work area indicates whether the work practices are effectively keeping fiber levels within limits set by regulations or job specifications.

Another reason for sampling the air is to alert personnel to a potentially dangerous atmosphere from escaping gases or other chemicals, or to warn of the uncontrolled release or leak of a material into the environment. The air may be sampled continuously or on a frequent and regular basis by an automated system connected to an alarm, which sounds at a predetermined level set by the user.

Air sampling is required by OSHA before workers enter a confined space, such as a tunnel, pit, tank, or other enclosed area, where the air may be unsafe. Specific tests must be performed to determine the amount of oxygen present, to measure the potential flammability of the air, and to find out if toxic materials are present at concentrations that pose a hazard to workers entering the space.

Finally, we arrive at what one might consider the most obvious reason for air sampling: to determine whether or not the levels of contaminants in the workplace air are within the OSHA-prescribed allowable levels, or PELs. For some contaminants, for example asbestos, OSHA regulations specify the sampling method to be used; use of a different method might yield results that are of no use in determining exposures relative to the regulatory limits. In addition to specification of a sampling and analysis method, OSHA may also specify the frequency with which sampling must occur.

For most regulated substances, the employer is required by OSHA to perform initial air sampling and may also need to implement a regular monitoring program. For example, for lead, if the results of the initial sampling reveal that airborne concentrations exceed the action level, sampling must be repeated every six months. If the samples exceed the PEL, sampling must be repeated on a quarterly basis. If periodic sampling data shows that the airborne concentrations of lead have decreased, the regulation requires that the decreased levels be demonstrated through repeated low results for samples taken at least a week apart – one low value is not enough to change the monitoring frequency. Other events that trigger additional sampling are changes in production, process, controls, or personnel that may result in actual or suspected increased exposure to lead. There are similar requirements for exposure monitoring in most OSHA substance-specific standards.

An effective industrial hygiene program will include a mechanism for identifying processes or activities for which exposure monitoring needs to be performed to 1) provide initial or baseline exposure data; 2) meet a regulatory requirement; or 3) provide historical data about exposures for company records. Monitoring programs can provide records of employee exposures over an extended time, allowing the industrial hygienist to alert management about conditions that may present unacceptable risks to employees. Monitoring is also useful for evaluating the adequacy of controls that are being used, such as ventilation or enclosed process systems. The use of respiratory protection also triggers a requirement for periodic sampling; OSHA's standard for respiratory protection, 29 CFR 1910.134, states that where respirators are used, the employer must monitor the workplace to verify that the concentrations of the contaminants of concern are within the use limitations of the respirators.

Thus, reasons for sampling the air in an occupational setting include:

— Identification of contaminants in an emergency situation (such as a hazardous material spill), or in an unknown atmosphere (as in a confined space);

— Identification of potential overexposure situations or high-exposure activities;

— Providing a historical record of employee exposures for company and employee records;

— Evaluating effectiveness of engineering controls;

— Verifying adequacy of respiratory protection;

— Initial determination of exposures;

— Periodic sampling to meet regulatory or company policy requirements;

— Evaluation of exposure status relative to an exposure limit.

No matter what the reason, it is important that the person taking the samples be familiar with the purpose and intended use of the results. The use of the results will affect: the selection of the sampling method, the type of sample(s) taken, the number of samples taken, the length of the sampling interval, the method of analysis, and the reporting of the results.

Sampling Approaches

There are two general methods for approaching the issue of air sampling. One is called **direct-reading** or real-time sampling. This type of sampling provides immediate or, at least, very fast feedback in terms of the sample results. The use of an oxygen meter to check the atmosphere inside a confined space is an example; once the sampling probe is placed in the air to be tested, the instrument will give a response within a few seconds or minutes. The second type of air sampling is **integrated sampling**. These samples are taken by drawing air through or across a collector, called the **sampling medium** (plural, media). The sampling medium is then analyzed by a laboratory to determine the amount of contaminant that is present. Sampling media commonly used for collecting integrated samples are listed in Table 5-1.

In order to collect samples for later analysis using an integrated sampling technique, the sample collector, which includes the holder or container that houses the sampling medium, is connected to the sampling pump using flexible tubing. The air enters the inlet and passes through the sampling medium, where the contaminant is absorbed or otherwise captured. The air then moves on through the pump and exits. The **sampling train** is composed of three elements: the sampling media in their holder, the tubing, and the sample pump. The components of

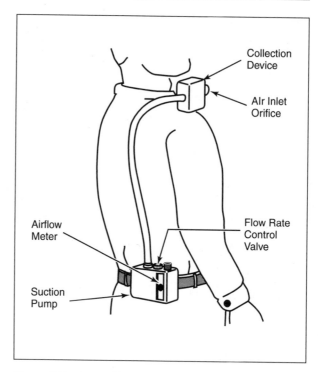

Figure 5-2: A sampling train consists of the inlet, collection device, and air sampling pump. The entire train is assembled, calibrated, and then attached to the worker as shown.

the sampling train are dictated by the requirements of the sampling method, which will specify the sampling medium to be used for collection, as well as the total volume of air that must be sampled to yield accurate analytical results. The rate at which the air is pulled through the sampling train by the pump can be adjusted and set at the flow rate that is specified in the sampling method, and must be verified with a representative sample collector connected to the pump. We will take a closer look at air sampling pump calibration later in the chapter. Figure 5-2 is a diagrammatic representation of a sampling train.

Sampling pumps come in a range of sizes with different flow capabilities. Pumps used for measuring individual exposures are generally small, weigh 1-2 pounds or less, and can pump air at rates ranging from as low as one cubic centimeter per minute (cc/min) to four liters per minute (lpm). The lower flow rates, ranging from .001 to a few hundred cc/min, are generally used for sampling gases and vapors, while flow rates of 1-4 lpm are used for many particulates. High flow pumps are capable of sampling rates of more than four lpm. These pumps are larger, heavier, and often noisier than those used for personal samples, so they are more apt to be used to take samples in a fixed location in a work area.

Contaminant/Aerosol Type	Sampling Media Commonly Used
Particulates – dusts, fibers, fumes	Filters
Dusts and mists	Filters, absorbing solutions
Gases and vapors	Absorbing solutions, solid sorbent tubes, passive samplers, evacuated containers, and sample bags

Table 5-1: The type of aerosol being collected influences the choice of collection media, which is usually indicated by the sampling method. In some cases, a combination of media is needed.

Figure 5-3: Personal sampling pumps such as this one are smaller and lighter than pumps commonly used for area sampling. Personal sampling pumps are used by the industrial hygienist to obtain data for evaluation against regulatory and other exposure limits.

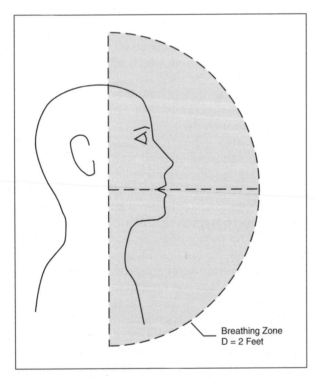

Figure 5-4: The breathing zone is defined as a two-foot diameter half-sphere around the worker's head and shoulders. This area contains the atmosphere that the worker is most likely to inhale. The inlet of the personal sample collector should be located within this region.

Personal samples are those that are obtained when a worker actually wears a sampling train during the work shift. The results of these samples are later compared against exposure limits. The sample collecting device is clipped or otherwise attached to the worker's shirt collar or lapel in order to collect (sample) the air that is representative of what the worker inhales. The region that is defined by a two-foot diameter half-sphere around the front of a worker's head and shoulders is the **breathing zone**; personal samples are therefore sometimes called breathing zone samples. In some instances, it is also appropriate to take samples of the air in a fixed location in the work area. These **area samples** provide the industrial hygienist with information about average levels of contamination and can be useful for evaluating overall air quality or the effectiveness of controls; however, they are generally not acceptable methods for evaluating worker exposure. Area sampling does have specific applications, such as monitoring the air outside of an asbestos or lead abatement (removal) area. Area samples are also used to evaluate the air following a hazardous material removal or cleanup action. These final area samples, called **clearance samples**, must show that levels of contaminant are at or below a specific concentration, which for asbestos is 0.01 fibers/cc, before the area can be released for normal occupation and work activities.

Checking Your Understanding

1. What are two reasons why OSHA might require more sampling, even if historical results show levels within regulatory limits?

2. Why is it necessary for the person doing the sampling to understand how the results will be used?

3. Explain the difference between engineering controls and exposure monitoring.

4. What is the difference between direct-reading and integrated sampling methods?

5. What type of sampling medium is used to collect particulates; dusts and mists; gases and vapors?

6. What is a personal sample? An area sample? A clearance sample?

7. Where is the breathing zone of a worker, and why is this the location for positioning a sampler when obtaining a personal sample?

5-3 Sampling Particulates

Filters

Filters are some of the most common sampling media used by industrial hygienists. Of the nearly 2,000 sampling methods described by NIOSH and OSHA, about one-fourth requires the use of some type of filter. Air sampling filters are made using different materials and methods. The type of filter to be used will depend upon the contaminant being sampled as well as on the sampling method chosen. Because the filters are quite thin, it is necessary to place them into rigid holders, called **cassettes**, during sample collection. Filter cassettes are commonly made of polystyrene, but other materials may also be used. A support pad is usually placed beneath the filter in the cassette. In some cases, the support pad is analyzed along with the filter. Like filters, support pads are made from various materials: some are cellulose, some are porous plastic; stainless steel mesh is also used.

Some methods specify the use of an **open face** cassette for sample collection. This means that the top portion of the cassette is removed during sampling. The filter and the support pad are held in place by a retaining ring that fits into the cassette base. Following sampling, the cassette is reassembled and returned to the laboratory for analysis. In a **closed face** sample, the small plug in the top of the cassette is removed and air enters the cassette through the small hole in the center; the plug is replaced following sampling, and the entire assembly is returned to the laboratory for analysis. Open face samples are used when distribution across the surface of the filter is important in the analytical result. For example, sampling for asbestos fibers requires an open face arrangement since analysis is performed on a section of the filter that is representative of the fiber loading across the entire filter. In contrast, evaluation of airborne metal fume involves analysis of the entire filter to determine the total amount of metal present. Since distribution across the surface of the filter is not a factor in metals analysis, a closed face arrangement is suitable for sampling. The sampling method will specify which arrangement is to be used.

It is logical to think that filters used for air sampling work like a screen or sieve to capture airborne contaminants. In fact, most filters used for industrial hygiene applications work much differently. Because of the way many filters are manufactured, there is no straight path through little holes in the filter. Instead, the openings or **pores** in the filter are more like a maze of many interconnected tunnels through which the air moves – turning, branching, speeding up and slowing down, as it passes through the thickness of the filter. This allows the contaminant to collide with surfaces in the filter (**impaction**), or stick to surfaces as it passes close by them (**interception**). Another possibility is that the particles might have an electric charge that causes them to be attracted to the filter material; this is called **electrostatic attraction**. The mechanism of collection will depend on the size of the particle, the electric charge of the particle and filter, the type of filter

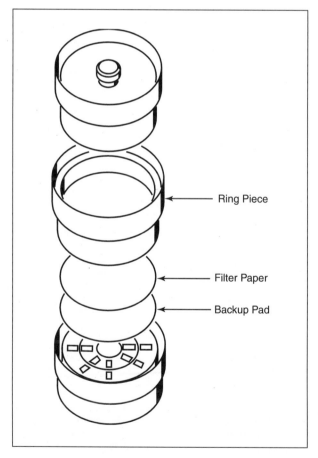

Figure 5-5: Filter and cassette assembly. Polystyrene cassettes such as this may be purchased preassembled, with the desired filter in place. Some analytical methods include the backup or support pad in the analysis.

Ring Piece

Filter Paper

Backup Pad

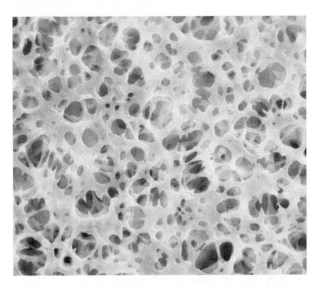

Figure 5-6: Photomicrograph of the surface of an air sampling filter. Filters used for collection of airborne particulates do not merely gather material on the surface; some particulate is actually deposited within the filter.

Mixed Cellulose Ester Filters (MCEF)

One of the most commonly used filters is the **mixed cellulose ester filter (MCEF)**. These filters are manufactured from a polymer that starts as a liquid, which is spread out in a thin layer to solidify or dry, like gelatin setting up. During the solidification process, the small openings or pores develop in the filter material. As mentioned above, the pores do not go straight through the surface of the filter (see Figure 5-7). This arrangement makes the filters very effective collectors of particles entrained in the air. The average size of the openings can be controlled in the manufacturing process, allowing the production of filters with pore sizes in a specific range suitable for collecting particulate of different sizes. MCEF filters are available with pore sizes ranging from 0.4 to 0.8 μm. Metal fume and asbestos fibers are collected on MCEF filters with pore sizes of 0.8 μm. Analysis of the filter is dependent upon the material that was collected. Filters used to sample

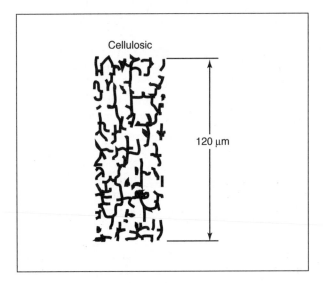

Figure 5-7: Cross-section through an MCEF filter. Filters are not sieves, but are more like mazes into which particles are pulled and become trapped.

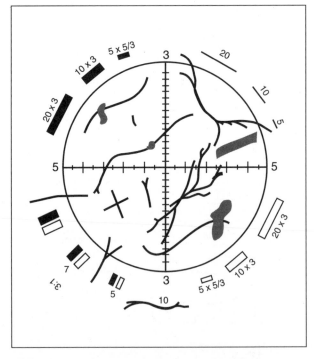

Figure 5-8: This is a diagrammatic representation of a filter being checked for airborne fibers, the method used for evaluating occupational exposure to asbestos. According to the counting requirements, all structures that are five microns or more in length and three times longer than they are wide are counted. The grid marks superimposed on the slide are from an optical insert – called a Walton-Beckett graticule – in the microscope.

being used, and the sample flow rate. It is possible that more than one mechanism will play a role in collection of a contaminant. As mentioned above, the pores make it possible to collect contaminants in the filters as well as on them. The capture efficiency of a filter is therefore not limited by the size of the pores.

for welding fumes, for example, are dissolved or digested in an acid solution, which is then analyzed for metals content. Fibers are quantified by removing a section of the filter and treating it with acetone vapor, so that it becomes clear. It is then possible to count the fibers while viewing the cleared filter through a microscope.

PVC Filters

Polyvinyl chloride (PVC) filters have good resistance to acidic and basic substances and do not absorb much water vapor from the air. Because of their **hydrophobic** properties (water resistance) and because they are lightweight, they are commonly used to collect dusts, such as silica-containing dusts. Before being used to collect dust, the filters are assigned identification numbers and placed into a sealed chamber, called a **desiccator**, which contains a water-absorbing substance like silica gel. The filters are left in the desiccator and allowed to dry completely. The filters are weighed using a sensitive electronic analytical balance, able to measure masses as small as ± 0.0001 grams. The filters are periodically weighed (e.g., every 24 hours) until they reach a constant weight within experimental error. These preweighed filters are then assembled in sampling cassettes and used for dust sampling.

After sampling, the filters are returned to the laboratory where the analyst carefully removes them from the cassettes and again places them into the desiccator. The weighing process is then repeated. This type of analysis, involving weighing of the sample, is called a **gravimetric analysis**. The presampling mass of the filter is subtracted from the after-sampling mass to determine how much dust was collected on the filter. The mass of collected dust and the sampled air volume are used to calculate the airborne concentration. These filters are sometimes further analyzed; for example if the sampling is to determine crystalline silica exposure, the filters might be examined after weighing using x-ray diffraction to identify which forms of crystalline structures are present – cristobalite, tridymite, or amorphous silica. This information can be used to determine the composition of the sand that was being used in the area where the samples were taken, which, in turn, is used to compute the PEL (see Chapter 4).

Sampling for dusts that might be inhaled requires that the industrial hygienist know what fraction of dust is of interest. As we saw in Chapter 4,

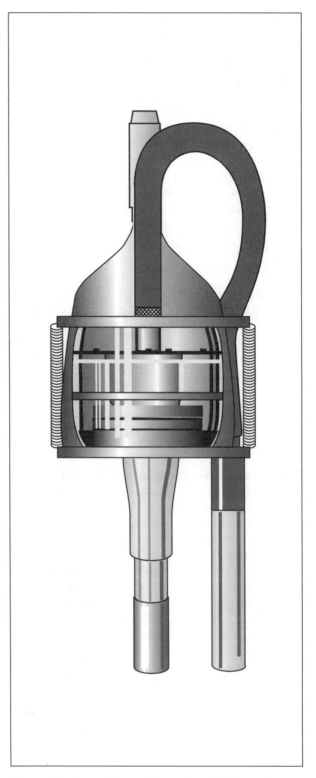

Figure 5-9: A sampling cyclone such as this is used to sample for the respirable fraction of airborne particulate. The air enters the inlet, then passes through the vortex-like separator, which removes larger particles from the air stream. The smaller particles are then collected on the filter inside the holder.

devices have been designed to separate and capture dust particles that fall into three specific size ranges: the inhalable, the thoracic, and the respirable fractions. The samplers, or **cyclones** as they are sometimes called, are placed on the inlet side of the filter cassette, where they separate the dust that enters, allowing only the desired fraction of dust to enter for collection on the filter. The two fractions most commonly sampled for are the inhalable fraction – which represents the total amount of dust that may enter the respiratory tract – and the respirable fraction – which contains the smallest particles that can penetrate into the deep lung.

Teflon Filters

Another polymer filter used for industrial hygiene sampling is the **Teflon® filter**, which is sometimes referred to as the PTFE (for polytetraflouroethylene) filter. These filters, like PVC filters, are chemical-resistant and hydrophobic. Aromatic hydrocarbons, such as benzo(a)pyrene that may be given off from hot tar or asphalt, are collected on PTFE filters.

Glass Fiber Filters

Glass fiber filters are composed of layers of fibers arranged in what appears to be a haphazard pattern to form a sort of tangled mat, similar to the filters used in air conditioners, furnaces, and automobiles. Glass fiber filters, sometimes referred to as AE filters, are used for collecting particulates and some droplets of contaminants, such as mercury and acid gases. Often these filters are used as a sort of upstream precollection device; they are placed in front of another collector so that contaminant particles of larger size do not enter the second collector. This can increase the efficiency of the sampling by increasing the percentage of the total amount of airborne contaminant that is captured or removed from the air as it moves through the media. This also allows simultaneous collection of contaminants that are present in the air in two different physical forms; for example, solid aerosols such as fumes are collected on the filter while vapors are collected on the sorbent.

Coated/Treated Filters

Some sampling methods require the use of filters that have been coated with a specific chemical: these are the **coated/treated filters**. These coatings vary, depending on the contaminant to be collected. The coatings enhance collection by chemically reacting with the contaminant as the air is drawn through the filter. An example is NIOSH Method 6004 for sampling sulfur dioxide. This method requires the use of a 37 mm diameter MCEF followed by a cellulose filter coated with potassium hydroxide. There are several OSHA methods for sampling organic amines that use two 37 mm glass fiber filters coated with sulfuric acid. Like other sampling media, these can be ordered from suppliers and arrive preloaded into cassette holders, ready for use.

Potential Problems Associated with Filter Collection

Collection on filter media does pose some potential problems to the occupational health professional. Among these problems are:

— Overloading – It is possible to overload filters. If the analysis requires examination of the filter, as in fiber sampling, the presence of too much particulate can make it difficult or impossible to accurately quantify the fibers collected on the filter. In the case of contaminants that are analyzed gravimetrically, including such contaminants as silica-containing dust or respirable dusts, too much particulate on the filter can make loss of material a source of error during handling and weighing of the filter.

— Static electricity – Static electricity can also be a problem for filters subject to gravimetric analysis. The filters pick up a charge and are then either repelled or attracted by the balance tray, causing errors in mass determination.

— Moisture or physical damage (tearing, bursting) – It is also possible for filters to become wet; they can tear or burst due to changes in flow or to failure of the IH to remove the inlet plug.

—Contamination with interfering substances – Filters can become contaminated with materials that interfere with the analysis and result in over- or under-estimates of the amount of contaminant that is present.

It is important that the professional who is performing the sampling be aware of the limitations and potential problems associated with the particular sample method so that the best sampling data possible may be obtained during the sampling event.

Checking Your Understanding

1. Explain the physical structure of the following filters used for air sampling: MCEF; glass fiber filters; coated filters.

2. Define/explain the following mechanisms for capture of contaminants by filters:

   a. Impaction;
   b. Interception;
   c. Electrostatic attraction.

3. Name the type of filter likely to be used for sampling the following:

   a. Silica-containing dust;
   b. Welding fume;
   c. Asbestos fibers;
   d. Aromatic hydrocarbons.

4. Explain the difference between open-face and closed-face sampling.

5. What is the purpose of adding chemical coatings to a filter?

6. What are the possible consequences of overloading a filter?

5-4 Sampling Gases and Vapors

Sorbent Tubes

Sorbent tubes are small glass tubes that contain sampling media. To use the tubes, the IH snaps off each end and inserts one end of the tube into a holder that is connected to the sampling pump. Most sorbent tubes have arrows on them to indicate the direction of airflow. If there is no directional arrow on the tube, the end nearer the small backup section should be placed into the sampling tube or holder. The tube holder has a clip for attachment to the worker's shirt, and most holders also cover the sharp inlet end of the tube to prevent the worker from possible cuts. The media or **sorbent** that is in the tube can be one of a number of materials, such as charcoal, silica granules (called silica gel), or one of several other granular polymers. As with filters, the sampling method will specify the type of sorbent

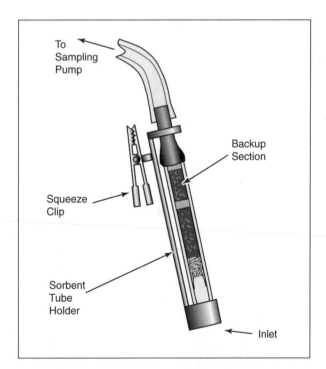

Figure 5-10: Sorbent tubes are made of glass; the ends must be broken off by the IH prior to sampling. Sorbent tube holders are constructed to cover the broken ends of the tube, thus protecting the worker from cuts. The tube must be oriented with the backup (smaller) section of sorbent closest to the sampling pump. The sample holder should be attached to the worker's lapel so that the sorbent tube is held in a vertical position.

tube that must be used for sample collection; the sorbent is also sometimes coated with a chemical compound to enhance collection efficiency for a specific contaminant. Collected compounds are removed, or **desorbed**, from the sorbent using a solvent, most often carbon disulfide. Some methods utilize heat to separate the contaminant from the sorbent.

Sorbent tubes collect contaminants through mechanisms that somewhat resemble those of filters. The most common types of sorbent used for occupational exposure sampling include charcoal and silica gel. Activated charcoal contains microscopic openings where molecules are trapped. Silica gel provides a surface on which contaminants may condense or to which they might simply adhere. Other sorbents have specific applications and are selected because of the affinity of the sorbent for certain kinds of compounds, or because of chemical coatings that have been added to the sorbent. Manufacturers have trademark names for some of these: Tenax, XAD2, Chromosorb, and others. We will not address these other sorbents; for more information, see the references at the end of the book.

Charcoal, or more accurately, **activated charcoal**, is used for a wide range of occupational exposure sampling applications. The charcoal, much of which comes from coconut shells, is activated through heating with 800-900°C steam. This causes the air spaces in the charcoal to expand, so that a cross-section looks more like a sponge or a layer cake; it is in these air spaces that the contaminant is trapped. Charcoal tubes are used to collect a variety of polar and non-polar molecules. Molecules that have a partial positive charge at one end and a partial negative charge at the other end are called **polar compounds**. Charcoal tubes are used in collection of organic solvents and vapors including many alcohols, ketones, and a variety of aromatic compounds. Another commonly used sorbent is silica gel, produced from the action of sulfuric acid on sodium silicate. Silica gel tubes are most often used to collect polar compounds; examples of such materials are amines and halogenated compounds such as hydrogen chloride, hydrogen bromide, and hydrogen fluoride. There are other sorbent tubes, some containing charcoal or silica gel with a coating of a chemical that increases the sorbent's ability to capture a specific contaminant. Again, the sampling method will specify the sorbent to be used.

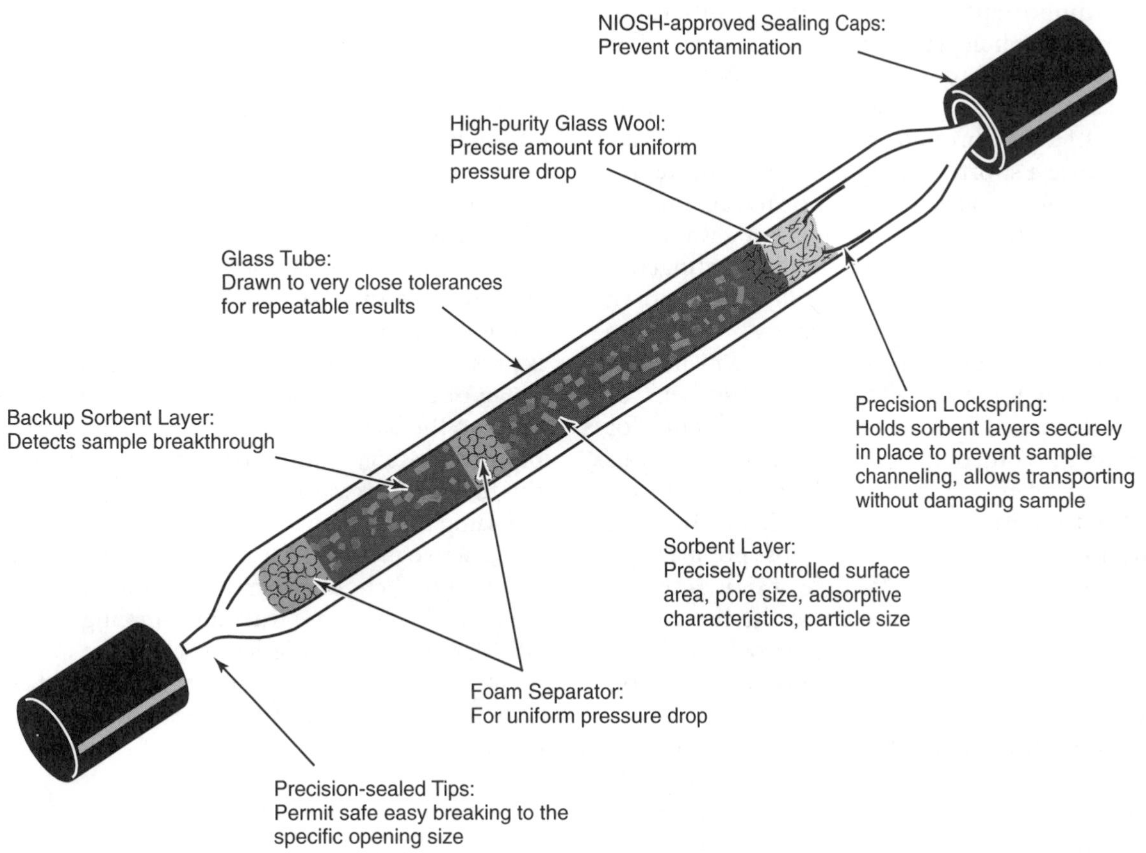

NIOSH-approved Sealing Caps:
Prevent contamination

High-purity Glass Wool:
Precise amount for uniform
pressure drop

Glass Tube:
Drawn to very close tolerances
for repeatable results

Precision Lockspring:
Holds sorbent layers securely
in place to prevent sample
channeling, allows transporting
without damaging sample

Backup Sorbent Layer:
Detects sample breakthrough

Sorbent Layer:
Precisely controlled surface
area, pore size, adsorptive
characteristics, particle size

Foam Separator:
For uniform pressure drop

Precision-sealed Tips:
Permit safe easy breaking to the
specific opening size

Figure 5-11: The internal assembly of sorbent tubes is based on an original specification from NIOSH. Larger tubes have proportionately more sorbent material in the sorbent and backup layers.

Many sorbent tubes in use today are assembled according to an original NIOSH specification that applied to charcoal tubes. The specification called for a 7 cm-long glass tube with an outside diameter of 6 mm and an inside diameters of 4 mm. The charcoal was to be small enough to fit through a 20/40 mesh screen. The main or front section was to hold 100 grams of media; a smaller section, called the back or backup section, was to contain 50 grams of media. Each section was separated by a plug of glass fibers, with additional glass fiber plugs at each end to hold the sorbent in place. Today, foam plugs are used instead of the glass fibers, since the foam presents less air resistance. Larger tubes are also available; they are larger in terms of overall dimensions and use proportionately greater volumes of sorbent. The larger tubes, with their increased volume of

sorbent, allow for longer sampling intervals and larger volumes of air to be sampled using a single tube. Figure 5-11 shows the construction of a typical sorbent tube.

The backup section of sorbent has an important role, which is capturing any contaminant that passes through the front or main section. The indicating arrow on the sorbent tube is to ensure that the tube is properly oriented, so that air enters the front section of sorbent first. If the laboratory analysis of the backup section shows that it contains an amount of contaminant equal to 20-25 percent of the amount of contaminant captured in the front section, it is an indication that the front section of sorbent was possibly completely saturated during the sample collection. This condition is called **breakthrough**. When this occurs, the sample results are at best an underestimate of the actual concentration of the contaminant.

It is important to consider the possibility of breakthrough when planning a sampling event. The presence of a high concentration of contaminant increases the chances of breakthrough. To minimize chances of breakthrough, the industrial hygienist needs to use a shorter sampling interval, make frequent changes of sorbent tubes, or use one of the larger sorbent tubes. Although it is not always possible to do so, failure to plan for this contingency can result in inaccurate sample results, which are of limited use to the industrial hygienist.

Another common problem associated with sorbent sampling is the use of the incorrect sorbent, or at least, the use of a sorbent that is not effective for capturing the contaminant of interest. This happens most often when the industrial hygienist is uncertain about the nature of the contaminant. In these situations, a charcoal tube is the most popular sorbent used; typically, a large volume of air is sampled, and the tube is sent for analysis in the hopes that something will show up. While charcoal is a useful and versatile sorbent, it is not a universal collector. Charcoal has its limits: it can be a good collector of polar molecules, like alcohols, as well as nonpolar compounds, like benzene. However, nonpolar compounds may displace polar compounds that have been adsorbed onto the charcoal. In situations where both polar and nonpolar compounds are being collected, it is common for a silica gel tube to be used in series or separately to collect the polar compound.

The temperature and relative humidity in the area during sampling can affect any solid sorbent's ability to capture and hold contaminants, usually adversely. Also, the presence of other compounds for which the sorbent might have a stronger attraction or capture ability can result in the interfering compound displacing contaminant that has already been collected. This will result in data that indicate a lower-than-actual concentration of the contaminant.

Passive Samplers

The term **passive sampling** is used to describe sampling that is done without an air sampling pump. The use of the term passive is a little misleading, since the contaminant does move toward the collection surface; however, its collection is accomplished through diffusion rather than through forcing air into a sampling device. Passive samplers – often called **sample badges** – are small clip-on devices that contain solid sorbent and can be used for collection of a wide variety of airborne materials. These devices are very easy to use: they simply clip to the worker's lapel and are worn throughout the shift. At the end of the day, the sampler is sealed in a container and sent to a laboratory for analysis. Other passive samplers provide user-interpreted results in the form of a color change in response to the presence of a particular contaminant. The response may indicate whether or not a contaminant is present, or it might vary in intensity to indicate a range of concentrations. There are obvious advantages to the use of such samplers: no pumps to calibrate, no battery failures to worry about, no putting the tube in backwards! Analytical results from passive sampling techniques are as good as samples taken using active methods for most applications, making them a good choice where ease of use is a primary concern. As is the case with any sampling event, specific data must be recorded, such as the identity of the person wearing the sampler, notes on their job activities, the length of the sampling interval, and other sampling data.

Impingers

Air contaminants that are nonreactive and highly soluble in a specific solution may be collected using a measured volume of that particular solution in an **impinger**, which is a glass container, also known as a bubbler or a gas-wash bottle. Impingers, which come in a range of sizes to fit a variety of applications, resemble a graduated cylinder with a long inlet tube fitted into a stopper. The inlet tube extends nearly to the bottom of the vial, which holds the solution. The sampling pump is connected so that a partial vacuum is created inside the impinger, drawing air through the inlet tube, where it exits into the solution. The air bubbles up through the solution, and this contact between air and liquid allows the contaminant to dissolve in the liquid. Some impingers have modified inlet tubes, designed to circulate air or otherwise increase contact between the air and the absorbing solutions. Some commonly used impingers have fritted inlets; the **frit** resembles a porous stone, similar to the ones used for aeration

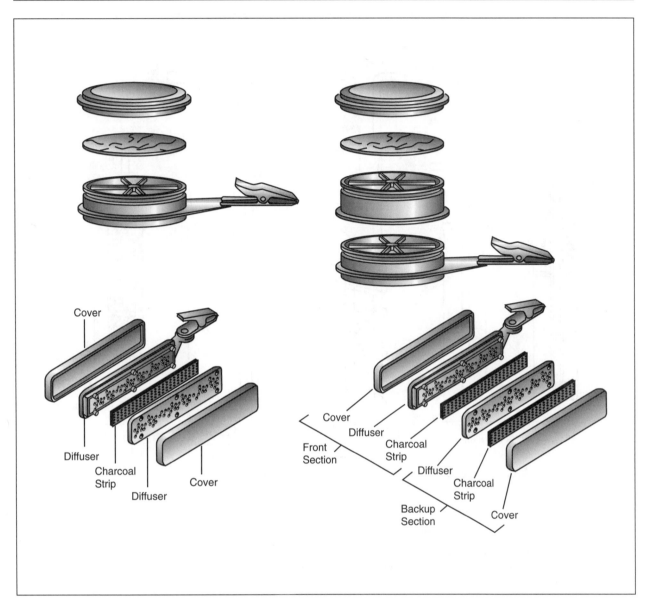

Figure 5-12: Some typical passive samplers. These devices are easy to use, since they require no calibration and no sampling pumps. Following sampling, a cover is snapped over the top of the sampler, and the entire assembly is sent to the laboratory for analysis.

of an aquarium. Frits may be coarse, medium, or fine depending on the number and size of the openings; selection of a particular grade of frit will usually be specified in the method. The frit breaks the airstream into small bubbles; this maximizes the surface area of air that contacts the solution, increasing the amount of contaminant that is dissolved or absorbed in the solution. The use of a fritted bubbler can increase the collection efficiency of hard-to-collect vapors and gases, but it also introduces

some flow losses due to pressure drops, which can be high for a fine frit.

Solutions used in impingers vary depending on the contaminant of interest. The solution may be a specific compound mixed at a certain concentration, for instance a sodium hydroxide solution used for ammonia sampling. At the other extreme, distilled water is a common absorbing solution for many materials that are water-soluble. After sampling, the solution may be sealed in the impinger cylinder, or

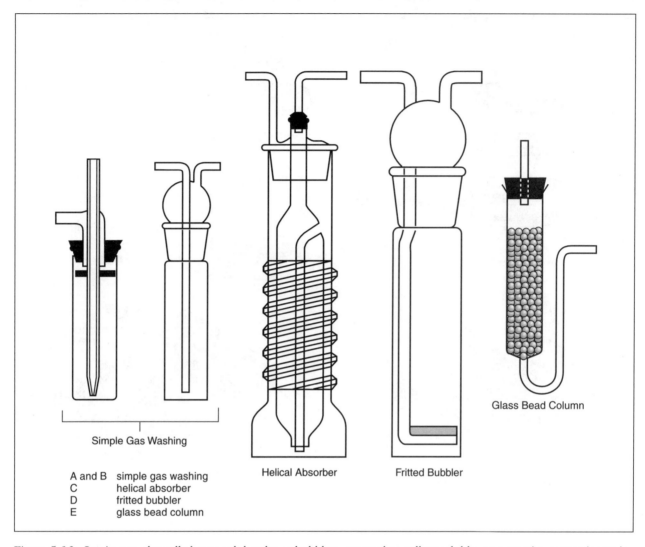

Simple Gas Washing

A and B   simple gas washing
C   helical absorber
D   fritted bubbler
E   glass bead column

Helical Absorber   Fritted Bubbler

Glass Bead Column

Figure 5-13: Impingers, also called gas-wash bottles or bubblers, are used to collect soluble, non-reactive contaminants in a solution. The solution is often transferred to a bottle or vial for transport to the laboratory for analysis.

placed into a different container, and sent to the laboratory for analysis. Again, the sampling method will specify the analytical steps for determining the amount of contaminant that has been collected in the solution.

Possible problems that can arise during use of impingers include loss of solution through spills or leaks; breakage of the cylinder; or evaporation during sampling. Also, some of the solutions used in impinger sampling are corrosive, which makes their use for personal sampling less desirable than a method that uses a sorbent tube or another collection medium.

Grab Samples – Sample Bags and Evacuated Cylinders

A **grab sample** is an integrated sample, but one that represents a very brief sampling period. Grab samples are like snapshots. The sample captures a moment in time, but is not necessarily an indication of constant or average conditions. Grab samples may be useful for evaluating compliance with a ceiling or peak limit. In many situations, the industrial hygiene professional may wish to obtain a grab sample for screening and identification of possible contami-

Figure 5-14: Sample bags made out of Teflon or another inert polymer are useful for obtaining grab samples. The bags are light and easy to use, and are returned to the laboratory for analysis.

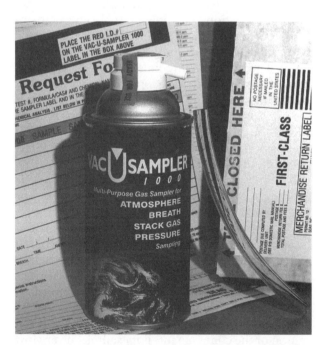

Figure 5-15: Evacuated containers – usually made of glass or metal – are used for grab samples. Air is drawn inside when they are opened; they are then sealed and returned to the laboratory for analysis.

nants, or as a first step in characterizing an atmosphere. These instances present opportunities for sampling gases and vapors using sample bags or evacuated containers.

Sample bags are often made of an inert polymer, such as Teflon® or Tedlar®. These bags have an attached fitting that allows connection to the pump, which pumps air into the bag along with the contaminants. The bag is then sealed and shipped to the laboratory for analysis. Figure 5-14 shows such a bag.

An **evacuated container**, also called an evacuated cylinder, is just what the name implies: a container from which air has been removed and the container has then been sealed. To sample using one of these, the seal is broken and the contaminated atmosphere is drawn into the evacuated container. The opening is then sealed, and the container is shipped to the laboratory for analysis of the contents. Figure 5-15 shows an evacuated container.

Checking Your Understanding

1. Name the two sorbents commonly used for industrial hygiene air sampling.

2. What is meant by activated charcoal?

3. What is meant by the term breakthrough, and how does this affect the results of an air sample?

4. Name two positive traits or aspects of passive samplers.

5. What is a fritted bubbler, and what does it do for sampling efficiency?

6. What is an evacuated cylinder and how is it used?

5-5 Standard Sampling and Analysis Methods

Sampling and analysis methods (also called "sampling methods" or simply "method" throughout the chapter) have been developed and validated for many airborne contaminants. Both OSHA and NIOSH have published sets of standard methods for sampling and analysis, the majority of which are for air samples. Our discussion will focus on those methods that are used for the sampling and analysis of air.

The OSHA sampling methods are collectively called the *OSHA Reference Methods*; NIOSH methods can be found in the *NIOSH Manual of Analytical Methods*, currently in its fourth edition. Sampling methods provide detailed instructions about collecting the sample and analyzing it for the contaminant, which is referred to as the **analyte**. Methods contain – among other things – the following information:

— The sampling media to be used;

— The sampling flow rate;

— The volume of air that is to be sampled;

— Instructions for sample preservation and handling;

— Detailed procedures to be followed for the analysis.

Methods also provide information about the statistical reliability of the sampling method, including an indication of the accuracy of the method. Before it is acceptable for use, a NIOSH method must be shown to be able to provide a result that is within 25 percent of the actual concentration, 95 times out of 100 tries. This means that we can be 95 percent sure that the method, when followed, will yield results that are within 25 percent of the actual concentration. The techniques used for sampling must allow for capture of the contaminant on the sample medium, where it will remain in a stable form. All or most of the contaminant must also be able to be removed from the medium on which it was collected, for analysis in the laboratory. To meet these stringent performance requirements, much planning and testing is involved in developing a sampling method. A detailed description of the criteria used in evaluating methods can be found in the introductory section of the *NIOSH Manual of Analytical Methods*.

Prior to performing sampling for a particular contaminant, the industrial hygienist will refer to the sampling method for the specifics of sample collection – such as selecting the sample medium and calibrating the pumps – and may take a copy along for reference on the day of the sampling. It will be helpful to refer to Table 5-2 to identify the various items or topics that are contained in an air sampling method as we describe them in this section.

One of the first skills learned by an industrial hygienist is the process for verifying the flow rate of air through an air sampling device. This process, called **calibration**, is done to make sure that the air passes through the collection or sampling medium slowly enough so that any contaminant present in the air is absorbed, dissolved, or otherwise collected, depending on the sample medium and the mechanism of collection. The flow must also be at an appropriate rate so that contaminant that has already been collected is not pulled off the sampling medium. The best sampling flow rate is determined as part of the method's development and validation, which involves many tests and trials to make sure the method will consistently provide results that are within the required 25 percent accuracy for the actual airborne concentration. NIOSH and OSHA sampling methods will often specify a range within which the sampling flow rate should be set; in the Ammonia by IC procedure, for example, the flow rate is given as 0.1 to 0.5 lpm.

The total volume of air to be sampled will also be specified in the method and may be given as a range as well; a number of different volumes may be specified for evaluating different concentrations of contaminant. In the latter case, the industrial hygienist performing the sampling must decide the appropriate volume, based on the expected concentration of the contaminant. For example, if very low concentrations are expected in the work area, a greater volume of air will need to be sampled. High concentrations of contaminant may require smaller volumes to avoid overloading the sample collector, which can invalidate the results. The method for ammonia sampling is quite flexible in this regard; the working range is given as 24 to 98 ppm for a 30-liter sample.

The sensitivity of the sampling and analytical method also plays a role in the volume of air that must be sampled. Laboratory techniques used to

AMMONIA by IC 6016

NH₃   **MW: 17.03**   **CAS: 7664-41-7**   **RTECS: BO0875000**

METHOD: 6016, Issue 1	**EVALUATION: FULL**	**ISSUE 1: 15 May 1996**

OSHA: 50 ppm
NIOSH: 25 ppm; STEL 35 ppm; Group III Pesticide
ACGIH: 25 PPM; STEL 35 PPM
(1ppm = 0.697 mg/m³ @ NTP)

PROPERTIES: gas: MP-77.7°C; BP -33.4°C; VP 888 kPa (8.76 atm) @ 21.1°C; vapor density 0.6 (air = 1); explosive range 16 to 25% v/v in air

SYNONYMS: none

SAMPLING

SAMPLER: SOLID SORBENT TUBE
(sulfuric acid-treated silica gel)
a 0.8 μm MCE prefilter may be used to remove particulate interferences.

FLOW RATE: 0.1 to 0.5 L/min

VOL-MIN: 0.1 L @ 50 ppm
-MAX: 96 L @ 50 ppm [1]

SHIPMENT: routine

SAMPLE STABILITY: at least 35 days @ 5°C [2]

BLANKS: 2 to 10 field blanks per set

SAMPLING

RANGE STUDIED: 17 to 68 mg/m³ [1]
(30-L samples)

BIAS: −2.4% [1]

OVERALL PRECISION ($\hat{S}_{rT}$): 0.071 [1]

ACCURACY: ± 14.5%

MEASUREMENT

TECHNIQUE: ION CHROMATOGRAPHY, CONDUCTIVITY DETECTION

ANALYTE: ammonium ion (NH₄⁺)

EXTRACTION: 10 mL deionized water

INJECTION VOLUME: 50 μL

ELUENT: 48 mM HCl/4mM DAP-HCl/4 mM L-histidine-HCl; 1 mL/min
alternate: 12 mM HCl/0.25 mM DAP-HCl/ 0.25 mM L-histidine-HCl; 1mL/min

COLUMNS: HPIC-CS3 cation separator; HPIC-CG3 cation guard; CMMS-1 cation micromembrane suppressor

CONDUCTIVITY SETTING: 30 μS full scale

CALIBRATION: standard solutions of NH₄⁺ in deionized water

RANGE: 4 to 100 μg per sample [3]

ESTIMATED LOD: 2 μg per sample [3]

PRECISION ($\bar{S}_r$): 0.038 [2]

APPLICABILITY: The working range is 24 to 98 ppm (17 to 68 mg/m³) for a 30-L sample. This method is applicable to STEL measurements when sampled at ≥ 0.2 L/min.

INTERFERENCES: Ethanolamines (monoethanolamine, isopropanolamine, and propanolamine) have retention times similar to NH₄⁺.
The use of the alternate (weak) eluent will aid in separating these peaks.

OTHER METHODS: This method combines the sampling procedure of methods S347 [4] and 6015 with an ion chromatographic analytical procedure similar to Method 6701 [5] and OSHA Method ID-188 [3].

NIOSH Manual of Analytical Methods (NMAM), Fourth Edition, 5/15/96

Table 5-2: NIOSH methods are the standard in the United States for sampling and analysis of airborne contaminants, although OSHA methods are also used. Analytical methods contain instruction in taking the sample, including sampling media and flow rate; they also provide instruction on how to conduct analysis of the samples in the laboratory.

analyze samples have boundaries or limits, beyond which the results are not reliable. These limits, which may be expressed as the **limit of detection (LOD)**, or the **limit of quantitation (LOQ)**, represent the smallest amount of contaminant that can be reliably detected and quantified, respectively, using the sampling and analytical techniques in the sampling method. These limits are usually listed on the first page of the NIOSH method; in the ammonia example, the estimated LOD is 2 μg per sample.

Some contaminants require special handling, such as refrigeration, or protection from light. If special handling is required, the method will provide this guidance; the ammonia method indicates routine shipment, which means there are no special handling requirements. Sample stability can be crucial if the collected contaminant is unstable or loses stability quickly following sampling; this can be the result of a chemical reaction or a weak chemical bond that holds the contaminant on the sample medium. Ammonia samples collected according to our example method are stable for at least 35 days if refrigerated.

The sampling method also contains detailed instructions to be followed in preparing the sample for analysis as well as setting up the equipment or instrument in the laboratory. We will examine some of the more common analytical techniques in the next section.

Calibration and Determination of Sample Volume

The process of determining the rate at which the sampling pump draws air through the sample collector is called calibration. In order to determine the flow rate, it is necessary to establish the time it takes for the pump to move a known volume of air. This is done by measuring the flow of air through the sampling train and comparing the measured flow to a reference or calibration standard. Most air sampling pumps used for industrial hygiene purposes have some method for adjusting the sampling flow rate – a valve, set screw, or an electronic controller. Many adjustment controls allow the user to approximate the desired flow rate, but most do not provide a precise setting and even if they do, it is wise to verify the flow rate. In order to obtain an accurate measure of the flow rate, the sampling train is assembled just as it would be used during the sampling event, using the pump that will be used for the sampling and the same type of collector – sorbent tube, filter, or impinger. The inlet side of the sampling train is then connected to a calibration standard.

The simplest example of a calibration standard used by industrial hygienists is an **inverted buret**, which resembles an upside-down graduated cylin-

---

Box 5-1 ■ Determining a Flow Rate

An example calculation for determining a flow rate follows. Imagine the sampling train in Figure 5-16 has been connected to an inverted buret for calibration. The industrial hygienist timed the movement of a soap bubble from 500 to 0 ml, with the following results:

Volume timed: 500 ml   Time #1: 13.42 seconds
                       Time #2: 13.50 seconds
                       Time #3: 13.49 seconds

The average time to move 500 ml of air is 13.47 seconds calculated by adding time intervals #1, 2, and 3,

and dividing the sum by 3. To determine the flow rate, we must convert ml/seconds to liters/minute, or lpm. To do this, conversion factors, or equivalencies, are used as follows:

$$\frac{500 \text{ ml}}{13.47 \text{ sec}} \times \frac{1 \text{ liter}}{1{,}000 \text{ ml}} \times \frac{60 \text{ seconds}}{1 \text{ minute}} = 2.22 \text{ lpm}$$

The calculated flow rate for the sampling train is 2.22 lpm.

der. Burets used for calibrating sampling pumps are typically marked with lines indicating various volumes – say, for a 1,000 milliliter (ml) or 1 liter buret, a line at zero, and then perhaps a line every 50 ml. The top of a buret used for calibration has a small glass fitting to which the inlet end of the sampling train is connected. To time the movement of the air, a soap bubble is generated inside the buret by momentarily submerging the bottom end of the buret in a weak soap solution. As the bubble moves up into the buret, its movement from one volume mark to another is timed with a stopwatch. This process is repeated using the same volume marks as reference points until at least three consecutive measurements are within a second or two of each other; the times are then averaged. The volume of air that was moved is equal to the volume represented by the distance between the lines that were used in timing. For example, say the watch was started at the 500 ml mark and stopped at the 0 ml mark; the volume is 500 ml. Once the time interval for this is measured, a flow rate – commonly expressed as liters per minute (lpm) – can be calculated. A volume of 500 or 1,000 ml is commonly used since the marks are far enough apart for the industrial hygienist to accurately time the bubble; also, the numbers are relatively convenient to work with. Figure 5-16 shows an inverted buret used to calibrate an air sampling pump.

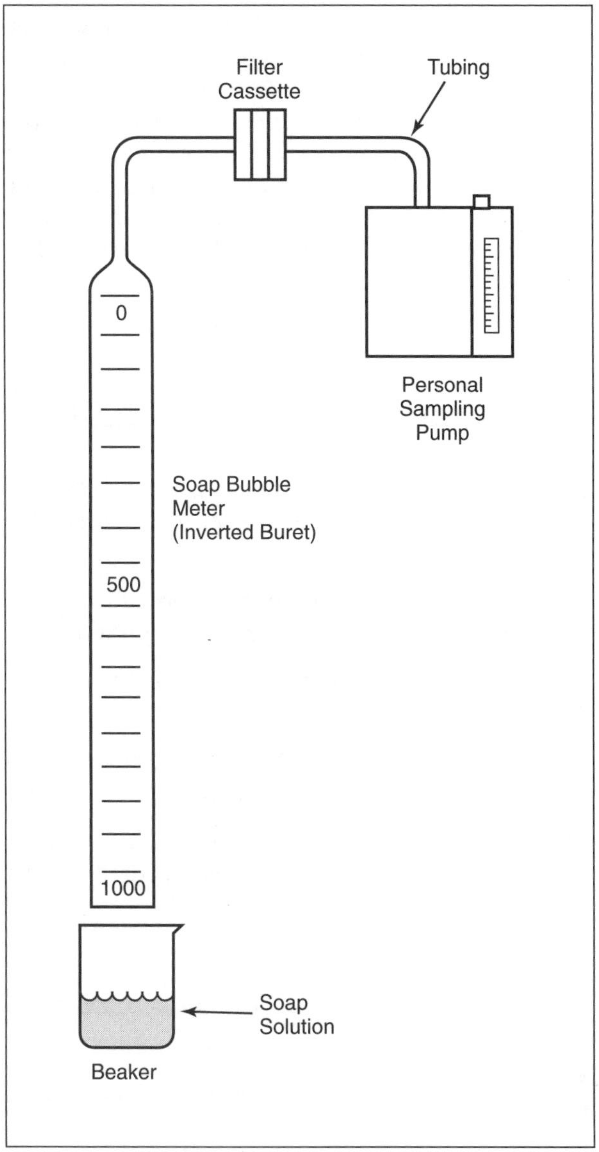

Figure 5-16: Calibration of the sampling train must be performed prior to the sampling event. The sampling train is assembled in the same configuration that will be used for sampling, with the inlet connected to the calibration device. This diagram illustrates a sampling train that is ready for calibration using a bubble meter, a primary calibration standard.

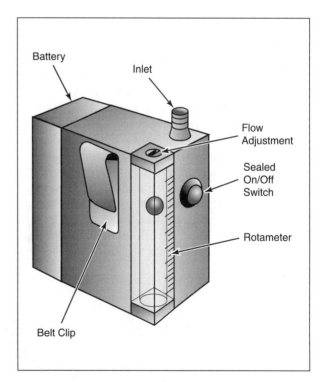

Figure 5-17: Some personal sampling pumps are equipped with a built-in rotameter. Rotameters – which are also available as free-standing devices – are secondary calibration standards. Once the rotameter has been calibrated against a primary standard, it can provide a convenient and accurate method for calibrations in the field.

Figure 5-18: Electronic calibrators such as this one operate using the same principle as the bubble meter. They offer the advantage of being faster; also, most can be purchased with options such as computer software, printers, and capabilities for calibrating a wide range of flow rates.

Although the inverted buret is a simple device, it is a **primary calibration standard**. Primary calibration standards are based on direct measures of a reference value, such as the known volume of the buret. **Secondary calibration standards** are calibration devices that are checked against primary calibration standards. While they do not provide the same high degree of accuracy as a primary standard, they do provide an acceptable level of precision and are often more convenient (and practical) to carry into the field for use during sampling. Flow meters and **rotameters**, such as those found on some sampling pumps, are examples of convenient secondary standards. A rotameter is a secondary calibration standard that consists of a clear tube with a metal or plastic float; the position of the float on a graduated scale along the length of the rotameter indicates the flow rate. Secondary calibration standards must be periodically checked against a primary standard; OSHA's technical manual recommends an interval of six months. If the secondary standard is damaged, it should be recalibrated before it is used again.

While the inverted buret is simple to use, other more convenient and more portable devices have been developed, such as automatic calibrators that use electronic eyes to detect the passing of the soap bubble within a glass cylinder of known volume. Although the volumes used in these instruments are much less than a liter, the devices are in fact variations on the inverted buret. Some are capable of being connected to printers or computers, enabling the production of a printed or electronic record of the calibration test. Figure 5-18 shows some electronic calibration devices.

Checking Your Understanding

1. What are the statistical criteria that NIOSH air sampling methods must meet to be acceptable for use?

2. List at least seven items or information topics that are included in an air sampling method.

3. Name three reasons why calibration of sampling pumps is necessary.

4. What is the difference between a primary and a secondary calibration standard?

5. Why do you think it is important to periodically check a secondary calibration standard against a primary one?

5-6 Laboratory Analytical Techniques

The analysis of the air sample in the laboratory is described in detail in the sampling method. Instruction is provided for preparation, setup, and calibration of laboratory instruments. The particular method of measurement will depend on the contaminant, or analyte, that is to be measured. As is the case with sampling methods, the analytical method that is used depends in part on the characteristics of the contaminant. This section will briefly address some of the methods used in the analysis of airborne contaminants.

Dusts and Fibers

Particulate contaminants such as dusts and fibers may be analyzed using gravimetric methods, as described earlier. The evaluation of dusts is the most common application of gravimetric analysis. Airborne dusts may also be evaluated in terms of particle size, while fibrous contaminants, such as asbestos, are counted. Both particle sizing and fiber counting techniques involve the use of a microscope fitted with an opti-cal insert, called a **graticule** or **reticle** (see Figure 5-8). The graticule may contain grid lines, circles, rectangles, or other patterns, which are of a known size and are precisely spaced at known distances, thus serving as reference. Because the graticule is part of the optics, it is superimposed over the field of view of the observed slide. This allows the analyst, called a **microscopist**, to use the grid marks to estimate particle sizes, or count the particles or fibers that fall within a specific area of the superimposed image of the grid. Analysts performing these types of evaluations receive specialized training, and take part in quality control programs to provide a periodic check on their application of technique, since a certain level of proficiency is necessary to ensure consistent, accurate results.

Gas and Vapor Analysis

Gases and vapors may be collected on sorbent tubes, filters, or in solutions. The method used for analysis of such samples depends in part on the collection

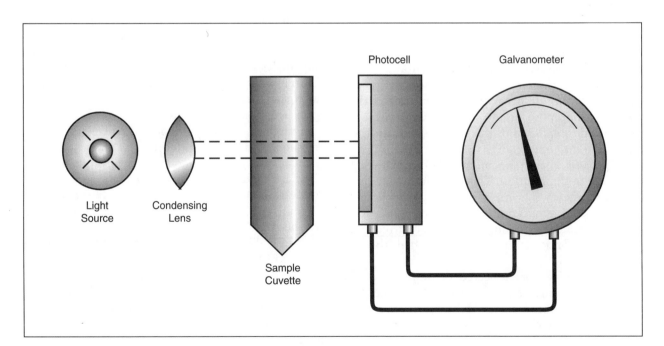

Figure 5-19: Spectrophotometers are used in analysis of some industrial hygiene samples. This method measures the amount of light transmitted through a solution, which is proportional to the degree to which a chemical reaction has taken place between the contaminant and the solution reagent.

medium as well as on the chemical properties of the contaminant.

Gases and vapors that are collected by dissolution or absorption in a solution may be analyzed by a **titration** method involving the use of chemical reagents that are added to the absorbing solution. The reagent is added gradually, in measured amounts, until the endpoint of a reaction is reached. The endpoint may be indicated by a change in the color of the solution; for instance, the solution may remain colorless as more reagent is added, and then change to a bright pink color. The intensity of the color, which is proportional to the concentration, is then measured in an instrument called a **spectrophotometer**. Spectrophotometers contain a light source on one side, and a light detector on the opposite side. The solution to be tested is placed into a small test tube called a **cuvette**, which is placed into the test cell in the spectrophotometer, between the light source and the detector. The amount of light that is absorbed by the solution is proportional to the concentration of the compound created by the reaction between the contaminant and the reagent present in the solution. The instrument is first set at zero, using unreacted absorbing solution; this provides a reference value for the amount of light that would pass through a solution containing no contaminant. The solution resulting from the titration is then tested. The reading is expressed in terms of **absorbance**, which is the amount of light absorbed in the solution, or, conversely, as the **percent transmittance**, which is the amount of light that passes through the solution, as compared to the solution used to zero the instrument.

Some reactions result in formation of an insoluble product or **precipitate**, which causes the solution to become cloudy. In such situations, the light beam is scattered by the particles that are suspended in the solution. The degree of light scattering is proportional to the amount of precipitate that is present in the solution. This analytical technique is **nephelometry**. The use of light-scattering detectors has some applications in laboratory analytical techniques and is also widely used in direct-reading instruments for detection of dusts and other airborne particulates.

Gases and vapors collected on solid sorbents such as filters and silica gel or charcoal must be **desorbed**, or removed from the sorbent, prior to analysis. In the case of sorbent tubes, this is done by passing a solvent through the tube to "wash" the contaminant off the solid. Carbon disulfide is one of the most common desorbing solvents; heated air is also sometimes used as a desorbing mechanism. Filters used as sorbents may be treated with a desorbing solvent, or the entire filter (and support pad) may be dissolved, or **digested**, in an acid solution, as is the case in analysis of some metals. The entire solution is then analyzed.

The instruments used to analyze industrial hygiene samples collected on solid sorbents employ a variety of principles. Instruments and detectors used to identify specific compounds utilize the chemical properties of the contaminant. Some of the analytical instruments used in the laboratory are described briefly in the next section.

Gas Chromatography

A **gas chromatograph**, or **GC**, operates something like a still. The contaminant is desorbed and then injected into the GC, where it passes through a long tube or column containing a solid sorbent. The column acts like a distillation tower, allowing some molecules to pass through more quickly than others, separating the compounds in the sample. Sorbents used in the column of a GC vary, depending on the analytes, and include materials such as silica gel. The particular sorbent used in the column is specified in the analytical method, along with other variables such as the length of the column and the temperature at which the column is maintained during analysis. The sample is moved through the column by an inert **carrier gas**, the identity of which is also specified in the method. The carrier gas is pumped through the column at the rate specified in the method. As each compound emerges from the column, a graph is produced. The amount of time needed for a compound to pass through the column is characteristic of a specific compound or class of compounds and is used to identify the analyte. The area under the peak of the graph is proportional to the amount of material. The identity of the compound is determined by comparing the sample results to those produced by known concentrations of specific compounds. Figure 5-20 shows the basic components of a gas chromatograph. Figure 5-21 shows a sample output from a GC analysis used to identify an unknown.

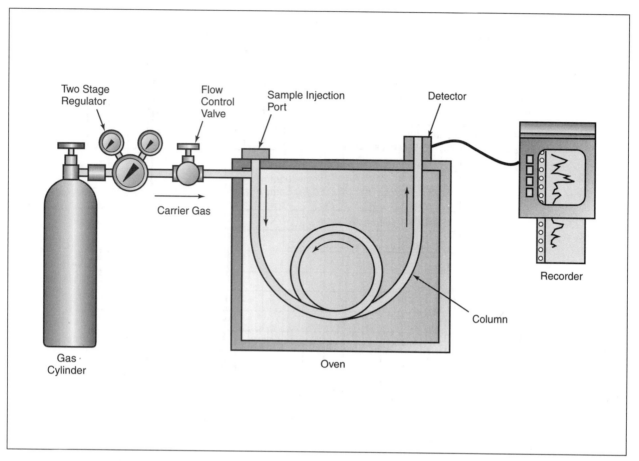

Figure 5-20: Gas chromatography is used for analysis of many gas and vapor contaminants. This analytical technique separates the molecular components of the sample in the same way as a still: each component moves through the column at a different rate. The time necessary for passage through the column is indicative of the compounds present in the sample.

Mass Spectrometry

In a **mass spectrometer**, also referred to as a **mass spec** or **MS**, the sample is desorbed from the collection medium and injected into the instrument, where it is bombarded with a beam of electrons. This causes the molecules in the sample to become **ionized**. Each charged particle, or **ion**, that is produced, has a specific mass. The relationship between the mass of an ion and its charge is unique and is expressed as a ratio of mass (m) to charge (e), or **m/e**. When the sample is bombarded with electrons, a number of ions may be produced, each with its own m/e value. The intensity of each m/e value is proportionate to the amount of the ion that is produced. The m/e value that has the highest intensity is called the base peak, and is assigned a value of 100; all the other m/e values are compared to the base peak. Because most ions have a +1 charge, the m/e ratio value is usually equal to the mass of the ion, making it easier to identify them. For example, in Figure 5-22, the m/e value for the $C_2H_5^+$ ion produced when neopentane is bombarded with electrons is equal to the mass of the $C_2H_5^+$ ion, which is equal to 24 (2 carbon atoms at 12 mass units each) plus 5 (5 hydrogen atoms at 1 mass unit each), or 24 + 5 = 29. The mass value is divided by the charge, 1, or 29/1 = 29. The m/e value for $C_2H_5^+$, then, is 29.

The MS is equipped with a detector that measures the intensity of each m/e value, producing a plot that shows the relative intensities of each. This

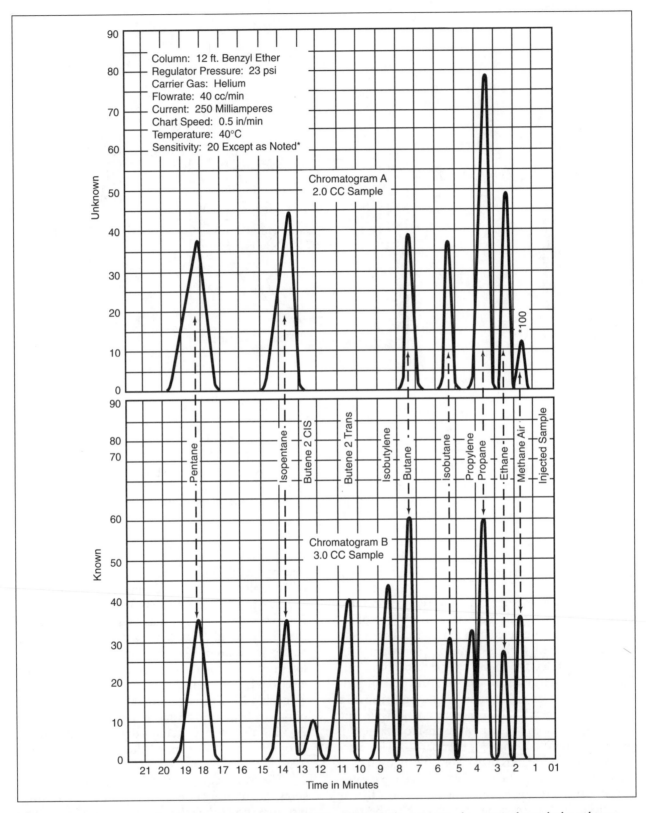

Figure 5-21: The output of a gas chromatograph shows how long it took the compound to move through the column; the area under the peak in the graph is proportional to the amount present. The output produced by analysis of an unknown can be compared to the output produced by analysis of a sample of known composition. In this example, the unknown compound contains pentane, isopentane, butane, isobutane, propane, ethane, and methane.

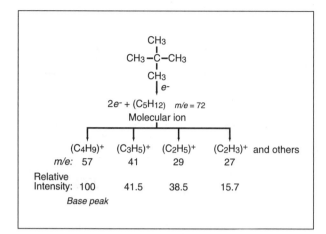

Figure 5-22: This mass spectrum for neopentane can be used as a standard to compare against unknowns. If an unknown material produced a mass spectrum with the same m/e ratios, it would be considered a match, and the unknown would be identified as neopentane.

plot is called a **mass spectrum** (see Figure 5-23). The mass spectrum of a specific compound is like a fingerprint, unique to a compound, and can therefore be used to identify the compound. An air sample analyzed by MS produces a mass spectrum, which is compared to a reference set. When we have found the spectrum that is a match for the sample spectrum, we have identified the compound in the air sample. Mass spectrometry is sometimes used in conjunction with gas chromatography to provide conclusive evidence of the identity of a sampled compound. The use of these combined analytical techniques is referred to as **GC/MS**.

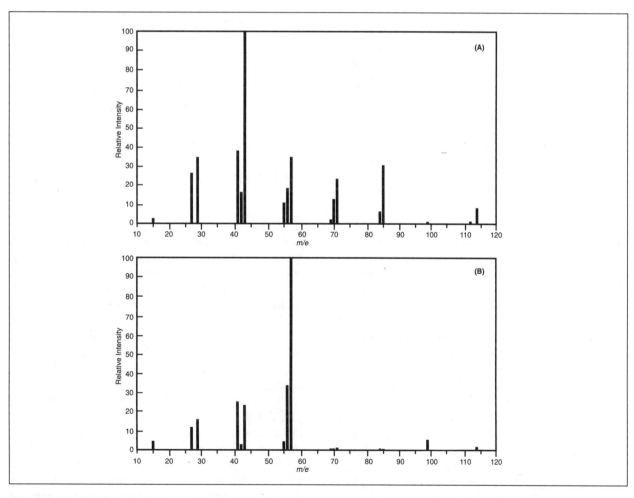

Figure 5-23: Similar to analysis by gas chromatography, the output produced by the mass spectrometer must be compared against a known or reference output to determine the compound that is present. These mass spectra are for n-octane (A) and 2,2,4-trimethylpentane (B).

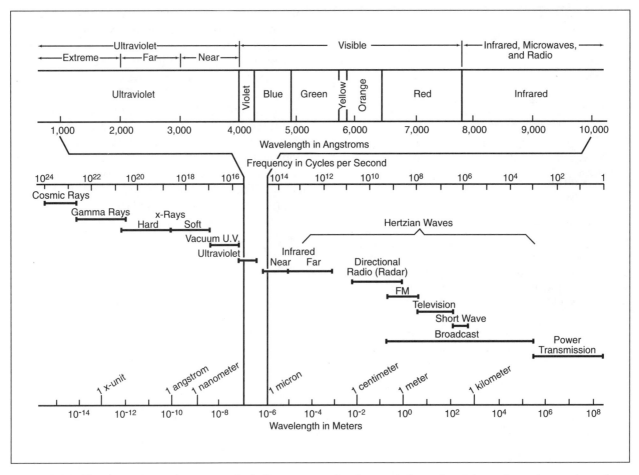

Figure 5-24: The electromagnetic spectrum includes wave energies that range from ultraviolet and cosmic rays through visible light, infrared, microwaves, and power transmission. Analytical techniques for identifying organic compounds utilize the absorbance of specific wavelengths in the ultraviolet and infrared regions.

Absorption Spectroscopy

Ultraviolet and Infrared Spectrometry – These two analytical techniques, as well as atomic absorption (described below), involve the measurement of the amount of energy that is absorbed by a compound. Such techniques are called **absorption spectroscopy**. The particular wavelength where energy is absorbed indicates the identity of the compound. The wavelengths used for analysis of many organic compounds are those in the ultraviolet to infrared regions of the electromagnetic spectrum (see Figure 5-24). For example, benzene absorbs energy at specific wavelengths in the ultraviolet region of the electromagnetic spectrum. Organic compounds exhibit energy absorbance at specific regions of the infrared region, depending on the types of chemical bonds that are present in the molecules. Ultraviolet (UV) and infrared (IR) spectrometry techniques produce a plot or spectrum similar to the mass spectrometer. These spectra are then compared against spectra in a reference set to determine which compounds are present in the air sample. Some ultraviolet and infrared spectra for various compounds are illustrated in Figures 5-25 and 5-26.

Atomic Absorption – **Atomic absorption**, or **AA**, is similar to some of the other analytical techniques we have already discussed in that the technique relies on a specific pattern, or spectrum, of energy absorption to identify the analyte compound. The AA is used most commonly to detect metals. During atomic absorption analysis, the sample passes through a flame or other heat source. This added energy results in an increase of the energy level of some of the electrons, while others remain in an unexcited, stable energy condition called **ground state**. Because most of the electrons remain in a ground state, it is better from an analytical stand-

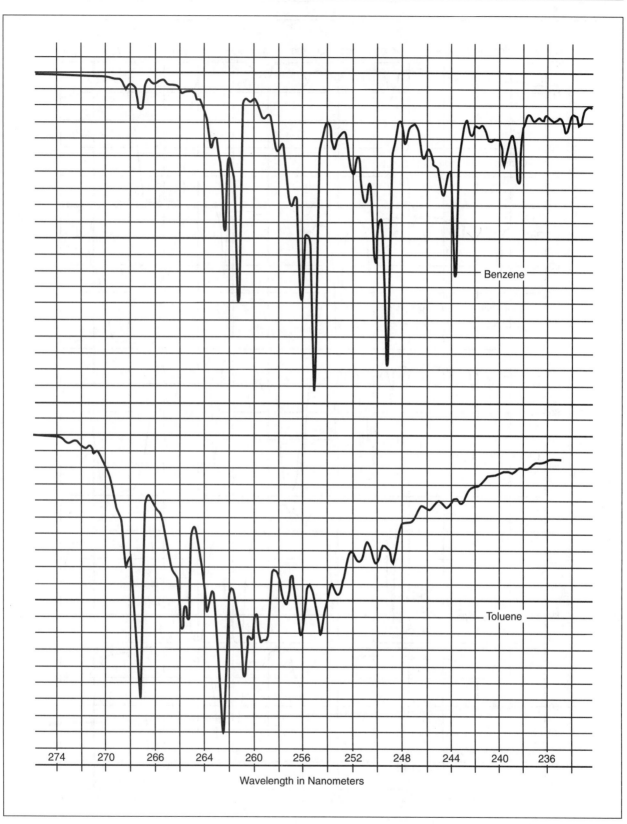

Figure 5-25: The absorbance of energy at specific wavelengths is a characteristic of organic compounds. These patterns of absorbance are like fingerprints, in the sense that they allow the specific compound to be identified. Shown here are the absorbance spectra of wavelengths in the ultraviolet region for benzene and toluene.

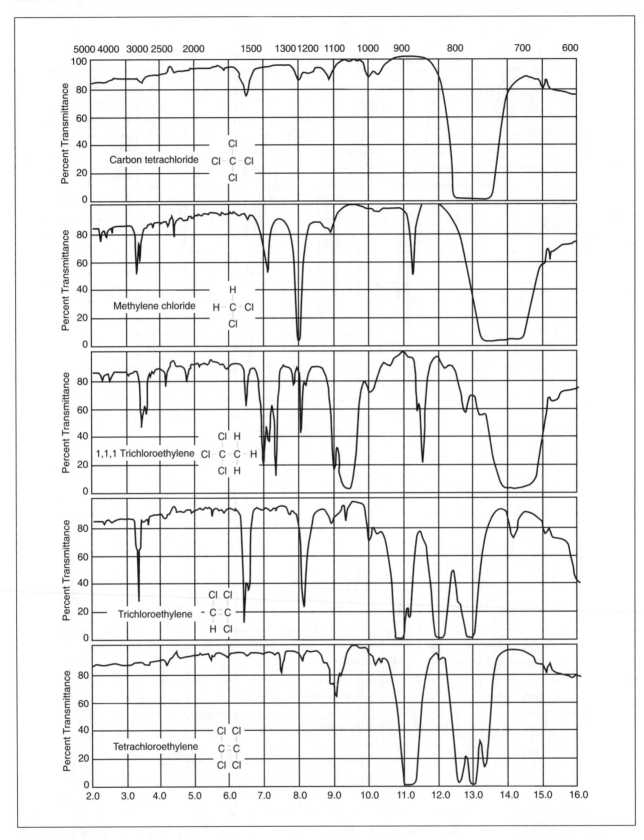

Figure 5-26: Chlorinated hydrocarbons tend to absorb energy in the infrared region. Shown here are IR spectra for five common industrial solvents (wavelength in micrometers).

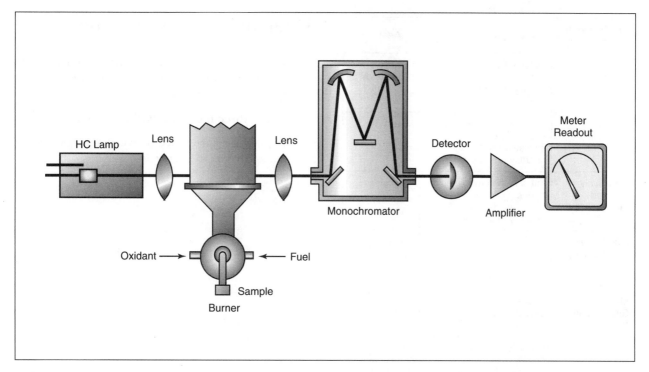

Figure 5-27: In atomic absorption analysis (AA) electrons in the sample absorb energy when passing through a cloud of atomic vapor. The wavelength of the energy absorbed by these electrons indicates the particular element that is present in the sample. This type of analysis is commonly used for metals.

point to take measurements on these. In an AA instrument, these unexcited electrons absorb energy when they pass through a cloud of atomic vapor. The AA detector indicates the wavelengths at which the strongest absorption of energy is occurring by the ground-state electrons. The absorption of energy at a particular wavelength indicates the presence of a specific element. The degree of absorption is proportional to the amount of the element that is present in the sample. Figure 5-27 shows the basic structure of the atomic absorption spectrometer.

Inductively Coupled Plasma and Fluorescence Spectrometry

An **inductively coupled plasma** or **ICP** analysis is an example of **emission spectroscopy**. Like absorption spectroscopy, emission spectroscopy utilizes the ability of electrons to absorb energy. In contrast to absorption spectroscopy, however, emission spectroscopy measures the energy loss of the excited electrons as they return to the ground state. As in the case with other methods of analysis, the spectrum

of emissions is specific to the analyte, and the intensity of the emissions is proportional to the amount that is present. This analytical technique is useful for metal scans, where a number of elements can be analyzed from the same sample; for example, an air sample taken during welding might be analyzed to determine which of a number of metals are present in the fume.

Fluorescence spectrometry is similar to ICP analysis in that the intensity and wavelength of the energy that is emitted from excited electrons is used to indicate the presence of certain compounds. Unlike ICP, fluorescence spectroscopy is used in the analysis of organic compounds, not metals. The method is most useful for organic compounds that contain aromatic rings, as in benzene, or conjugated double bonds, such as the aromatic hydrocarbons present in coal tar emissions. The source of energy used to excite the sample is a lamp capable of producing light energy through either a portion or all of the ultraviolet region of the electromagnetic spectrum. Lamp selection depends on the analytes of interest; common lamps include the xenon-arc, tungsten, and mercury lamps, each of which produces light in a specific range of the spectrum.

Instrument/ Technique	Principle of Analysis	Analytes
Gas Chromatograph (GC)	Column travel time and area under peak indicates identity and quantity of analyte.	Organic compounds
Atomic Absorption (AA)	Intensity of absorbance at specific wavelength indicates identity and quantity of analyte.	Metals
Inductively Coupled Plasma (ICP)	Intensity and wavelength of energy emissions indicates identity and quantity of analyte.	Metals
Fluorescence Spectrometry	Intensity and wavelength of energy emissions indicates identity and quantity of analyte.	Aromatic hydrocarbons
Mass Spectrometry (MS)	Intensity of ions formed indicates identity of analyte.	Metals, organics
Infrared Spectrometry	Intensity and pattern of energy absorbance indicates identity and quantity of analyte.	Organic compounds
Ultraviolet Spectrometry	Intensity and pattern of energy absorbance indicates identity and quantity of analyte.	Organic compounds

Table 5-3: Industrial hygiene samples are analyzed using a variety of instruments and techniques, depending on the contaminants. This table summarizes some of the more common methods used for laboratory analysis of air samples.

Checking Your Understanding

1. What properties of a contaminant determine the best method to be used for analysis to detect a specific contaminant?

2. Name three methods used for metals analysis.

3. Explain the difference between absorption and emission spectrometry analyses.

5-7 Direct-reading Methods

So far we have looked at integrated sampling methods; we will now take a look at some of the ways that samples can be taken and analyzed right at the site. As a group, these methods are referred to as direct-reading methods. They include the use of instruments like combustible gas meters and oxygen meters, photoionization detectors and flame ionization detectors, and length-of-stain or detector tubes.

Probably the most important issue surrounding the use of gas meters and other direct-reading instruments is that the user must be familiar with the instrument. Operation, including alarms, indicators, sensors, and limitations, must be understood so that the instrument can be used appropriately and effectively. Calibration of direct-reading instruments is critical and must be performed prior to use and again at intervals specified by the manufacturer or dictated by use conditions; some field instruments are calibrated several times a day. The process of calibrating a direct-reading instrument is different from calibrating a sampling pump; instead of verifying a flow rate, the calibration of a direct-reading instrument is a process to verify that the instrument is responding to the contaminant and providing an accurate reading or indication of the detected concentration. Such calibrations require the use of specialty gases that contain a known mixture or concentration and to which the instrument produces a specific response. Instrument responses and alarm level settings can usually be adjusted, following the manufacturer's instructions. In addition to calibration, many instruments require a functional check, which is a sort of "go/no-go" test to make sure the instrument is working properly.

Records of calibration and maintenance actions, such as sensor replacement, must also be maintained. If direct-reading area samples are used as an indication of employee exposure, the records should be treated as other records that are generated using standard air sampling methods, with detailed notes taken regarding who, what, when, where, why, and how. Many direct-reading instruments available today have a hygiene function; this usually includes the ability to perform such functions as computing time-weighted average concentrations and producing a record that can be transferred to a computer or printer to show actual measurements taken over the sampling interval. These records can help identify peaks in concentration that occurred over the shift, allowing more detailed scrutiny of the pattern of exposure and therefore better controls for worker protection. Often direct-reading instruments with a hygiene function are small and lightweight, which allows them to be used for obtaining personal samples.

The following sections provide a brief introduction on several direct-reading methods available to the industrial hygienist. For more detailed information, refer to the bibliography at the end of the book.

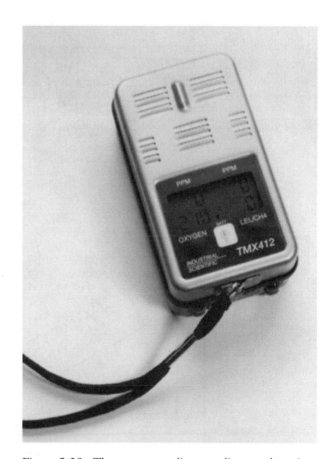

Figure 5-28: There are many direct-reading gas detection instruments available to the industrial hygiene professional today. The instrument shown here is built to withstand regular use in a variety of industrial settings. Common applications include evaluation of confined spaces and continuous monitoring of areas to detect leaks or unsafe levels of airborne contaminants.

Gas Meters

A wide variety of gas meters is available to the industrial hygiene professional for field use. These units are typically relatively small and portable, most weighing less than two pounds, and many newer models weighing in at just a few ounces. They contain a small pump that draws air into the instrument, where it passes into a sampling chamber containing the sensor unit, which is connected to a readout device. Meters can have specific applications such as oxygen detection, or they may contain several sensors, which makes them multi-gas or multi-function units. For example, many meters used for atmospheric testing of confined spaces contain a sensor for oxygen, another sensor that detects combustible gas, and a third sensor that detects the amount of toxics – usually total hydrocarbons – that are present. The "toxic" function on many of these meters is **nonspecific**, that is, it does not identify the specific contaminant that is present, but it does indicate that a hazardous substance is present. Nonspecific instruments do have useful applications in screening for contaminants, initial characterization of unknown sites, and leak detection.

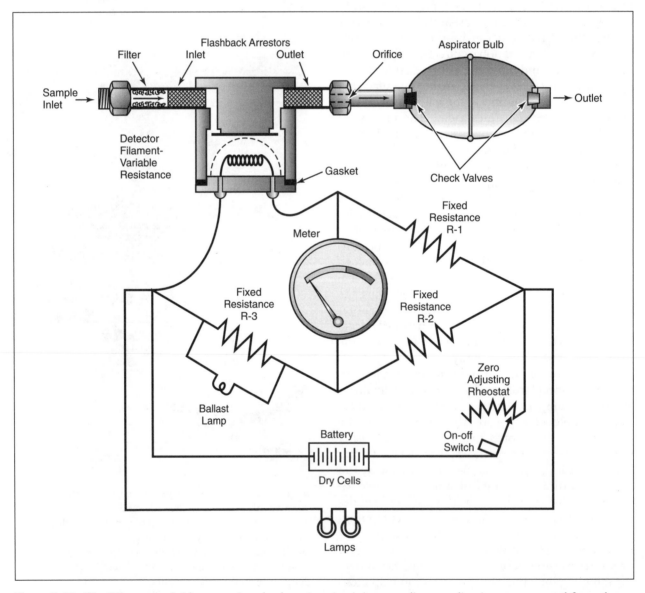

Figure 5-29: The Wheatstone bridge comprises the detection circuit in many direct-reading instruments used for evaluating the presence of combustible gases.

For many gases, the sensors consist of a gel or solid matrix in which the sampled air dissolves. The resulting reaction is detected and translated into an electronic signal, which is interpreted on a digital or analog (meter) readout. Sensors are available to detect a variety of gases, including hydrogen sulfide, chlorine, and others.

Combustible gases, such as carbon monoxide, can be detected using a fairly simple sensor device that contains an electric circuit called a **Wheatstone bridge circuit**, shown in Figure 5-29. These sensors, called catalytic combustion sensors, contain a heated filament that makes up one leg of the Wheatstone bridge circuit. The filament is coated with a catalyst that reacts with a combustible gas and generates heat. The heated filament's electrical resistance changes, causing an imbalance in the circuit. The change in conductivity of the filament produces a degree of imbalance in the circuit, which is proportional to the amount of combustible gas present. The readout device on the meter indicates the imbalance as a concentration of combustible gas, typically expressed as a percent.

Another sensor used to measure combustibility also relies on detecting a change in electrical conductivity; in this sensor, the change is produced when the combustible gas is adsorbed onto the solid surface of a **metal-oxide semiconductor (MOS)**. These sensors are named for the detector device, and are referred to as MOS sensors.

A third sensor used in combustible gas meters involves measurement of **thermal conductivity**, which is the ability of the tested air to conduct heat. Like the catalytic combustion sensor, these sensors contain a filament that is part of a Wheatstone bridge circuit. The atmosphere containing the combustible gas enters the test chamber and passes over the filament, which is heated; the filament responds to the surrounding air, which, if it contains a combustible gas, will cause a change in the temperature of the filament and thus its electrical resistance. The change in resistance is translated by the meter and shown on the indicator.

All three of the sensors used to detect combustible gases rely on specific properties: heat of combustion, thermal conductivity, or adsorption onto an MOS. These properties are unique to each gas; therefore the sensor's responses are also unique to each gas, and some gases will generate a stronger or weaker response from the sensor. Different gases may be used for calibration, which will produce a different response from the instrument. Often the instrument response can be adjusted by the user during the calibration process. Also, many manufacturers

GAS BEING TESTED	CALIBRATION GAS							
	ACETONE	ACETYLENE	BUTANE	HEXANE	HYDROGEN	METHANE	PENTANE	PROPANE
Acetone	1.0	1.3	1.0	0.7	1.7	1.7	0.9	1.1
Acetylene	0.8	1.0	0.7	0.6	1.3	1.3	0.7	0.8
Benzene	1.1	1.5	1.1	0.8	1.9	1.9	1.0	1.2
Butane	1.0	1.4	1.0	0.8	1.8	1.7	0.9	1.1
Ethane	0.8	1.0	0.8	0.6	1.3	1.3	0.7	0.8
Ethanol	0.9	1.1	0.8	0.6	1.5	1.5	0.8	0.9
Ethylene	0.8	1.1	0.8	0.6	1.4	1.3	0.7	0.9
Hexane	1.4	1.8	1.3	1.0	2.4	2.3	1.2	1.4
Hydrogen	0.6	0.8	0.6	0.4	1.0	1.0	0.5	0.6
Isopropanol	1.2	1.5	1.1	0.9	2.0	1.9	1.0	1.2
Methane	0.6	0.8	0.6	0.4	1.0	1.0	0.5	0.6
Methanol	0.6	0.8	0.6	0.5	1.1	1.1	0.6	0.7
Pentane	1.2	1.5	1.1	0.9	2.0	1.9	1.0	1.2
Propane	1.0	1.2	0.9	0.7	1.6	1.6	0.8	1.0
Styrene	1.3	1.7	1.3	1.0	2.2	2.2	1.1	1.4
Toluene	1.3	1.6	1.2	0.9	2.1	2.1	1.1	1.3
Xylene	1.5	2.0	1.5	1.1	2.6	2.5	1.3	1.6

Example: The instrument has been calibrated on methane and is now reading 10% LEL in a pentane atmosphere. To find actual % LEL pentane, please multiply by the number found at the intersection of the methane column (calibration gas) and the pentane row (gas being sampled)... in this case, 1.9. Therefore, the actual % LEL pentane is 19% (10 × 1.9).

Multiplier accuracy is ± 25%, subject to change without notice pending additional testing.

If the sensor is used in atmospheres containing unknown contaminants (silicone, sulfur, lead, or halogen compound vapors) methane is the recommended calibration gas. Periodic comparison of methane and pentane readings is recommended when using this chart.

Table 5-4: The response of a direct-reading instrument to a particular gas depends on what gas was used to calibrate the instrument. This chart illustrates how to correctly interpret the response of the instrument shown in Figure 5-28, based on the gas being tested and the gas used for calibration.

Combustible gases such as methane are displayed as a percent of the **lower explosive limit** (or level), **LEL**, sometimes called the **lower flammable limit**, or **LFL**. Most gas-air mixtures that support combustion will only do so within a certain range of concentrations, which has both lower and upper limits. The LEL is the lowest gas-air mixture that will support combustion. The acronym used to refer to the upper limit of combustibility is the **UEL**, for **upper explosive limit**, or sometimes, the **UFL**, for **upper flammable limit**. Most combustible gas meters come with the alarm level preset at 10 percent of the LEL. Since the LEL is itself a percent value, the alarm setting is a percent of a percent. This provides a fairly wide margin of safety. For example, the LEL of ethylene oxide is about three percent. A combustible gas meter that had been calibrated using the same gas, with an alarm set at 10 percent of the LEL, would alarm at a concentration of 0.3 percent of ethylene oxide.

the location being tested. Meters equipped with these sensors are equipped with a flashback arrestor in order to prevent the ignition of the combustible atmosphere that they are being used to test. Similar possibilities of ignition may apply to the circuitry and electronic components of any direct-reading instrument. Instruments that can be safely operated in a possibly combustible atmosphere are identified as **intrinsically safe**. Again, any questions about instrument limitations should be clarified by the manufacturer.

Photoionization and Flame Ionization Detectors

Photoionization detectors or **PIDs** (see Figure 5-30), as they are sometimes called, are an example of a nonspecific instrument that is most often used to

provide calibration curves for different gases, which can be used to cross-reference the meter's response to other gases. For instance, a meter calibrated using methane gas will still respond to atmospheres containing pentane or butane, but the response will not be the same as to methane. Table 5-4 shows some correction or correlation factors for an instrument calibrated using methane.

Some words of caution: The presence of a combustible gas just above the UEL is potentially worse than a concentration approaching the LEL. This is because a little dilution of the mixture will push the concentration into the combustible range. Any reading above the UEL should be considered potentially very hazardous!

Another caveat in the use of a combustible gas or other direct-reading meter is that the sensors can become less sensitive over time. Depending on their design, they may have a fixed capacity for absorbing combustible materials; some have a recommended useful life, beyond which they should be replaced. Sensors can also be contaminated by interfering compounds that cause the sensor to respond incorrectly or inaccurately; silicone vapors, for example, can poison the sensor of a combustible gas meter. It is important to know the manufacturer's specifications for acceptable use conditions, including temperature, humidity, and whether the instrument may be used in a hazardous or combustible atmosphere. Since some direct-reading instruments use a sensor with a heated filament or flame, there is a risk of igniting an explosive or combustible atmosphere in

Figure 5-30: Photoionization detectors (PIDs) are direct-reading instruments that are used a great deal at hazardous waste sites. They are nonspecific, but useful for screening purposes during initial entries at sites where specific contaminants are unknown. The most common application is for screening to detect organic compounds.

Figure 5-31: The flame ionization detector (FID) is another type of non-specific instrument often used at hazardous waste sites during investigation and cleanup. It is generally better than a PID at detecting hydrocarbon compounds, making them useful for field screening at sites where leaking underground fuel tanks have been removed.

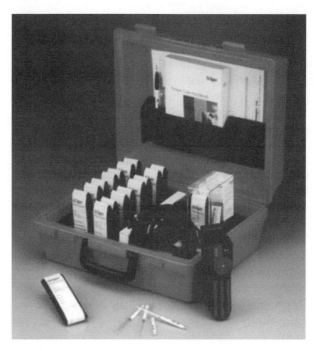

Figure 5-32: Detector or length-of-stain tubes are among the most portable and versatile direct-reading methods for chemical detection. Tubes are available for detecting a wide variety of materials within a wide range of concentrations.

detect organic vapors, such as alcohols, ketones, ethers, and others. The mechanism of operation utilizes the light energy from a tiny ultraviolet lamp. The air pump in the instrument draws air into the test chamber, where the light energy given off by the lamp is absorbed by the contaminant molecules. These become charged particles or ions, which are collected by an electrode; this produces a current that is proportional to the concentration of the contaminant ions (molecules) in the test chamber. The signal that results is translated by the meter into a readout. One limitation of these instruments is that they are nonspecific; on the other hand, this makes them most useful for screening an atmosphere for a set of suspected, or known, contaminants.

Another limitation is that its functioning is based on the molecules being ionized by the lamp in the test chamber. Some molecules require more energy to become ionized; the amount of energy required to ionize a molecule is a physical property called the **ionization potential**, or **IP**, which is expressed in electron volts using the symbol eV. (The ionization potential of ethyl benzene, for instance, is 8.76 eV.)

The energy output levels of lamps used in PIDs are given in eV also. In order to detect a contaminant using a PID, the ionization potential of the contaminant molecules must be equal to or lower than the energy output of the lamp in the instrument. Because ionization potentials of common airborne vapors vary widely, most PIDs can be fitted with lamps that have different energy outputs. This allows the use of a lamp in the energy range that is needed to detect the presence of the contaminants of concern. Using a high-energy lamp may seem to provide the widest range of applications; however, high-energy lamps do not last as long, and they may not generate the same response as a lower-energy lamp for a specific compound. The use of the higher energy lamp does nothing to increase the specificity of the instrument, which is limited by the physics inherent in the sensor.

Flame ionization detectors, or **FIDs** (see Figure 5-31), operate on a principle similar to that of PIDs in that the contaminant molecules are ionized through absorption of energy. In the case of the FID, this energy comes from a heat source specifically, a hydrogen flame. FIDs are commonly used to detect hydrocarbons, to which they are most sensitive, but they also can detect most organic compounds. During sampling, air is drawn into the instrument, and carbon atoms present in the molecules of the contaminant become positively charged when they pass through the flame. The resulting ions are collected on a platinum loop, creating an electric current that is proportionate to the rate at which the ions are collected. This current is translated into a readout value on the instrument. Because it is nonspecific, the FID is subject to limitations similar to that of the PID. However, both are useful for screening and initial characterization of unknown atmospheres as well as for monitoring known contaminants that are within the detection abilities of the instrument being used. As with any direct-reading instrument, calibration and familiarization with the operation are important factors in providing the best and the safest utilization of the equipment.

Detector or Length-of-Stain Tubes

Detector tubes are sometimes referred to as **length-of-stain tubes**, which is a reference to the way they are used to indicate the presence of a contaminant. These tubes – which resemble an oversized sorbent tube – are sealed at each end and contain a solid sorbent coated with a reagent that reacts with a contaminant and causes a color change. The tubes are marked with graduations along their length; generally, the length of the colored area or stain is indicative of the concentration. Once the ends are snapped off and the tube is in place, air is drawn into the tube by a hand pump or an electric pump. These pumps, like other samplers, must be calibrated so that the volume of air drawn through the tubes is known. Higher volumes of air, useful for detecting low levels of contamination, are sampled by drawing additional pump strokes of air through the tube. These tubes are versatile, portable, and useful for field activities such as screening, leak detection, initial site characterizations, and other applications. They are available for a wide variety of contaminants; the results are typically accurate within 25 percent. Because of this, they are a good choice when the contaminants of concern are known, when simple "go/no go" kinds of results are adequate, and when the situation calls for a simple but reliable sampling method.

Checking Your Understanding

1. What is meant by the term nonspecific when talking about a direct-reading instrument?

2. Name three applications for nonspecific instruments.

3. What is the importance of intrinsic safety as it applies to direct-reading instruments?

4. What is a length-of-stain tube?

5-8 Air Sampling Strategies

Performing the actual air sampling event can be compared to completing a series of steps. Each step must be completed before continuing the process. The steps are:

1. Determine the agents to be sampled. This may be based on a regulatory requirement, baseline or initial sampling, employee complaints, or a review of MSDSs for materials being used in the area.

2. Select the sampling method based on the expected concentration of contaminant, interfering compounds, laboratory recommendations, or other factors.

3. Obtain and review a copy of the sampling method, which will contain information about sample media, flow rates, analysis, etc.

4. Obtain adequate sampling media and equipment, based on type and number of samples, including blanks.

5. Calibrate sampling pumps or instruments to be used for sampling.

6. Perform sampling.

7. Ship samples to laboratory for analysis.

8. Interpret analytical results.

9. Prepare a report to employees and management.

The entire process should be methodical and documented, usually on a standardized form, to assure that all the required information is recorded. Some steps, such as calibration of a direct-reading instrument or sampling pump, may generate a separate record.

The number of samples that are to be obtained is usually determined according to one of two approaches. One approach is to attempt to identify the workers who are most likely to be exposed to the highest levels of contaminants – these are the **maximum risk employees**. Another approach is to identify populations or groups that have similar exposures. Such a group, called a **homogenous exposure group** or **HEG**, can often be defined by job title or the function of the workers – material handlers, hopper fillers, appliers of spray coatings, and so on. Once the HEGs have been defined, the industrial hygienist can employ a random sampling scheme to select persons for sampling out of the group. These approaches are described in more detail in the NIOSH publication, *An Occupational Exposure Sampling Strategy Manual* and in the AIHA publication *A Strategy for Occupational Exposure Assessment*. The importance of selecting who is to wear the sampler cannot be stressed too much, since these few measurements may be used to represent the exposure of the entire group.

Blanks

If the number of workers is small, it may be possible to obtain sampling data for all. More often, due to limited resources, the industrial hygienist must choose who will wear the sampling train or where the area sample will be placed. As a general guideline, at least three samples should be obtained. **Field blanks** should also be prepared and submitted with the samples for analysis. A field blank consists of sample media that are exposed to the same conditions as the media used for the actual sampling, but are not connected to a sampling pump. For example, during a solvent sampling activity involving five samples, two charcoal tubes are opened and immediately capped, sealed, and labeled. These are the field blanks. The accepted number of field blanks is 10 percent of the total number of samples, or at least two regardless of the total number of samples. The field blanks are subject to the same **chain of custody** and other sample handling procedures as the rest of the samples. This is a quality control step intended for detection of any potential contamination that could be introduced into the process through sample handling. Field blanks that analysis reveals to contain measurable levels of contaminant could indicate problems with sampling or handling of the media, or also a previously unsuspected source of interference or contamination.

Laboratory blanks are sample media that are not sampled on, but are prepared and analyzed by the laboratory. This is another quality control step

Air Sampling Worksheet

U.S. Department of Labor
Occupational Safety and Health Administration

1. Reporting ID	2. Inspection Number	3. Sampling Number ▶ 91151697 9

4. Establishment Name	5. Sampling Date	6. Shipping Date

7. Person Performing Sampling (Signature)	8. Print Last Name	9. CSHO ID

11. Employee (Name, Address, Telephone Number)	14. Exposure Information	a. Number	b. Duration

c. Frequency

15 Weather Conditions	16 Photo(s) Y N

11. Job Title	12. Occupation Code

13. PPE (Type and effectiveness)	17. Pump Checks and Adjustments

18. Job Description, Operation, Work Location(s), Ventilation, and Controls

Cont'd

19. Pump Number:

Sampling Data

20. Lab Sample Number

21. Sample Submission Number

22. Sample Type

23. Sample Media

24. Filter/Tube Number

25. Time On/Off

26. Total Time *(in minutes)*

27. Flow Rate ☐ l/min ☐ cc/min

28. Volume *(in liters)*

29. Net Sample Weight *(in mg)*

30. Analyze Samples for:	31. Indicate Which Samples To Include in TWA, Ceiling, etc. Calculations

32. Interferences and IH Comments to Lab	33. Supporting Samples	34. Chain of Custody	Initials	Date
	a. Blanks:	a. Seals Intact:	Y N	
		b. Rec'd in Lab		
	b. Bulks:	c. Rec'd by Anal.		
		d. Anal. Completed		
		e. Calc. Checked		
		f. Supr. OK'd		

Case File Page ____ / ____ of

OSHA-91A (Rev 1/84)

Table 5-5: This sampling data form is used by OSHA officers when they perform air sampling. When completed, the form provides a legal record of the sampling event and includes many important details that are useful for preparing the final report. There is no required format for such a form; however, this form is a good example of the type of information and data that should be recorded for any sampling event.

Pre-Sampling Calibration Records

P r e	**35.** Pump Mfg & SN	**38.** Flow Rate Calculations
	36. Voltage Checked? ☐ Yes ☐ No	
	37. Location/T & Alt	

39. Flow Rate	**40.** Method ☐ Bubble ☐ PR	**41.** Initials **42.** Date/Time

Post-Sampling Calibration Records

P o s t	**43.** Location/T & Alt	**44.** Flow Rate Calculations

45. Flow Rate	**46.** Initials	**47.** Date/Time

Sample Weight Calculations

48. Filter No.						
49. Final Weight *(mg)*						
50. Initial Weight *(mg)*						
51. Weight Gained *(mg)*						
52. Blank Adjustment						
52. Net Sample Weight *(mg)*						

53. Calculations and Notes:

Table 5-5: Continued.

that is taken to detect problems with preparation and analysis of the samples. Blank values are sometimes deducted from sample values – this produces data that are said to be **blank-corrected**. If both field blanks prepared in the example above had contamination at a level of two micrograms, it would be assumed that all the samples contained this extra amount. The analytical data for the samples would then be blank-corrected by subtracting two micrograms from the total amount of contaminant found on each of the tubes.

Sources of Error in Air Sampling

The results of air sampling events can be affected by many things, some of which cannot be controlled. Uncontrollable sources of error are usually assumed to be self-correcting or self-limiting through randomness. It is also possible to introduce error into any sampling process through actions or through the failure to take certain actions. These sources of error are not random and – to a certain extent at least – they are controllable. Standard sampling and analytical methods are developed to include steps that, if followed, will assure sampling accuracy, such as the calibration of sampling pumps. Direct-reading methods are also subject to operator-introduced error. Some examples of sampling error or of **bias** that can be introduced (and therefore controlled) by the industrial hygiene professional performing the sampling are:

—Using inappropriate sample collection media;

—Using a direct-reading instrument outside of its limitations and applications;

—Use of incorrect flow rates, resulting in a sample volume that is too large or too small to allow accurate quantitation of the contaminant;

—Failing to perform calibrations or functional checks on direct-reading instruments;

—Overloading the sample collector;

—Errors in calculations – flow rates, sample volumes, resulting concentrations;

—Sampling in the presence of interfering compounds;

—Failure to follow special handling procedures such as protecting samples from light or heat;

—Placement of sample collectors by the industrial hygienist in hopes of detecting higher or lower measurements;

—Neglecting to keep complete and accurate records.

Once samples have been received at the laboratory, they are again subject to possible mishandling or other events that can affect the results. Some events that can happen at the laboratory include:

—Improper storage/handling of samples while awaiting analysis;

—Delays in analysis that exceed time limits for stability of samples (holding times);

—Use of incorrect analytical techniques, including use of instruments that are not set up according to the parameters specified in the analytical method;

—Failure to properly prepare the samples for analysis;

—Contamination of samples with other samples or laboratory chemicals;

—Errors in calculations;

—Mix-ups such as incorrect labeling of samples and transposing numbers;

—Loss of samples due to breakage, spillage, or other events.

Most analytical methods include a numerical factor that accounts for the uncontrollable portion of **sampling and analytical errors**, or **SAEs**. This numerical value is used to evaluate whether the sample results indicate exposures within acceptable limits such as an OSHA PEL, for example. We will look at the application of the SAE near the end of the chapter.

Laboratory Accreditation

The American Industrial Hygiene Association administers a laboratory accreditation program; facilities that meet the criteria of the program are said to be AIHA Accredited. Laboratories selected to perform analysis of industrial hygiene samples should be AIHA-Accredited. This provides the industrial hygienist with assurance that quality control procedures are in place, and that the laboratory consistently reports accurate results. The process for obtaining accreditation includes completion of a written application that describes the laboratory's operational procedures, followed by a site visit to verify that conditions at the laboratory are as described in the application. The application and review process comprehensively evaluates the laboratory, addressing areas such as internal quality assurance, sampling materials and procedures, qualifications of laboratory personnel, chain-of-custody and sample handling, calibration and maintenance of analytical instruments, and health and safety programs.

Laboratories seeking accreditation must also demonstrate analytical proficiency by participating in the AIHA's Proficiency Analytical Testing, or PAT, program. This program consists of periodic analysis of unknown samples, the results of which are reported to the AIHA by the laboratory. Areas of accreditation for industrial hygiene laboratories include metals, solvents, asbestos, and lead; laboratories may be accredited one, or all, areas. The AIHA also administers an accreditation program that meets EPA's requirements for environmental lead analysis - that is, analysis of samples of soil, paint chips and similar items. This program is known as the Environmental Lead Laboratory Accreditation Program, or ELLAP. A listing of accredited laboratories may be obtained by contacting the AIHA.

Limits of Air Sampling

Air sampling does have its limitations. Some understanding of these is necessary so that the industrial hygienist can appropriately interpret sampling results, and make decisions and/or recommendations to management regarding the use of methods for controlling or eliminating the hazardous exposure. This section will discuss the issues associated with reliance on air samples as an indication of the absorbed dose of the exposed worker and some related topics.

As already mentioned, air sampling data are often accepted as an indication of the actual level of exposure of the workers and are assumed to be an indication of the absorbed dose. This is a faulty assumption. Airborne concentrations probably are best thought of as a representation of a potential level of exposure. There are many variables that can affect the actual amount of the contaminant that is absorbed by the exposed worker population. These include:

—Age and gender;

—Health status/level of fitness;

—Nature of work (sedentary or strenuous);

—Breathing rate;

—Pre-existing health conditions, medications, allergies;

—Concurrent exposures, including hobbies and nonoccupational sources.

Besides the variety of influencing factors, each factor will vary from person to person. This makes the estimation of absorbed dose an educated guess at best. This is where the "art" part of industrial hygiene comes into play. The industrial hygienist must consider the entire set of variables, not just the numbers, and make a recommendation that will protect the health of the worker accordingly.

The problems associated with reliance on the air concentration as an indicator of absorbed dose have been addressed to some degree through the use of the biological exposure indices, or BEIs presented in Chapter 3. The BEIs are levels of contaminants, or of metabolic by-products associated with their absorption, that can be measured in blood, urine, exhaled breath and sometimes other tissues. Like air samples, concentrations of contaminants or indicators in body tissues are not a definitive and final answer; and like TLVs, BEIs do not represent black-and-white lines between what is safe and what is not safe. However, they do contribute to the overall exposure picture and can be quite useful when considered along with other factors such as air sampling measurements, work practices, engineering controls being used, and other observations made by the industrial hygienist. Some regulatory standards require the use of biological samples, in addition to air sampling. For example, lead levels are measured in the blood of lead-exposed workers. These levels are monitored and used to determine how long a worker must be away from lead exposure to allow the body to eliminate some of the accumulated lead. Table 5-6 shows an excerpt from the ACGIH TLV booklet, which contains a list of BEIs as well.

Another issue associated with air sampling data is one of **representativeness**, that is, whether or not the data are truly representative of the air concentrations in the work location(s) where the sampling took place. Most sampling data are obtained over a one-day or two-day time period, often following a

ADOPTED BIOLOGICAL EXPOSURE DETERMINANTS			
Chemical [CAS #] Determinant	Sampling Time	BEI	Notation
ACETONE [67-43-1] (1994)			
Acetone in urine	End of shift	100 mg/L	B, Ns
ANILINE [62-53-3] (1991)			
Total p-aminophenol in urine	End of shift	50 mg/g creatinine	Ns
Methemoglobin in blood	During or end of shift	1.5% of hemoglobin	B, Ns, Sq
ARSENIC AND SOLUBLE COMPOUNDS INCLUDING ARSINE [7784-42-1] (1993)			
Inorganic arsenic metabolites in urine	End of workweek	50 µg/g creatinine	B
BENZENE [71-43-2] (1987) [See Note Below]			
Total phenol in urine	End of shift	50 mg/g creatinine	B, Ns

Benzene in exhaled air:
      mixed-exhaled
      end-exhaled
	Prior to next shift	
		0.08 ppm
		0.12 ppm
			
			Sq
			Sq

CADMIUM AND INORGANIC COMPOUNDS (1993)			
Cadmium in urine	Not critical	5 µg/g creatinine	B
Cadmium in blood	Not critical	5 µg/L	B

Note: The Chemical Substances TLV Committee has proposed a revision of the TLV for benzene. See Chemical Substances Notice of Intended Changes and TLV Documentation.

Notations provide the following information:

"Sc" Notation	This notation indicates that an identifiable population group might have an increased susceptibility to the effect of the chemical, thus leaving it unprotected by the recommended BEI. The specific BEI documentation should be consulted for information.
"B" Notation	This notation indicates that the determinant is usually present in a significant amount in biological specimens collected from subjects who have not been occupationally exposed. Such background levels are included in the BEI value. For information on background levels, consult the specific documentation.
† **"Nq" Notation**	Biological monitoring should be considered for this compound based on the Committee's review of the literature; however, a specific BEI could not be determined due to insufficient data.
"Ns" Notation	This notation indicates that the determinant is nonspecific, since it is observed after exposure to some other chemicals. These nonspecific tests are preferred because they are easy to use and usually offer a better correlation with exposure than specific tests. In such instances, a BEI for a specific, less quantitative biological determinant is recommended as a confirmatory test. The documentation should be consulted for information on factors affecting interpretation of these BEIs.
"Sq" Notation	This notation indicates that the biological determinant is an indicator of exposure to the chemical, but the quantitative interpretation of the measurement is ambiguous (semiquantitative). These biological determinants should be used as a screening test if a quantitative test is not practical or as a confirmatory test if the quantitative test is not specific and the origin of the determinant is in question.

In some instances, BEIs for confirmatory and screening tests are not listed, but the pertinent documentation provides information for estimation of the reference value. Examples are measurements of inhaled chemicals (which are extensively metabolized) in exhaled air. BEIs for some screening tests are derived as an upper (or lower) limit of the levels observed in unexposed populations. Examples are measurements of cholinesterase or methemoglobin.

Table 5-6: This excerpt from the ACGIH TLV booklet shows biological exposure indices for some common industrial materials. While not a substitute for TLVs or PELs, the BEIs are useful as complementary data points in evaluating a worker's absorption of a hazardous material.

triggering event such as a spill, a leak, a system failure or breakdown of engineering controls, or an employee complaint. As such, the sample may represent a worst-case situation or at least something that is not "normal" for the operation. One of the most common remarks to the industrial hygienist performing exposure sampling is: "You should have been here yesterday (or last week, or last month), it was really (bad/strong/thick/heavy) then!" Repeated sampling over a long period will provide statistically more accurate data regarding the actual concentrations.

One of the larger issues associated with air sampling is that the results are inevitably compared to a predetermined exposure limit. We looked at exposure limits in a previous chapter. Whether the limit is a statutory (regulatory) limit or a recommended limit from NIOSH or the ACGIH, or an internal company standard, there are some questions to consider in comparing measured air concentrations to such a limit. These include:

—What health effect is the limit intended to protect or prevent against? Acute effects, such as irritation or a nuisance, or a chronic effect such as cancer?

—What types of toxicological studies have been done? Have some areas been neglected, such as reproductive effects?

—Are the toxicological data applicable to the exposed population? (Most of the current data for human exposures are based on studies that involved white males.)

—Does the limit account for exposure to a combination of toxic materials? What about cumulative and synergistic effects?

—Does the limit adequately protect hypersensitive workers, cigarette smokers, and others who might be more susceptible to the negative effects associated with exposure? Examination of the ACGIH *Documentation of TLVs* will reveal that some TLVs have been established at levels known to cause some individuals to experience an adverse effect. (Recall the disclaimer by the ACGIH that TLVs are to protect most, not all, workers.)

There are no easy answers to the above questions, but they should be considered by the industrial hygienist who is interpreting the air sampling results and applying exposure limits.

Calculating and Interpreting Results

Once data are obtained from the laboratory, it may be necessary to calculate the airborne concentrations that correspond to each sampling interval. Many laboratories calculate the concentration as a service to the industrial hygienist; however, their calculations assume that the IH has provided all the correct information.

The concentration of particulates and fumes is typically reported in terms that express a mass per unit volume, as in micrograms (μg), or milligrams (mg), per cubic meter (m^3), as: $\mu g/m^3$ or mg/m^3. The convention is to report concentrations of most gases and vapors as parts per million, or ppm, which is a volume-to-volume ratio. It may be necessary to convert concentrations from one ratio to the other. The conversion is easy to make, but it is necessary to have certain information about the contaminant, such as its molecular weight and the concentration of the contaminant in either a mass per unit volume or a volume-to-volume ratio. The equations for making these conversions are:

$$ppm = \frac{C\ (mg/m^3) \times 24.45}{MW\ of\ contaminant}$$

$$mg/m^3 = \frac{C\ (ppm) \times MW\ of\ contaminant}{24.45}$$

Where

C   = the concentration of the contaminant;

MW = molecular weight; and

24.45 = a constant that represents the volume, in liters, of one mole of a gas at standard temperature (25°C) and pressure (1 atmosphere).

The conversion factors above may also be used to convert an exposure limit from one unit to another. Here are a couple of examples of conversions:

From ppm to mg/m³:

50 ppm n-butyl acetate, which has a molecular weight of 116.2; converted to mg/m³:

$$mg/m^3 = \frac{50 \text{ ppm} \times 116.2}{24.45}$$

$$= \frac{5810}{24.45}$$

$$= 237.6 \text{ mg/m}^3$$

From mg/m³ to ppm:

14 mg/m³ of diphenyl, which has a molecular weight of 169; converted to ppm:

$$ppm = \frac{14 \text{ mg/m}^3 \times 24.45}{169.2}$$

$$= \frac{342.3}{169.2}$$

$$= 2.02 \text{ ppm}$$

Laboratories generally assume the concentration will be compared against an 8-hour time weighted average limit, but this is not always the case. We saw in an earlier chapter how to calculate time-weighted averages; we will use another example here to illustrate how air sampling data are compared to exposure limits.

Consider a case where sample data obtained over three sampling intervals indicated an 8-hour TWA exposure to 2-heptanone was 105 ppm. The OSHA PEL is 100 ppm. The compliance officer must determine whether this measurement indicates an exposure within regulatory limits.

Remember that the NIOSH sampling methods have been validated to provide reliable results 95 times out of 100. The analytical results are evaluated statistically to determine whether the results are at this 95 percent confidence level. The analysis looks at the results from two positions: one of underestimation of exposure, where the result is corrected for error using the SAE and compared to an upper limit; and one where the results are compared to a lower limit. The statistical terms for these limits are the 95 percent **upper confidence limits (UCL)** and **lower confidence limits (LCL)**, or 95 percent UCL and 95 percent LCL. These limits are calculated and used by the OSHA compliance officer as follows:

1. First, a standardized concentration, represented by Y, is calculated. This is done by dividing the full-shift sampling result by the PEL value. In our example, it would be the 105 ppm 8-hour TWA sample result, divided by the PEL of 100 ppm; the result is 1.05, a unitless number since the ppms canceled out:

    $$Y = \text{full-shift sampling result}/PEL$$

    $$Y = 105 \text{ ppm}/100 \text{ ppm} = 1.05$$

2. Second, the 95 percent upper confidence limit or UCL is calculated. This limit is equal to the standard concentration Y, plus a factor that accounts for the sampling and analytical error associated with the method. For our example, the SAE value listed in the method is 0.10. To determine the 95 percent UCL, then:

    $$95 \text{ percent UCL} = Y + SAE$$
    $$95 \text{ percent UCL} = 1.05 + 0.10 = 1.15$$

3. The lower confidence limit is calculated similarly to the UCL, only the SAE factor is subtracted from, rather than added to, the standard concentration Y:

    $$95 \text{ percent LCL} = Y - SAE$$
    $$95 \text{ percent LCL} = 1.05 - 0.10 = 0.95$$

4. OSHA interprets the results of the above calculations as follows:

    —If the 95 percent UCL is less than or equal to 1, a violation of the allowed exposure limit does not exist.
    —If the 95 percent LCL is greater than 1, a violation of the allowed exposure does exist.
    —If the 95 percent LCL is less than or equal to 1, and the 95 percent UCL is greater than one, there is a possible overexposure. This is the case in our example.

Employers can use these same tests to determine whether or not they are violating an exposure limit; however, in the case of the possible overexposure, as in our theoretical example above, it is better for the employer to assume that an overexposure does exist. This is especially critical when dealing with agents for which there are specific control measures, training, medical surveillance, and ongoing air sampling requirements that apply to exposures at or above a PEL or action level.

Checking Your Understanding

1. Explain how air sampling results that are below the PEL can indicate a possible overexposure situation.

2. What are BEIs and how are they used by an industrial hygienist?

3. Name five factors that can affect a worker's absorbed dose of a hazardous material.

4. Name and discuss three issues or problems related to use of air concentrations as safe levels of worker exposure.

Summary

Sampling the air may be done for many reasons:

— Identification of contaminants in an emergency situation (such as a hazardous material spill) or in an unknown atmosphere (as in a confined space);

— Identification of potential overexposure situations or high-exposure activities;

— Providing a historical record of employee exposures for company records;

— Evaluating effectiveness of engineering controls;

— Verifying adequacy of respiratory protection;

— Initial determination of exposures;

— Periodic sampling to meet regulatory or company policy requirements;

— Evaluation of exposure status relative to an exposure limit.

The purpose of the sampling may affect the method chosen for sample collection and analysis. There are publications available from NIOSH and the AIHA to assist the industrial hygienist in designing a sampling scheme to ensure that the samples provide representative data that are statistically defensible.

There are two basic approaches to air sampling for industrial hygiene applications: 1) direct-reading, which produces immediate feedback about contaminants that are being measured, and 2) integrated sampling, where the sample is sent to a laboratory for analysis. The type of approach that is used will depend on the purpose of the sampling, the regulatory requirements, and the availability of methods and equipment. Each approach has its advantages; direct-reading methods provide fast feedback and involve simple and straightforward methods. Integrated sampling requires more preparation and must be performed by a trained professional following established and validated methods. Because the analysis is done in the laboratory, there is a delay between the sampling event and receipt of results. In both approaches, results may be affected by errors or bias introduced during the sampling event or – for integrated sampling – during sample handling or analysis.

Direct-reading methods involve the use of portable instruments such as gas meters and detector tubes. The instruments may be specific or nonspecific, depending on which sensor or detector is used. Many sensors are specifically designed to detect a particular substance – such as oxygen or hydrogen sulfide – or they may respond to a certain type or class of compounds – such as combustible gas mixtures. It is important to understand the capabilities and limitations of detection devices to allow for appropriate interpretation of the readings they provide. Other direct-reading methods involve colorimetric or length-of-stain tubes, which undergo a color change if the contaminant tested for is present. The length of the stain is read against a scale on the tube and is an indication of the concentration. Other methods involving a color change come in the form of badges or clip-on devices that resemble a credit card. These change color in response to the presence of a particular contaminant and may provide a response that varies in intensity through a range of concentrations. Other direct-reading instruments include the photoionization and flame ionization detectors, which are used for detecting organic vapors.

Integrated sampling methods require a medium for collecting the contaminants present in the air. The air may be drawn through the sample medium by a pump; the contaminant may also be collected through diffusion onto the surface of the medium, as in passive sampling. Integrated methods for collecting air contaminants utilize the physical and chemical properties of the contaminants themselves: impaction, interception, filtration, and dissolution are among the mechanisms used to remove contaminants from the air and collect them on the sample medium. Sorbent tubes and impinger solutions are commonly used for collecting gases and vapors, while filters are generally used for collecting particulates. Passive samplers in common use for gas and vapor sampling contain activated charcoal or another solid sorbent for collecting the contaminant.

Sampling methods for integrated sampling undergo a rigorous testing process to validate the steps used for sample collection and analysis. Validated methods are available from NIOSH and OSHA. The methods contain specific instructions about collecting the samples, including sampling equipment, sample collection media, flow rates and sample volumes, shipping, and analysis. The methods also contain information about the sensitivity of the analytical technique for measuring the contaminant. Following the method will minimize the amount of error that is introduced into the system by the industrial hygienist and the laboratory.

Several laboratory techniques and instruments are used to analyze the media used for collection of air samples. The method of analysis depends on the medium with which the sample has been collected as well as on the physical properties of the contaminant; it may also utilize the chemical properties of the contaminant, or certain characteristics that are inherent in the compound due to the arrangement of the molecules and the type of chemical bonds that are present. Analytical techniques vary in sophistication from gravimetric and counting techniques to more sophisticated methods such as gas chromatography, atomic absorption, and others.

Interpretation of sampling results requires an understanding of the accuracy of the sampling method. To determine compliance with OSHA limits, data are compared to upper and lower confidence limits to test the possibility that an overexposure condition exists. It is possible that data may be shown to be inconclusive at proving whether or not an overexposure exists. In this situation, the employer should assume the worst and proceed as if there were an overexposure or over the PEL condition.

The industrial hygiene professional must be aware that there are many issues associated with the use of airborne concentrations as representing the absorbed dose of the exposed worker. Age, gender, fitness level, pre-existing medical conditions, exposure to a combination of materials, and the physical demands of the work are all factors that can influence the body's absorption of a hazardous material. The industrial hygienist must also consider the representativeness of the data that was obtained during sampling. The use of air concentrations as defining safe levels of exposure also raises questions about the appropriateness of the limit as it applies to the conditions of exposure: what is the condition and susceptibility/sensitivity of the exposed population, what effect(s) does the limit protect against, and how complete are the toxicological data? These are some of the issues that comprise the "art" part of the art and science of industrial hygiene.

Critical Thinking Questions

1. Convert the following concentrations from ppm to mg/m³:

   a. 1,200 ppm toluene (MW: 92)
   b. 2,500 ppm ethanol (MW: 46)

2. Convert the following concentrations from mg/m³ to ppm:

   a. 0.050 mg/m³ lead (MW: 207)
   b. 2,000 mg/m³ cyclopentane (MW: 70.2)

3. Explain what the 95 percent confidence limits are and how they are used to compare sampling data to an exposure limit value.

4. What are the BEIs and what sort of information do they provide? Are they better than PELs or TLVs? Explain your answer.

5. Name and discuss several factors that can affect a worker's absorbed dose of a hazardous material to which they are exposed.

141

6. A worker is exposed to a hazardous material at a level that is about 2/3 of the allowable limit, yet is reporting symptoms consistent with ex-

143

posure to the material. Explain this in terms of exposure limits, absorbed dose, and other relevant factors.

7. List the issues associated with the use of air concentrations as safe levels of worker exposure.

143

6

Indoor Air Quality

Chapter Objectives

Upon completing this chapter, the student will be able to:

1. **List** some of the common complaints associated with poor indoor air quality (IAQ).

2. **Describe** basic HVAC systems and their components.

3. **Apply** some simple testing and troubleshooting to evaluate IAQ problems.

4. **Use** checklists and standardized questionnaires designed for simple IAQ investigations.

5. **List** the contents of some of the consensus standards for IAQ.

6. **Explain** how microorganism contamination as well as radon and asbestos can affect IAQ.

Chapter Sections

6-1 Introduction

The occupational hazards associated with **indoor air quality (IAQ)** are a recent phenomenon, being related in many cases to buildings that were constructed or modified with energy conservation in mind, starting in the 1970s and 1980s. Energy-conserving features often include windows that cannot be opened and thermostat controls that are not adjustable or accessible to building occupants. These modern buildings often are self-contained environments, with temperature, humidity, and airflow being controlled by one or more air-handling units. Air is drawn into the ventilation system and circulated through a network of ducts, louvers, and grates, which may or may not be adjustable by the building occupants. The occurrence of illnesses among occupants of these modern structures has given rise to the terms **sick building syndrome** and **tight building syndrome**. These terms refer to the indoor building environment as the cause for the symptoms that plague its occupants. While sick buildings may be recipients of much attention, sometimes with a good deal of publicity, there are other contributors to poor air quality. Among them are gases emitted by cleaning chemicals, building materials, and office furniture; radon; asbestos; **pathogenic** organisms, such as the bacterium *Legionella* that causes Legionnaire's disease; and plant pollens, mold spores, and animal dander that can cause illnesses or allergic reactions.

In this chapter, we will examine some common causes of poor indoor air quality as well as some simple methods for investigating air quality issues, including basic troubleshooting of the building's heating, ventilation, and air conditioning (**HVAC**) system. Our discussion will concentrate on nonindustrial situations, such as offices, schools, or other work locations where the source of the problem is not as obvious as it is, for example, for a welding process. However, many of the principles are applicable to ventilation-related problems in any setting. We will also look briefly at some other issues related to indoor air quality, such as radon and asbestos contamination.

6-2 Indoor Air Quality as a Public Health Concern

While buildings with less than optimum air quality have certainly existed for many years, the energy crisis of the 1970s spawned thousands of buildings designed and constructed to meet targets for conservation of energy. Heating, cooling, and filtering air, as well as delivering it throughout the building requires consumption of energy. A poorly designed HVAC system, or one that is not properly maintained, will require more energy to operate than one that is installed and operated correctly. Indoor air quality, or more accurately, the lack of it, is typically associated with a set of common complaints among building occupants. The most common complaints are reports of odors, uncomfortable temperatures (too hot or too cold), and physical symptoms, such as headaches and respiratory irritation, which are attributed to chemicals or biological agents. As building occupancy levels increase, so does the number of complaints. The most common types of complaints include those associated with the following parameters:

— Temperature – Too hot, too cold, drafts, and similar complaints are often heard. The specific problems may vary with seasonal changes in temperature and HVAC operation.

— Humidity – The air being too dry is a common complaint, contributing to irritation of the respiratory tract and eyes. Too much humidity contributes to the growth of microorganisms, and encourages odors and mustiness.

— Stuffiness or lack of air circulation – These can be related to location of the diffusers or outlets relative to the occupants. There may also be a lack of air movement attributed to a less than adequate supply of air from the HVAC system. Poor circulation may lead to stratification of the building air, with some areas receiving plenty of air and others little or none. These "dead zones" might allow odors and carbon dioxide (CO_2) to accumulate to unacceptable levels.

— Odors – Objectionable odors are a common complaint and are as varied as flavored coffee smells to body odor, vehicle exhaust, or chemical smells. Indoor sources of odors may include newly painted surfaces, off gassing of furniture or carpet, or smells produced by operation of copy machines and other office equipment. Odors may also be drawn into the HVAC system from outside sources if the outdoor air intakes are located near areas such as loading docks, trash dumpsters, incinerators, or exhaust stacks from a chemical process.

— Physical symptoms – Occupants may report an array of symptoms ranging from dryness of the eyes or respiratory tract to headaches, tiredness, upset stomach, runny noses and congestion, and others. Some of these symptoms may be attributable to contaminants in the air; for example, CO_2 levels above 1,000 ppm may cause headaches or drowsiness in some people. Unfortunately, most of these symptoms are also nonspecific enough that a cause-and-effect relationship between the symptoms and one or more air contaminants is often not apparent, or difficult to establish, at best. The exception to this would be an outbreak, such as Legionnaire's disease, in which a set of severe symptoms would appear among a group of people occupying the same building, making it easier to establish the cause. Aside from these clear-cut situations, is important for the investigator to understand that the lack of a clear link between the reported symptoms and the building's air supply does not diminish the possibility of a correlation between the two.

The **olf** was first used by researchers in Denmark to measure the bioeffluent odor load produced by a "standard" building occupant, defined as a person who bathes about 0.7 times/day, changes their underwear daily, has a skin surface area of 1.8 square meters, and spends their day in sedentary or seated tasks. Some studies of IAQ use the olf to quantify odor levels, and sources of undesirable odor are assigned an olf equivalency. For example, a wet and moldy filter in an air conditioner might produce an undesirable odor at a level equivalent to 10 standard persons or olfs.

NIOSH indoor air quality investigations performed in the United States in the last 25 years have revealed an interesting set of statistics. For one, a large percentage (about half in the NIOSH studies)

Box 6-1 ■ Two Case Studies

These case studies illustrate the difficulties that can be encountered in identifying the specific offending contaminants causing poor indoor air quality. In some situations, the culprit is clear, firm conclusions can be reached, and concrete recommendations can be made. In others, there are no clear answers, and an educated guess is the best conclusion that can be made from the available data.

Case 1

An industrial hygienist was called upon to help identify the cause of periodic incidences of respiratory irritation that occurred among some workers at an offset printing location. Without exception, the staff that was in the area at the time of each event experienced an irritating, sulfur-like odor, and some persons reported more severe reactions involving the eyes, throat, and respiratory tract. The affected area included an office adjacent to an open work area where printed materials were sorted and bundled. Both areas were serviced by a roof-mounted HVAC unit that investigation revealed also serviced the photo processing room. This room was found to be under positive pressure relative to the rest of the plant.

Air sampling found measurable levels of sulfuric acid in the affected areas some 48-60 hours after each event. Inspection of the HVAC unit revealed that a minimal amount of outside air was being drawn into the unit, resulting in mostly recirculated air being supplied to the two areas. The indoor intake grille for the HVAC unit was located near the photo processing area. It was deduced that sulfuric acid – and possibly other irritating compounds present in the photo processing chemicals– were produced in the photo processing area, then drawn into the HVAC unit and circulated to the part of the building where the symptoms were experienced. Recommended corrective actions included increasing the amount of outside air being drawn into the HVAC unit, relocating the indoor intake grille, and installing a local exhaust ventilation system in the photo processing area to remove chemical vapors.

Case 2

An industrial hygienist was called upon to investigate the air quality of an office building as the result of occupant complaints of respiratory tract and other irritation. The building had windows that opened, high ceilings, personal cooling fans, and plenty of room for each occupant. There was no central HVAC unit, nor any individual units, servicing any part of the building. Some offices were equipped with window-mounted air conditioners, and some occupants ran these units on the fan setting, even during the winter, to draw some outside air into their work area. Heat was provided by a boiler, and the steam remained in the pipes and radiators when the boiler was operating. As an added precaution by the building maintenance staff, vents on each radiator had been installed and routed directly to the outside of the building through windows. The boiler had been cleaned using a chemical agent 18-24 months before the investigation.

The investigation included discussions with occupants, completion of questionnaires, and air sampling. Symptoms experienced by the occupants included irritation of eyes, nose, and throat, as well as the upper respiratory tract. Odors resembling urine and dead fish were also reported. Most questionnaire respondents indicated relatively high job satisfaction.

Air sampling was performed in a closed room directly above an open vent in the radiator, in efforts to simulate a worst-case situation. During sampling, the hygienist experienced symptoms of respiratory and eye irritation similar to those reported by the building occupants. These symptoms disappeared almost immediately upon leaving the sampling area. Laboratory analysis revealed no detectable levels of any of the suspected contaminants. For at least one of the suspected chemical contaminants, the limit of detection for the analytical method was well above the estimated odor threshold.

It was concluded that some building occupants might be sensitized to the chemicals used in the boiler, which would account for their reporting of symptoms even at the very low levels of chemicals that could have been present in the building. The hypothetical route for the chemicals entering the building was that chemical-containing steam was escaping out of the radiator vents, only to be drawn back into the building through open windows, or through the window-mounted air conditioners. However, a definitive solution could not be found, and specific corrective measures could not be offered.

of the IAQ problems were attributable to the HVAC system. These problems ranged from poor HVAC system design to improper maintenance. Other causes of IAQ problems were identified as chemical contaminants or microbes, either inside the building – i.e., the HVAC system itself – or drawn into the building from outside sources. In about 10 percent of the cases, the investigation failed to reveal an obvious source for the reported problem.

It should be noted that indoor air quality problems are sometimes linked to other factors that have no bearing on air quality, but may be linked to the occupants' perception of the air quality. One of these factors is job satisfaction, which encompasses a broad range of issues related to personal and professional relationships, level of experience, and other social and economic issues that are beyond the scope of our discussion. Still, it is worth mentioning here since the industrial hygiene professional who investigates IAQ complaints will find many questions related to this on the questionnaires that are a standard part of the IAQ investigator's toolbox (see Appendix 3, NIOSH questionnaire). It is worth noting that job satisfaction, or rather the lack of it, has been statistically linked to other occupational health issues, such as ergonomic injuries. Another significant issue is the degree of control that the building occupants have over their environment. Examples of this would be access to adjustable thermostats and temperature controls as well as to functional windows, which are windows that can be opened and that are near the desk or assigned work area.

Standards for Indoor Air Quality

Although OSHA has proposed development of a standard for IAQ in places of employment, a final standard has yet to be issued. The proposed standard contains provisions for 1) keeping CO_2 levels below 800 ppm; 2) maintaining relative humidity at or below 60 percent; 3) maintaining records on HVAC systems, including the original design specifications, cleaning, and repairs; 4) exhausting designated smoking areas to the outside and keeping them under negative pressure relative to the rest of the building; and 5) locating air intakes of systems to prevent capturing outside air contaminants. OSHA also proposed to require that HVAC systems be maintained and operated in a manner consistent with the building codes that were in force when the

building was constructed and the HVAC unit was installed.

The lack of OSHA regulation, however, has not prevented the establishment of other consensus standards for minimum levels of performance for HVAC systems and the quality of air they supply to building inhabitants. Listed below are summaries of the more significant sources that are used for design and evaluation of HVAC systems. They consist primarily of standards developed by the American Society of Heating, Refrigeration and Air Conditioning Engineers (ASHRAE). As is the case with other consensus standards, these are not regulations, but guidelines that are followed by most professionals in the absence of regulatory standards. Because they are consensus standards and not OSHA regulations, changes and updates are easier. Market pressures as well as changes in building requirements also influence consensus standards. In some locations, the ASHRAE standards may be enforceable through incorporation into local building codes.

ASHRAE 62-1989, Ventilation for Acceptable Air Quality

This recommended standard describes some of the minimum features and performance capabilities for HVAC systems in 100 different types of occupied buildings, including specific recommendations for offices, classrooms, laboratories, smoking lounges, and others. Among the provisions contained in this standard:

— There should be a method for verifying that the system is providing an adequate volume of airflow to the occupied space;

— The system should deliver air to the areas in the space where the occupants are located;

— The system should be designed to prevent the growth of microorganisms through design features such as self-draining condensate pans, steam humidifiers, and unlined ducts;

— Relative humidity should be maintained below 60 percent; CO_2 levels below 1,000 ppm;

— Air intakes and outlets should be located to avoid drawing contaminants into the system;

— Filters, scrubbers, and other treatment methods should be used to remove air contaminants and maintain acceptable air quality;

—The HVAC unit should be located so that it is accessible for cleaning and maintenance.

ASHRAE 55-1992, Thermal Environmental Conditions for Human Occupancy

This recommended standard contains a description of the temperature and humidity conditions that should be acceptable to most occupants of a space. Factors such as air movement, humidity, clothing, and activity level are considered.

ASHRAE 52-1992, Methods of Testing Air Cleaning Devices Used in General Ventilation for Removing Particulate Matter

This recommended standard contains two methods that can be used to test the effectiveness of air filters: 1) the ASHRAE Arrestance test, for filters used to remove larger particles, and 2) the ASHRAE Dust Spot Efficiency Test for filters that are used to remove fine dusts and smaller particles from the air stream. These tests result in assignment of values for the percentage of the test agent that a specific filter removes. Foam-type filters may be found 70-80 percent efficient in the arrestor test but only 15-30 percent efficient for the dust spot test. Fibrous mat-type filters, by comparison, can be rated as high as 95 percent or more for the arrestor test and 90 percent or better for the dust spot test. Specific information on the rating of a filter can usually be obtained from the manufacturer.

Checking Your Understanding

1. Name four common areas of complaint related to IAQ.

2. Explain why IAQ problems are often linked to structures built or modified in the 1970s or later.

3. Aside from problems with the HVAC system, what other factors may contribute to poor IAQ (or the perception of poor IAQ)?

4. What are the ASHRAE-recommended levels for humidity and CO_2?

5. Name three significant ASHRAE standards and briefly describe what they address.

6. What are some of the proposed contents of the OSHA standard for IAQ?

6-3 Heating, Ventilating, and Air Conditioning Systems (HVAC)

It is no accident that HVAC systems are so often linked with problems of indoor air quality. The complexity of some of these systems is enough to require a regimented maintenance program; failure to follow schedules for cleaning and other general maintenance can lead to poor air quality. One of the first steps in performing an investigation into an IAQ complaint is to determine the type of HVAC system involved. Before we discuss the most common aspects of HVAC operation and IAQ problems, we must become familiar with the terms used to describe HVAC systems.

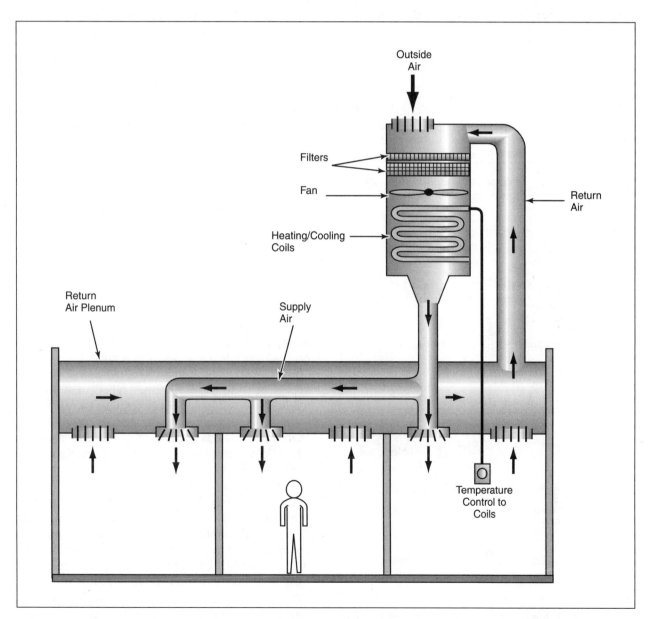

Figure 6-1: Typical single zone, constant volume HVAC unit. The temperature controls affect the heating/cooling of supply air by the unit.

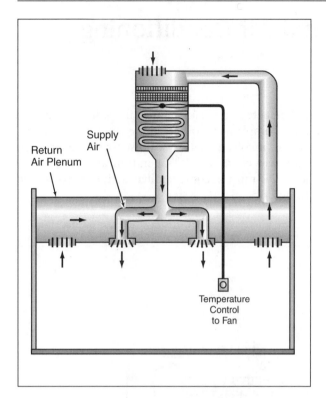

Figure 6-2a: Single zone, variable volume system. The temperature controls regulate the fan in the HVAC unit.

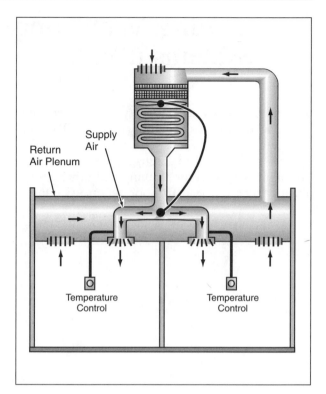

Figure 6-2c: Multiple zone, variable volume system. The temperature controls affect airflow at the point of distribution. The fan in the HVAC unit adjusts in response to pressure changes in the supply duct.

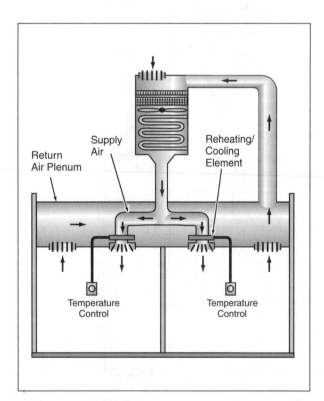

Figure 6-2b: Multiple zone, constant volume system. The temperature controls regulate reheating/cooling of supply air at the point of distribution.

A building is typically serviced by one or more HVAC units. In larger buildings, the HVAC units may operate in unison to service the entire building. As an alternative to that, the building may also be separated into various air spaces, or **zones**, with each air space being serviced by a single HVAC unit. Some HVAC systems may introduce a large percentage of outside air, sometimes called **makeup air**, while other systems recirculate all of the air within the space. Many HVAC units are comprised of more than mechanical fan and heating/cooling sections. The HVAC may also contain humidifiers, dehumidifiers, and filters.

In the typical HVAC system, air from within the zone (**return air**) is drawn through one or more **registers** (adjustable flow) or **grills** (nonadjustable flow) into the system. As the air travels through the return air duct or **plenum**, it approaches the HVAC unit for conditioning. Makeup air from outside the building may be added to the air stream, usually just before or at the entrance to the HVAC unit. Outside air is introduced into the system to assure that the air contains an adequate amount of oxygen, as well as to dilute odors, carbon dioxide, and other contaminants. The air then enters the HVAC unit, where it is tempered (heated or cooled), filtered, humidified, or dehumidified. The air then travels via the **supply air ducts** back to the zone served by the unit.

The volume of makeup air introduced into the system is typically determined by the outside weather conditions. If the outdoor air temperature is moderate, more air will be allowed to enter the system. If the temperature of the outdoor air is very hot or cold, little or no makeup air will be drawn into the system. This is necessary to conserve energy that would be needed to cool or heat the air prior to delivery to the zone.

Supply and return ducts may be insulated on the inside or outside; ducts lined with insulation may present problems associated with cleanliness and erosion of the insulative lining. In some systems, there are no return air ducts. The air is drawn into the HVAC system from an area such as the space between a suspended ceiling and the roof. In these systems, the entire space functions as the return air plenum.

Most HVAC systems fall into one of four main types: 1) single zone, constant volume; 2) single zone, variable volume; 3) multiple zone, constant volume; or 4) multiple zone, variable volume. Some HVAC systems may be hybrids, combining aspects of one or more of these systems.

A single-zone, constant volume HVAC unit provides tempered and cleaned air to a single zone at a preset volume of flow. Similarly, multiple zone systems provide air to more than one zone in the building. Variable-volume systems are adjustable; this means that the volume of airflow provided by the system can be adjusted to suit the needs of the occupants. This adjustment capability presents an advantage during periods of low or no occupancy and during seasonal variations in temperature and humidity. Single zone systems are usually installed in or very near the zone they serve. Multi-zone units may be located some distance from the areas they serve; for example, they may be roof-mounted or installed in special rooms, such as mechanical rooms or fan rooms, or other dedicated areas in the building.

IAQ problems related to the operation of the HVAC system are often attributable to poor or irregular maintenance. Table 6-1 summarizes some of the more common maintenance-related contributors to poor IAQ. Many of the potential problems can show up in the HVAC unit itself, as well as at

Potential Problems	Possible Cause
Mold, mildew, other microbes growing on filter media. Clogged filters preventing airflow. Dirt/contaminants recirculated in building air.	Wet/dirty filters
Spores, microscopic fragments of molds, etc. being circulated in building air. Odors from decaying matter circulated by HVAC.	Wet/decaying organic matter
Growth of mold, mildew; growth of microbial organisms including bacteria, algae; odors.	Standing water in drip/condensate pans
Odors; particles recirculated in building air.	Dirty cooling/heating coils
Odors; irritation or sensitivity from chemicals.	Residue from chemicals/cleaners
Little or no airflow.	Fan belts slipping/broken
Little or no airflow.	Drive motors inoperative
Growth of molds, mildew, bacteria; odors; erosion of damaged insulation resulting in airborne particles circulated in building air.	Wet, dirty, or damaged duct insulation
Encourages microbial growth; odors.	Settled water/signs of water damage (at any location in the HVAC system, unit, or ducts)
Dirt/stains on ceiling tiles around diffusers, grills, or registers.	Entrainment of dirt in system; poor filter maintenance; dirty duct interiors.

Table 6-1: Many operational HVAC characteristics are associated with poor IAQ. Problems with IAQ can arise during lapses in maintenance.

locations inside the zone. A simple checklist for inspecting an HVAC system is included as Appendix 4.

Aside from operations-related problems, other contributors to poor quality of HVAC-supplied air may be related to one or more aspects of airflow. These problems may be due to:

—Little or no outside air being drawn into the system;

—Poor system design, or installation of the system not as designed;

—Failure to maintain and operate the system as it was designed;

—Alterations and modifications to the system after installation, or alterations in the zones served by the unit;

—Placement of air intakes near outside sources of pollutants.

Inadequate outside air may be linked to buildup of noticeable odors as well as increased levels of carbon dioxide and other undesirable gases. Because humans produce carbon dioxide, its levels are often used as an indicator of the effectiveness with which the HVAC unit is able to supply fresh air to the building or occupied space. Some special IAQ instruments are now available that include sensors for measuring air velocity, temperature, relative humidity, and CO_2, all in one instrument. Less sophisticated methods, such as detector tubes, can also be used.

The recommended amount of outside air has varied over the years. In 1905, the recommended rate was 30 cubic feet per minute (cfm) of outside air/person; in 1936, the recommendation was 10 cfm of outside air/person; in 1973, ASHRAE recommended 5 cfm of outside air/person, probably a reduction for energy conservation purposes. More recent ASHRAE recommendations – revised in 1989 – set the level at 20 cfm/person for an office area. This upward trend is the result of increased concerns with indoor air quality issues such as secondhand tobacco smoke. ASHRAE recommendations for outside air are minimums and should be considered such. If the occupied space contains sources of aerosol contaminants or odors, or combustion sources, additional outside air may be required.

To determine whether there is adequate intake of outside air, a number of causes or potential situations should be ruled out before more costly investigative measures are employed. The system design should be reviewed and compared to the present installation. It may be that the zone's requirements for air volume and flow exceed the capabilities of the unit. If the unit is capable of providing the necessary volume of flow, the unit should be checked to see if the outside air dampers are stuck or rusted closed. Sometimes these types of complaints are seasonal, becoming more common when air is being cooled in summer, or heated in winter months. Mechanical problems should also be ruled out. Fans, motors, and drive belts should be inspected to determine if there is need for repair or replacement. The supply ducts should also be checked to rule out blockage, sharp turns, and "user" modifications that interfere with airflow.

If the unit is supplying a volume of air (mixed with adequate outside air) that is appropriate for the occupied area, the problem may be one of inadequate air distribution. This can be linked to causes as simple as a blocked room diffuser or dampers being closed. Direct measurement of airflow is useful for determining whether the air supplied by the system meets design specifications. A **velometer** is an instrument commonly used to measure airflow. The reading is taken at the face of the diffuser or outlet, and the instrument provides a reading in feet per minute (fpm). Measurement of the length and width of the opening allows calculation of the area (measured in square feet). The velocity (fpm) multiplied by the area of the duct (ft^2) provides the volumetric airflow in cubic feet per minute (cfm). Airflow through a duct is defined by the following equation:

$$Q = VA$$

Where

Q = the volumetric rate of airflow in cfm,

V = the velocity of the air in fpm, and

A = the area of the duct.

Let us illustrate with an example. The IH is evaluating airflow to an office and measures the flow of air coming out of a 2 ft × 2 ft opening with a velometer. The reading is 10 fpm. The application of the equation, $Q = VA$, results in the following:

$$Q = VA$$
$$Q = (10 \text{ fpm}) (2 \text{ ft} \times 2 \text{ ft})$$
$$Q = (10 \text{ fpm}) (4 \text{ ft}^2)$$
$$Q = 40 \text{ cfm}$$

The amount of air that is being supplied to an area by the ventilation system is an easy measurement to take and can be useful for basic troubleshooting, as we will see later.

The addition or rearrangement of walls, partitions, or major furniture within the space, after the installation of the HVAC unit, may cause problems by creating a configuration of supply and return volumes that were not considered in the original design. This may give the impression of an apparently malfunctioning HVAC system. In order to ensure that mixing of air occurs in an occupied area, supply and return ducts should be placed as close as possible to the occupants. If this is not possible, persons who work in areas with poor mixing can be provided with pedestal or small personal fans that can be used as needed. Adding, blocking off, or removing ducts can also create an imbalanced system. It is sometimes possible to rebalance or optimize systems if a limited amount of changes have taken place; however, these changes should be made by a competent HVAC engineer.

Inspecting the ducts and intakes is a sometimes overlooked but important aspect of an IAQ evaluation. Ducts may be located in areas that are hard to inspect; above ceilings, in crawl spaces, under floors, and in walls. Nevertheless, with some effort, it is usually possible to locate and inspect most ducts. Worn insulation located inside ducts can lead to fibers or particles from the insulation being drawn into the HVAC system and circulated through the building. Wet insulation, or insulation with stains and other evidence of water damage, can be a breeding ground for microbes. Some of these cause disease, others are capable of causing allergic reactions among sensitized individuals, while yet others may cause building occupants to develop sensitivities.

Intakes are also areas that need to be checked visually. Like ducts, the intakes for a system may be difficult to locate or to inspect; they may be on roofs, on sides of buildings, or between floors. When inspecting intakes, the IH needs to check for blockage, including leaves, bird nests, and debris on screens; stuck or malfunctioning louvers, and other barriers to airflow. Potential sources of pollutants that could be drawn into the system need to be investigated. Garbage dumpsters, parking structures, loading docks, incinerators, and chemical storage areas are a few examples of potential sources of odors and contaminants. The presence of processes or odor-producing equipment must be considered, including machines that use solvents or fluids such as blueprint machines. Fresh air intakes should not be located near such machines, or near the exhaust of a ventilation system used for hazardous contaminant control. Another possibility to consider when odor problems develop suddenly is the relocation of a process, a machine, or any other possible source of the odor. It is possible that rearrangement plans did not take into account the ventilation requirements or the presence of air intakes. Ventilation equipment that has been disassembled, moved, and reinstalled, may not have been reassembled properly. Fan motors that are wired incorrectly will result in blades that turn in the opposite direction. In this case, some fans are still capable of moving air, but the volume of air that is moved is probably going to be less. Incorrect wiring is a common cause of less-than-optimal fan performance.

Checking Your Understanding

1. Name and describe the four basic types of HVAC systems.

2. Name five functions performed by the HVAC system.

3. List at least six operational aspects of an HVAC system, which can cause or contribute to poor indoor air quality.

4. Name four things that can affect the flow of air in a supply zone, including two items that are not related to changes in the HVAC system.

5. What is the volumetric rate of airflow from an opening measuring 12 × 24 inches, with a velometer reading of 8 fpm?

6-4 Basic Instruments for Use in IAQ Studies

In addition to inspecting systems visually, many simple instruments can be used to help the occupational health professional evaluate air quality. Table 6-2 describes some of them.

Evaluating the Amount of Makeup Air

The function of makeup air is to ensure adequate oxygen content and to dilute indoor air contami-

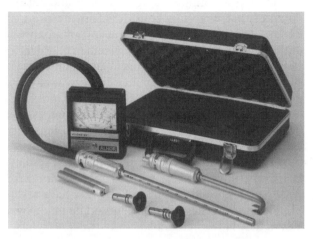

Figure 6-3: A velometer is a common instrument for measurement of air flow across the face of a duct or air diffuser. This measurement can be used to calculate the amount of air being supplied to the diffuser by the HVAC unit, which can then be compared to a recommended airflow.

Instrument	Use/Purpose
Thermometer	Determine temperature of outside, supplied, and return air; helps in determining percentage of outside air being drawn into system.
Velometer; rotating vane anemometer; hot-wire anemometer	Direct measurement of velocity of airflow; can be used to evaluate actual performance of HVAC vs. design specifications.
Gas-detection instruments; detector tubes	Direct measurement of contaminant concentrations in ducts and areas of the building; carbon dioxide is commonly used as an indicator gas for IAQ benchmarking.
Psychrometer	Relative humidity
Smoke tubes	Generate small amounts of visible smoke used to observe airflow patterns.
IAQ multi-function instruments	These instruments are capable of providing direct-readout measurements such as: air temperature; humidity; CO_2; air velocity; and dewpoint. Many of them have computer-interface capabilities for downloading numerous stored measurements taken during an investigation.

Table 6-2: These are just a few of the instruments that are used to evaluate indoor air quality. Although many of the instruments are simple, the data they provide can assist in identifying most air quality problems associated with HVAC system operation.

nants to levels that are acceptable and healthful for most of the occupants of the space. Possible contaminants include carbon dioxide as well as odor- or symptom-producing gases, vapors, dusts, molds, and other airborne substances.

Carbon dioxide is relatively easy to measure and is widely used as an indicator of the effectiveness with which the HVAC system is diluting contaminants and mixing the air. The levels of CO_2 present in a space will vary depending on how many people occupy the area; consequently, the levels will tend to increase in the course of a workday or shift. After

When using outside air conditions such as temperature or CO_2 concentration to evaluate flow into a space, the IH should try to obtain the data for outside air as near as possible to the air intake location. For example, if the CO_2 measurement is taken outside the main entry, it may be different from the concentration of CO_2 that might be obtained on the roof, or in the alley, where the intake is located.

the workers have gone home and the area is vacant, levels of CO_2 will drop if the HVAC system continues to provide outside air to the space. The predictability of this cycle allows for some simple evaluations of airflow in an occupied space, using CO_2 as the indicator gas. The basis for these tests is that a volume of outside air brought into a space will reduce the concentration of the indicator gas (in this case, CO_2) present by an amount proportional to the volume of outside air introduced.

A simple method for estimating the volume of outside air flowing into a space involves taking measurements of the CO_2 levels both inside and outside the space. The levels can be compared using the following equation:

$$Q_{OA} \approx \frac{13,000\,x}{C_{in} - C_{out}}$$

Where

Q_{OA} = the volumetric flow rate of outside air in cfm,

C_{in} = the concentration of CO_2 inside the space; and

C_{out} = the concentration of CO_2 outside the space.

To illustrate, let's assume the IH has done some preliminary IAQ investigative work and suspects that the HVAC system is not delivering enough outside air to a corner office. She counts 10 people occupying the space. By noon, the CO_2 concentration has leveled off at about 1,100 ppm, as indicated by a direct-reading instrument. The outdoor CO_2 concentration is 335 ppm. The IH inserts her data into the equation:

$$Q_{OA} \approx \frac{13,000\,x}{C_{in} - C_{out}}$$

$$Q_{OA} \approx \frac{13,000\,(10)}{1,100 - 335}$$

$Q_{OA} \approx 170$ cfm, or about 17 cfm/person.

This volume of airflow is lower than the current flow of 20 cfm/person recommended in ASHRAE Standard 62-1989.

An instrument as simple as a thermometer can provide a great deal of information about the efficiency of an HVAC in bringing outside air to the occupied space. There is a method for estimating the percentage of outside air that involves the use of temperature along with the following equation:

$$\%OA \approx \frac{T_{RA} - T_{SA}}{T_{RA} - T_{OA}} \times 100\%$$

Where

T_{RA} = the temperature of the return air,

T_{SA} = the temperature of the air supplied to the occupied area, and

T_{OA} = the temperature of the outside air.

An example of the use of this equation follows.

The following temperatures were measured at the designated locations in the HVAC system:

T_{RA} = 72°F

T_{SA} = 65°F; and

T_{OA} = 50°F

What is the percentage of outside air being drawn into the system?

Using the equation:

$$\%OA \approx \frac{T_{RA} - T_{SA}}{T_{RA} - T_{OA}} \times 100\%$$

$$\%OA \approx \frac{72 - 65}{72 - 50} \times 100\%$$

$\%OA \approx 31.8$, or 32 %

The percentage of outside air must be compared to the total amount of air supplied by the HVAC unit to determine whether an adequate volume of outside air is being circulated by the unit. For example, if the 32 percent calculated above is applied to an HVAC unit supplying 1,000 cfm to an area occupied by 10 people, or 100 cfm/person, the system is providing approximately 32 cfm/person of outside air. This exceeds the minimum ASHRAE recommendation of 20 cfm/person.

A similar method involves the use of measured levels of CO_2 for estimating the percentage of outside air, using the equation:

$$\%OA \approx \frac{C_{RA} - C_{SA}}{C_{RA} - C_{OA}} \times 100\%$$

Where

C_{RA} = the concentration of CO_2 in the return air,

C_{SA} = the concentration of CO_2 in the air supplied to the occupied area, and

C_{OA} = the background concentration of CO_2, in outside air.

Let's assume the IH has taken some CO_2 measurements in the HVAC system, with the following results:

C_{RA} = 800 ppm

C_{SA} = 700 ppm

C_{OA} = 350 ppm

We can use the equation to estimate the amount of outside air as follows:

$$\%OA \approx \frac{C_{RA} - C_{SA}}{C_{RA} - C_{OA}} \times 100\%$$

$$\%OA \approx \frac{800 - 700}{800 - 350} \times 100\%$$

$$\%OA \approx 22\%$$

The percentage of outside air must be applied to the total volume being supplied to the occupied area as described above. If we assume that this example is for the same HVAC unit and occupancy levels, the amount of outside air is approximately 22 cfm/person. This is slightly above the ASHRAE minimum recommendation for outside air of 20 cfm/person.

Another simple but useful evaluation involves taking CO_2 measurements in the occupied space throughout the day, at specific times; for example, in the morning near the start of the day, at midday, after quitting time, and one hour or so later. These measurements will provide an indication as to how well the system is diluting buildup of CO_2 levels after the occupants leave. It is important that most of the people have gone from the building before taking the final two measurements. If the concentration of CO_2 does not drop, or if it drops slowly, it could be an indication that little or no outside air is being drawn into the system. Beware, too, of some systems that automatically dial back the amount of outside air after a preset quitting time.

Other equations are used to evaluate the performance of an HVAC system. Additional information about the use of other equations and/or more complex IAQ studies is available in the references at the end of the book.

Checking Your Understanding

1. Name four instruments commonly used for IAQ studies. What types of measurements do they provide?

2. What types of measurements can be used to estimate the amount of outside air drawn into an HVAC system? Where are the measurements taken?

3. In the course of conducting an IAQ investigation, the IH measures and records the following CO_2 levels inside the building: 8 am (500 ppm), 1 pm (1,200 ppm), and at 6 pm (1,000 ppm). Outside the building near the HVAC intake, the levels are measured and found to be 300 ppm. The workers have gone home at 5:00 pm. What conclusions might you draw from these data?

6-5 Microorganism Contamination and IAQ

Microorganisms in an HVAC system can lead to odors and in some cases, disease or allergic reactions. These problems arise when the numbers of bacteria, fungi, or viruses increase due to conditions that favor their growth. The presence of these types of contaminants might be suspected if the inspection of the HVAC system revealed wet filters, wet duct insulation, or standing water in any part of the system, such as condensate pans or ducts. The presence of greenish or other slimy growth in the wet areas is another obvious clue. The fact that intakes are located adjacent to cooling towers is also a potential clue. The solution in all these cases is to identify and remedy the uncontrolled source of the moisture and to clean or replace the affected areas of the system. However, strong-smelling cleaning solutions used in the HVAC system may produce undesired odors or may be irritating.

Among the notable incidents involving disease-causing organisms in HVAC systems, Legionnaire's disease is one of the best known. This pneumonia-like disease is caused by bacteria that have been found in hot water systems, air conditioners, cooling towers, and condensate pans. The bacteria are probably transmitted through air and are inhaled, causing high fever, chills, general aches and pains, headache, and digestive tract discomfort within two days to a week after exposure. Although Legionnaire's disease has achieved much notoriety, it causes illness to only five percent or less of the exposed population. The notoriety is more likely related to the approximate 15 percent death rate among those that do become ill.

Small mammals and birds might gain access to HVAC systems, building nests at intakes and traveling through ducts. They may leave droppings or they might die in the ducts, resulting in contaminants being drawn into the system and circulated in the building air. The **Hanta virus** has been isolated in the droppings of deer mice in the western United States, but is suspected to be present in other parts of the country as well. Infections occur when disturbance of the droppings or nest materials generates airborne dust containing the virus, which is then inhaled. The illness produces a series of flu-like respiratory symptoms, which may be mistaken for other diseases, such as flu or bronchitis, sometimes resulting in death. The Hanta virus can be easily neutralized using a three percent hypochlorite (bleach) solution, which should be applied to the potentially infected materials and surfaces, and allowed to soak for 30 minutes or so. The creation of dust during cleanup should be avoided. Respirators with high efficiency particulate air filters may be advisable if the cleanup involves a large amount of material, or if personnel routinely perform such cleanup activities in the course of their job. Once such materials are removed from an HVAC system, it is advisable to try to prevent re-entry of the rodents (a task sometimes more easily approached than solved!).

Histoplasmosis is a fungal infection that usually affects the lungs, resulting in an allergic response, or a serious illness, sometimes causing death. The fungus is found in bat guano and bird droppings, which is why it is also known as bird-fanciers disease. Chickens, pigeons, and starlings are among some of the common carriers of the fungus. The discovery of nests and/or droppings in an HVAC system could warrant installation of wire mesh barriers and cleaning of the area using a sterilizing solution of three percent hypochlorite. Again, the creation of dust should be avoided, and appropriate PPE should be worn.

Allergic reactions to airborne allergens can result from hair, dander, droppings, and other materials being present in the HVAC system and circulated in the air supply. Symptoms may include fever, cough, and chest tightness, often occurring within a few minutes or hours after exposure. Identifying and removing allergens can be a tricky and costly business.

If a microbial or allergen contamination of the HVAC system is suspected and is causing severe reactions, it is wise to consult with a professional who specializes in microorganism sampling and detection. The methods and equipment used for these types of sampling are specialized and beyond the price range of most building maintenance budgets. The cost of hiring a consultant to perform such an investigation might be justified by liability concerns on the part of the employer or building owner.

Checking Your Understanding

1. What are some of the visible clues or signs that odors or other complaints might be related to microorganism contamination of an HVAC system?

2. Name some of the methods that can be used to prevent animals and birds from entering and contaminating the HVAC system.

3. Why should one avoid creating dust when cleaning out nests or droppings?

4. Why is PPE recommended for cleaning operations where Hanta virus or other pathogenic microorganisms are suspected?

6–6 Radon and Asbestos

Radon

Radon is a **decay product** of radium-226, which itself is a product of the decay of uranium-238. Radioactive materials give off energy in the form of **alpha particles**, **beta particles**, or **gamma rays** – depending on the isotope – as they change to a more stable isotope. The process by which they give off this energy is called **decay**, and the intermediate steps in the process are the decay products; the decay products of radon are called **radon daughters**. Both radium-226 and uranium-238 occur naturally in rocks and soil, with higher concentrations occurring in some rocky geographic locations. Buildings

made with brick or stone may therefore contain some radon. The U. S. Geological Survey has produced a map that shows the general locations of geologic formations with potential to contain radon across the United States.

Radon is a gas that enters a building by diffusion through cracks and pores in the building foundation. It may also enter through the water supply, though this is less common. The amount of radon that enters a building is directly related to its location, the porosity of the soils, the porosity of the building foundation, the presence of cracks or openings in the building foundation, and the air pressure of the building relative to the surrounding

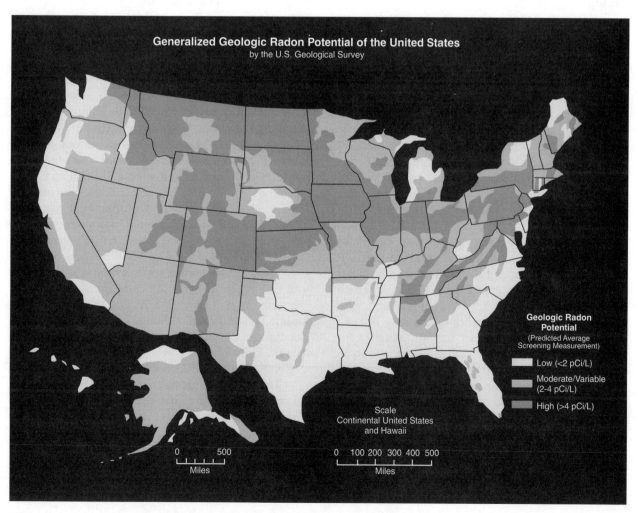

Figure 6-4: Map from USGS showing radon potential across the United States.

atmosphere. Buildings under low or negative pressure relative to the outdoors will tend to draw in air (and radon) from the outside. Because the basements of large buildings comprise a smaller proportion of the entire building volume, radon is of more concern in a single-family dwelling where the basement may comprise as much as 50 percent of the building volume.

The health concerns associated with radon are linked to two of its decay products, polonium-218 and polonium-214, both of which emit alpha particles as they decay to reach more stable forms. Alpha particles are composed of two protons and two neutrons, but have no electrons, resulting in a charge of +2. Alpha particles are primarily an internal hazard. They are not capable of traveling far through air – about two to four inches is the maximum range – or of penetrating simple barriers – skin or paper are effective shields – but if they are inhaled, they deposit their energy in tissues. This is suspected of causing lung cancer.

Related to decay is the **half-life**, which is the amount of time it takes for the radioactive material to decay to one-half of its original level of activity. Most radon decay products have very short half-lives and decay to stable forms quickly, in seconds or just a few minutes. However, for the two polonium isotopes mentioned above, the half-lives are about 30 minutes each, thus providing enough time to be inhaled. Other than inhalation of the particles themselves, building occupants can also inhale dust particles to which decay products of radon have become attached. This is why radon is considered a significant health risk where it occurs in high concentrations.

Like other radioactive particles, it is possible to describe radon and its radon daughters in terms of their **activity**, or rate of decay. The unit used for this is the **pico Curie**, expressed as **pCi**, and named for the physicists Marie Curie and her husband, who were pioneers in the field of radioactivity. One pCi is equal to 0.037 decays per second. Samples for radon and its decay products quantify the activity in terms of pCi per liter of air sampled, expressed as pCi/L. EPA has recommended that radon concentrations indoors should not exceed 4 pCi/L. Sampling for radon can be fairly simple and involves placement of canisters of activated charcoal in the space to be tested, leaving them for up to a week, and returning them to a laboratory for analysis. The most reliable measurements of radon are those taken over several months, since the concentration can vary on an hourly, daily, or even seasonal basis.

Since radon and its decay products pose an inhalation hazard, most control methods are aimed at reducing, removing, or preventing entry of radon gas into occupied areas of the building. Methods for controlling radon involve actions such as application of non-porous coatings on foundations or increasing ventilation of basements or crawl spaces. Each EPA regional office also has one or more persons assigned to radon issues, and state health departments provide local resources for information and testing.

Asbestos

The presence of asbestos is often perceived as an indoor air quality issue in many buildings. This is due to the general assumption that the mere presence of asbestos poses a health hazard, even though this has been proven not to be accurate. The EPA regulations of the 1980s and early 1990s resulted in many school districts launching full-scale removal of all asbestos-containing building materials – often by unskilled contractors – although removal of asbestos was never recommended by EPA. In fact, the EPA's "green book" (*Managing Asbestos in Place, a Building Owner's Guide to Operations and Maintenance Programs for Asbestos-Containing Materials*) on management of asbestos-containing materials in buildings, issued in 1990, states that asbestos-containing building materials are often best managed in place. Where removal work was done improperly, the removal actions created, rather than abated, an asbestos hazard.

The primary concern with asbestos is the carcinogenic potential of the airborne fibers, not its mere presence in a building. EPA has evaluated many public and commercial buildings containing asbestos building materials, with the conclusion that in the majority of cases, airborne asbestos fiber levels are well below current occupational limits.

The 1995 revisions to the OSHA regulations for asbestos require that building owners identify the asbestos-containing materials present in their buildings. They are also required to notify the occupants of the building that the materials are present and what actions, if any, personnel should take to avoid creating a hazard. Locations that contain large amounts of asbestos, such as mechanical rooms or boiler rooms, must be posted to alert persons to the presence of asbestos. Other provisions are included in the regulation to ensure that construction and maintenance contractors are informed about any asbestos that might be disturbed by their work.

The OSHA regulation does require air monitoring during an abatement action. Neither OSHA nor EPA require air sampling as part of the hazard assessment process once asbestos has been identified in the building. This is significant since recent studies have shown little correlation between levels of airborne asbestos fibers – which present a health risk – and the physical condition of asbestos. An assessment based on the physical condition of the asbestos-containing materials, their location and accessibility, likelihood and source of disturbances, as well as airborne fiber levels, is probably the best approach. Additional information about asbestos hazards and managing asbestos materials in place can be obtained from regional EPA offices, or from state environmental quality agencies.

Checking Your Understanding

1. What is the primary health concern associated with radon? What level does EPA consider to be safe?

2. How does radon get into buildings, and what are some ways for remediating a building with an identified radon problem?

3. What is the primary health concern associated with asbestos?

4. List four provisions of the OSHA regulations for asbestos in buildings.

Summary

Concern about indoor air quality is a relatively recent phenomenon that is increasingly encountered by occupational health professionals. Potential sources of air quality problems and contaminants include chemicals from cleaners and building materials such as furniture and carpet; odors from office equipment; plant pollen, microbes, and mold spores; and odors, chemicals, and various types of foreign materials that are drawn into the HVAC system from outside of the building. About half of NIOSH studies performed in the past 25 years found IAQ problems attributable to the HVAC system, specifically in terms of poor design or improper maintenance. Other problems associated with poor IAQ included chemical contaminants or microbes in the HVAC itself or drawn in from outdoors. In some situations, a low level of job satisfaction played a role, while in some others no obvious cause was identifiable. The possibility for poor IAQ is directly proportional to the density of the building population. Common complaints heard by occupational health professionals include problems with temperature extremes, humidity, stuffiness or lack of air circulation, odors, as well as physical symptoms such as headaches, dry eyes, respiratory irritation, and fatigue.

Although there is no OSHA standard for IAQ, consensus standards are used for design and evaluation of HVAC systems. These include ASHRAE standards for ventilation for acceptable air quality, thermal environmental conditions, and methods for testing air cleaning devices. These standards contain recommendations for minimum amounts of airflow in occupied areas, design requirements for location and maintenance of HVAC systems, recommended levels of humidity and CO_2, and others.

Most HVAC systems fall into one of four main types: 1) single-zone, constant volume; 2) single-zone, variable volume; 3) multiple zone, constant volume; or 4) multiple zone, variable volume. HVAC systems may also combine aspects of one or more of these four types. Each HVAC unit serves a portion of the building space, called a zone. Typically, the HVAC unit heats/cools, humidifies/dehumidifies, and filters air that it supplies to the zone it services. Many IAQ complaints can be attributed to some aspect of HVAC maintenance or operation. Inspection of the HVAC unit and the supply and return ducts is necessary to identify and correct HVAC-related IAQ problems. Accumulation of moisture, dirt, organic matter, animal droppings, or carcasses may all contribute to microbial growth and odors, which result in poor IAQ. Improper or blocked airflow is another possible HVAC problem. Checklists are provided in the Appendix to aid in performing HVAC inspections. The use of simple instruments to gather data on air temperature and CO_2 concentrations is explained, and sample problems are worked to show how these data can be used to evaluate HVAC system performance in practical situations.

Radon and asbestos are two of the more common IAQ concerns that may pose serious health hazards at high concentrations. Radon is a radioactive gas that is linked to lung cancer. It can be measured using simple passive detectors, which are analyzed by a laboratory following the sampling period. Effective methods for controlling radon are 1) ventilation, and 2) sealing cracks and other openings in basements and foundations to prevent the entry of radon gas.

Asbestos typically does not pose a hazard to building occupants, but may be of concern to maintenance and custodial workers whose job duties involve disturbance of installed asbestos-containing materials. OSHA regulations require building owners to identify asbestos-containing materials and inform building occupants, contractors, and others who perform work that might disturb the asbestos materials as to the location of such materials. Evaluation of the hazard posed by asbestos-containing materials present in an occupied building is probably best done through an assessment of the physical condition of the materials, their potential for disturbance, and through air sampling to evaluate levels of airborne asbestos fibers.

Critical Thinking Questions

1. Explain why IAQ studies may result in data that, while interesting, do not identify the source or cause of the complaints.

2. Why do you suppose that OSHA has not issued a regulation that dictates minimum air quality requirements for non-industrial settings? Are there any advantages to relying on consensus standards, and if so, what are they?

3. You are called upon to investigate complaints of moldy odors and stuffiness in the offices located at the front of the plant. Describe how you would complete the investigation, including your strategy for performing inspections and tests.

4. During a safety meeting, someone expresses concern about the presence of asbestos insulation in the penthouse mechanical room and then states that they have noticed a lot of people in the building suffering from respiratory illnesses. How would you respond?

7

Controlling Airborne Hazards

Chapter Objectives

Upon completing this chapter, the student will be able to:

1. **List** several options that can be used for the control or elimination of different types of airborne hazards.

2. **Recognize** whether dilution or local exhaust ventilation is most appropriate for contaminant control for a given operation or process.

3. **Explain** some basic principles of dilution ventilation.

4. **Explain** some basic principles of local exhaust ventilation systems.

5. **Take** simple measurements and make observations to evaluate ventilation system performance.

6. **Locate** regulations and other standards that contain information about ventilation system design and performance.

7. **Calculate** the amount of dilution ventilation required to reduce contaminants to acceptable levels.

7-1 Introduction

In earlier chapters, we discussed the role of the industrial hygienist as the occupational health professional who anticipates, identifies, evaluates, and recommends controls for chemical and physical agents that may cause harm to the health and well-being of workers. This chapter will address engineered controls – primarily ventilation – that can be used to control airborne hazards that have been identified or are anticipated to occur in the workplace. We will look at local exhaust ventilation systems, used for removing contaminants from the work area by capturing them at the point of emission, and dilution ventilation systems that add air to dilute contaminants in the work environment. The appropriate conditions and good general practices for use of each type of system are addressed.

Airflow through a ventilation system is traced from the intake through the ducts, air cleaner, and finally out through the fan and exhaust. The importance and function of each part of the system is explained. Some basic principles of airflow physics are presented along the way, and examples are given as to how these principles can be applied in system design and for evaluating ventilation system performance. The most common types of fans and air cleaning devices are presented, along with examples of their appropriate application in ventilation systems.

7-2 Hazard Control Options

Once an airborne hazard has been identified, it must be quantified in a manner that allows the industrial hygiene professional to determine whether the material poses a health hazard. Methods for identification and evaluation of airborne hazards are addressed in Chapters 4 and 5. This chapter deals primarily with the use of engineered controls to protect workers.

Let's assume that we have two industrial processes that involve the use of toxic materials. One process requires workers to manually empty bags of material into a large hopper. About 25 percent of the dust particles released during this activity are in the respirable range (<10 μm in diameter). The second process involves a volatile solvent with irritating properties. Workers are required to periodically open an access door and check the pH of the solvent. If the solvent levels are low, the workers must reach in with their entire arm in order to reach the solvent. This sometimes results in the worker's clothing becoming wet with solvent, as the gloves do not cover the entire arm. Also, when the workers reach inside, their eyes often become irritated by the solvent vapors. We will use these two situations to illustrate the use of various control methods.

There are several actions that can be taken to reduce or control airborne hazards; they include engineered and administrative controls as well as PPE. These approaches – presented in Chapter 1 – are listed and will be discussed in the order of their preferred use.

Engineered controls are incorporated into the process itself, sometimes as part of the equipment. These types of controls greatly reduce the amount of contaminants or prevent their release into the air. One of the most common and easiest engineered controls to employ is substitution, which is the use of another material that poses a less significant health hazard. To prevent the trading of one set of problems for another, the substitution must be carefully considered by both the process engineer and the industrial hygiene professional. The substitute material must meet the process requirements as well as protect worker health. This method could be applied to either of our previous example processes.

Process alterations are another way to reduce or eliminate the release of contaminants into the work environment. Examples of changes that can be made in processes include automation and enclosure. In our second sample process, the task of checking the pH of the solvent could be automated through installation of a sensor monitoring pH and providing a readout or record of each measurement. This would eliminate the need for the workers to reach into the process vessel. For the first sample process, manual bag-emptying could possibly be replaced by large hoppers of material that are drawn into the process vessel using an automated dumper, or perhaps a pneumatic material transfer system.

Enclosure of the process machinery separates the process environment from the one inhabited by the workers. In the case of our first sample process, the use of an enclosure around the process machine would prevent the dust from escaping into the work area where it could potentially be inhaled by the workers. However, enclosure is not always practical, as it might not be possible to isolate all of the workers if the tasks that pose an exposure hazard are not also automated or somehow eliminated.

Work practices alteration can aid in the control of airborne contaminants. In the first sample process the use of a water mist might be a feasible alternative. Misting the area where the bags are emptied could significantly reduce the airborne dust levels. Of course, the process has to be able to tolerate the added moisture. Housekeeping practices can also be an aid, especially for processes that generate dust. Regular cleaning using vacuums – or **HEPA** vacuums if respirable dust is an issue – as well as wet sweeping and wiping of surfaces will prevent stirring settled dust back into the air. If the dust is combustible, housekeeping will be an important part of a fire and explosion prevention program.

Administrative controls that may be applied to hazardous materials include establishment of controlled or limited access areas, usually through posting signs and erecting some type of physical barrier. OSHA standards define these as **regulated areas**. By controlling access to regulated areas, the employer can limit the number of workers that are potentially exposed to the hazard. Additional administrative measures used to control exposures are worker rotation and worker training. Generally, these are appropriate only in situations where engineered controls are not effective or feasible. Although worker rotation will keep exposures to levels that do not exceed regulatory limits, the end result will be a larger number of exposed workers. Also, OSHA prohibits the use of worker rotation as a method for reducing exposure in some instances; cadmium is one such example.

Worker training can be an effective control method if the principles learned are applied consistently and in the manner that is needed to provide the intended protection. Employees should receive training to address any special work practices and procedures, such as use of water sprays or special handling techniques to minimize dust; operation of emission controls such as fans or dust collectors; the proper use and limitations of any PPE that is assigned; and how to respond to situations where controls do not function properly, or when unusual conditions arise. Significant results are usually obvious immediately following training, but these results decrease over time if the training is not periodically reinforced. Workers should also receive training in how to respond to emergencies, including spills or uncontrolled releases of contaminants, if their job duties include such activities. Awareness training, so that workers can recognize a potentially unsafe situation and take prompt measures to escape and warn others, is the minimum OSHA-required training for personnel who do not actually respond to and contain a spill or release.

The use of PPE is typically considered the last resort to be used only 1) when engineered and administrative controls are not feasible, 2) while these controls are being instituted, or 3) if already existing controls are not sufficient to reduce contaminant concentrations to acceptable levels. Additional measures might include the installation of monitors that will automatically check the atmosphere to verify that airborne concentrations do not exceed preset alarm limits. Employees should know what to do if the alarm sounds. Passive exposure badges or dosimeters are another precaution. Personnel who wear such devices should understand how to read and interpret the results and should know what to do if an overexposure or other unsafe condition is indicated by the device.

An additional but possibly overlooked method for controlling contaminants is the proper maintenance of the process equipment. Leaking pipes and lines; worn gaskets, seals, and/or valve packings; cracks in ducts; problems with fans, such as worn or slipping drive belts; and broken or missing guards and enclosures are just a few examples of maintenance items that, if neglected, contribute to airborne releases. Workers who perform the necessary repairs and preventive maintenance on equipment are at increased risk compared to individuals who operate the process, since they are the ones that will be exposed to materials normally contained within the process machinery. If the process involves highly toxic or sensitizing materials, special PPE or work procedures may be required to assure maintenance and repair workers are protected adequately.

Finally, we should not overlook the importance of personal hygiene. Workers whose duties require their presence in atmospheres that contain airborne contaminants should have the opportunity to cleanse exposed hair and skin prior to eating or taking breaks, and again before leaving work. They should also change out of work clothing that may contain hazardous dust, fumes, or other contaminants before leaving work. For some substances, such as lead and asbestos, workers must be provided with a change of clothes for work so that they do not carry contaminants home on their personal clothing. Showers may be required. There have been numerous instances of occupational disease among family members of workers whose only exposure was to the traces of material that were carried home on the clothing and body of the employee. OSHA requires employers to provide clean lunch and break areas, in locations where levels of airborne contaminants are below regulatory limits. A personal exposure monitoring program that includes area sampling can be helpful in tracking the effective performance of work practices and engineering controls, as can occasional area inspections by supervision or safety committees.

Checking Your Understanding

1. What are the three primary approaches to hazard control and how might they be applied to airborne contaminants? Give specific examples of each.

2. Explain the importance of the maintenance of process machinery in controlling airborne contaminants.

3. Other than OSHA requiring it, why is showering and changing clothes before going home a good idea for workers whose jobs involve handling or working in atmospheres that contain hazardous materials?

4. What types of topics should be covered in employee training when it is used to help control airborne emissions into the work area? How often should the training be repeated?

7-3 Local Exhaust Ventilation Systems

The use of ventilation is among the more common engineered controls. Ventilation systems are generally of two types: 1) **local exhaust ventilation (LEV)** systems designed to capture the contaminants at the point of generation or release; and 2) **dilution ventilation** systems, which are based on the use of an added volume of air to dilute contaminants without removing them from the work area atmosphere.

The primary advantage of a local exhaust ventilation system compared to dilution ventilation is the ability to effectively remove the contaminant from the work area. Because they are typically designed for a specific application, LEV systems tend to be more efficient in terms of energy consumption. An LEV system is also less likely to impact the overall heating and cooling requirements of a facility. Depending on the contaminant, the LEV system may incorporate cleaners, scrubbers, or a device for capturing raw materials for reuse in the process. LEV systems are appropriate for use in situations where the airborne contaminants:

— Pose a health, environmental, or fire/explosion hazard;

— Are irritating, or create an unacceptable nuisance such as impaired visibility;

— Create significant housekeeping problems;

— Are released at irregular times and in irregular volumes, or occur in or near the breathing zones of workers; or when

— State, local, or federal regulations require that an LEV system be used.

A typical LEV system has five basic components: 1) an intake or hood, which captures and draws the contaminant into the system; 2) ducts, which carry the contaminant from the work area toward the cleaner and exhaust; 3) an air cleaner, which removes the contaminant from the air before it is released to

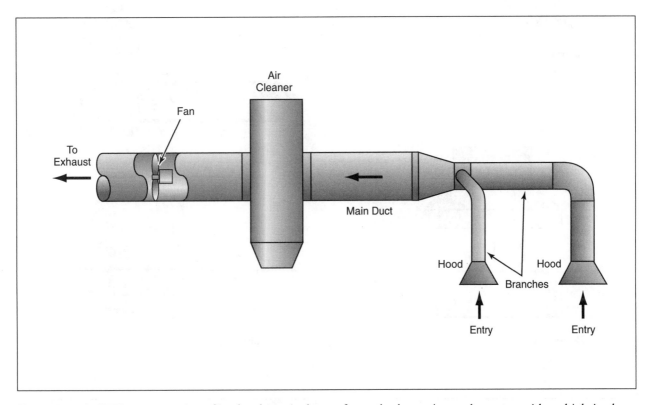

Figure 7-1: An LEV system consists of intake, duct, air cleaner, fan, and exhaust. A complex system with multiple intake locations requires careful design to ensure that required airflow is maintained throughout the system.

the environment; 4) a fan, which provides the necessary movement of air through the system; and 5) the outlet or exhaust, where the air is released from the system.

We will now trace the path of air as it enters an LEV system at the hood and travels through the various components to the exhaust. Some basic equations that describe how air moves through the system will be presented along the way so that system evaluation and measurements can be discussed later in the chapter.

Hoods

The hood is the part of the LEV system that cap-tures the contaminant. The design and placement of the hood are crucial to the effectiveness of the LEV system: if the hood does not capture the contaminant, the entire system is not able to perform its intended function. Three basic types of hood design are 1) the **capture hood**, exemplified by a nozzle or hose-type intake at the point of emission; 2) the **enclosing hood**, which surrounds the point of emission or generation, e.g., a hood around a grinding wheel; and 3) the **receiving hood** or **canopy hood**, best illustrated by a hood over a hot process tank (see Figure 7-2).

When air enters the hood, some turbulence is created, which results in some loss of the kinetic energy of the air. This is called **entry loss**. Entry losses vary with hood design; a sharp-edged entry has the highest losses. The use of a flange around

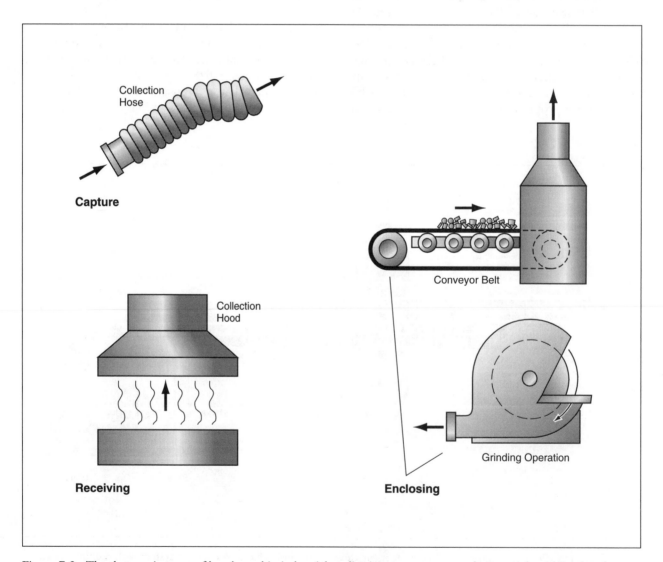

Figure 7-2: The three main types of hoods used in industrial applications are capture, enclosing, and receiving hoods.

the opening reduces these losses and also causes more air mass to enter from the front, where the contaminants are likely to be. A volume of air must be drawn into the hood at a high enough flow rate to capture the contaminants of concern. The velocity of air at the point of capture, called the **capture velocity**, is related to the volumetric flow rate of air that enters the hood according to the following equation:

$$Q = AV$$

Where

Q = volumetric flow rate, in cubic feet per minute (cfm);

A = the cross-sectional area through which the air flows;

V = the velocity of the air in feet per minute (fpm).

We looked at some applications of this relationship in Chapter 6, where we computed the volumetric flow rate being provided by an HVAC unit to an occupied area of a building. The use of the formula for ventilation systems is the same; the industrial hygienist can check the performance of the system against design specifications. The formula may also be used for design of a ventilation system, to determine the volume and/or velocity of air that the system needs to be able to move in order to control the contaminants.

The required capture velocity that should be designed into the system will depend on the contaminants that must be captured. Vapors and gases can usually be drawn into the system at relatively low velocities such as 1,000 fpm. Heavier contaminants require greater velocities; for example, wood or cotton dusts require 2,000-2,500 fpm, and lead and cement dust require 4,500 fpm or more. The ACGIH publication *Industrial Ventilation, A Manual of Recommended Practice* – often referred to by industrial hygienists as the "Vent Manual" – contains tables with recommended capture velocities for a variety of contaminants. The higher the capture velocity, the greater the volume of air that must be drawn into the system. The ACGIH "Vent Manual" also contains many diagrams and descriptions of recommended hood designs for specific operations.

The placement of the hood is also important in the ability of the LEV system to capture contaminants. Inlets more than two feet from the source of

contaminant will capture little or no emissions. It is possible to use process characteristics to increase the efficiency of capture. Using baffles and enclosing the point of emission can reduce the effects of drafts and crosscurrents that interfere with effective capture. Some good general practices for placement of hoods include:

—Enclose the process as much as possible with the hood, while not interfering with the worker;

—Place the hood so that contaminants are drawn away from the breathing zone of the worker;

—Take advantage of process features, such as heat and grinding, which can provide some initial movement of contaminants toward the intake;

—Locate the hood as close as possible to the point of generation or emission to minimize the volume of air needed for effective capture of contaminants;

—Do not fall prey to the fallacy that heavier-than-air vapors or gases should be captured near the floor; capture all contaminants that pose health concerns as near the point of generation as possible!

Ducts

Ducts carry the air and contaminants toward the outlet end of the LEV system. The movement of air through ducts requires the overcoming of friction between the air and the sides of the duct. Some of the kinetic energy of the air will be lost to this process; this loss is called **friction loss**. The longer the duct, the greater the total friction loss. Other duct features will contribute to loss of kinetic energy: turns, tapers, and the connection of multiple ducts to the system. The size of the duct is related to energy loss as well, as smaller diameter ducts tend to have higher losses. Values for these types of losses have been determined experimentally and can be looked up in a ventilation design handbook or text, such as the ACGIH publication cited above. Air movement through the LEV system creates pressure inside the system, called **velocity pressure (VP)**. Velocity pressure is related to the velocity, V, of air movement, as defined in the following equation:

$$V = 4,005 \times \sqrt{VP}$$

Where

V = the velocity of air in the duct, in fpm;

VP = the velocity pressure of air in the duct, in inches of water gage (inches w.g.)

The equation to calculate the velocity of air in a duct is based on work by the 18th century physicist Daniel Bernoulli (1700-1782) who studied fluid dynamics. Bernoulli's theorem, as it is sometimes called, is based on the basic physical principle of conservation of energy per unit volume. It takes into account the relationships between the energies of the fluid (air) that is flowing through the LEV system – kinetic, gravitational, and total pressure. The gravitational potential energy of air typically does not contribute to the velocity of airflow in an LEV system. Therefore, the equation used by industrial hygienists for LEV system design does not include it; the equation used here, then, is Bernoulli's equation, in a simplified form. It contains a constant, 4,005, which relates inches of water gage (pressure) to feet per minute (kinetic energy), eliminating the many steps used for conversion of units. For more information on Bernoulli's theorem, please refer to a college physics textbook.

Air moving inside the ducts of an LEV system actually exerts pressure in all directions. This is called **static pressure (SP)**. The sum of velocity pressure (VP) and static pressure is referred to as the **total pressure (TP)**. These three are related in the following way:

$$TP = SP + VP$$

The units for each of these pressures are normally inches of water gage.

The relationships between the various velocities, pressures, energy losses, and volumes of air in an LEV system are important in the design of the system. As hoods, turns, and duct lengths are added, the volume and velocity of air at a given point in the system must be maintained under conditions that will keep the contaminants moving through the system while **entrained** in the air, which means the contaminants must remain airborne and not settle out and collect in the duct. This is especially important if the contaminant is a solid, such as lead fume or dust, which will deposit in the ducts and eventually require removal or cleaning.

Periodic measurement of velocities or flow rate can help in troubleshooting problems with system performance while also providing confirmation of proper system function. The maintenance of the LEV system at optimum flow conditions will also minimize the amount of energy required to operate the system. There are two approaches to maintaining this balance. One involves sizing the ducts and fittings to maintain the desired velocity in each part of the system. This method – sometimes referred to as the velocity-pressure method – will prevent clogging of the system, and is required for LEV systems used to control explosive or radioactive materials. Balanced systems designed using the velocity-pressure method are also less susceptible to alterations or tampering by workers, since changes in one location will be noticeable at another location in the system. The other method for balancing an LEV system is one that uses dampers – sometimes called slide gates – to block parts of the system in order to achieve the desired airflow. Drawbacks to using this method include erosion of the slide gates and the tendency for dead spaces to occur behind the gates throughout the system; contaminants will accumulate in these locations. This method is generally not recommended for contaminant control for these reasons.

Air Cleaners

An air cleaning device of some kind is recommended – and in some cases dictated by regulations – for all air contaminants that pose a human or environmental hazard. Common air cleaners include filters, collectors, settling chambers, and precipitators. Air cleaners are available to remove most types of contaminants, including dusts, fumes, fibers, radioactive particles, and gases or vapors. Selection of the cleaner will depend upon the physical and chemical properties of the contaminants and the degree of cleaning that is desired. The possibility of fire or explosion may also be a concern and should be evaluated for all LEV systems. Another possible concern is the recirculation of highly toxic materials. In most instances, this practice is not recommended, and it may even be prohibited by regulation.

Air cleaners include industrially rated devices as well as the ones intended for light-duty situations. Light-duty applications usually involve dust or particulate levels typically found in uncontaminated air treated, for example, by HVAC systems and air conditioners. Industrial-type cleaners are capable of removing large amounts of contaminants from air and can operate under higher-pressure conditions that might exist in an LEV system. Industrial processes tend to produce large, localized amounts of materials; the air cleaning systems that are used for these applications must be capable of handling these potentially heavy loads.

When selecting the cleaner, the properties of the contaminant must be considered. These include its physical and chemical characteristics – particle size, density, toxicity, solubility, and its concentration in the air stream. Is the material light, heavy, wet, or dry? Is the air stream that is moving the contaminant hot or humid? Are gases or vapors highly toxic, corrosive, odorous? All of these characteristics affect the choice of collector.

Another important factor is how clean the exiting air must be. This may be dictated by 1) federal, state, or local regulatory limits; 2) public nuisance issues; 3) protection of surrounding vegetation and/or property; or 4) the desire to capture and reuse raw materials that are present in the exhaust air. The use of a cleaner capable of removing the majority of large particles may be removing most of the potential pollutant. However, the finer particles may still be visible and alarm the public.

The costs associated with operation and maintenance of any air cleaner cannot be overlooked. Usually there is more than one device that will satisfy the criteria for the specific application, but energy costs, design and space limitations, maintenance and cleaning considerations, and disposal of the collected contaminant must also be considered in the final selection. Table 7-1 compares some of the more common air cleaners and their uses in industrial applications.

Air Cleaner Type and Mechanism of Operation	Applications	Advantages	Disadvantages
Centrifugal collectors Uses centrifugal force to remove dust from air stream; also called cyclone.	Moderate-sized particles such as wood dust, shavings, chips; particles 10-40 μm diameter.	Low initial cost. Low maintenance. Little pressure drop. May use in series to increase efficiency.	Not an effective cleaner for removing finer, respirable dusts, i.e., those < 5 μm diameter. Cyclones used in series may have large pressure drops.
Wet collectors Uses water spray to increase particle mass to facilitate collection or to trap particles directly in water. Types include spray chambers, packed towers, venturi and centrifugal collectors.	Collection of particles 10-70 μm in diameter; also used for some gases and toxic dusts.	Pressure drop varies over wide range, from 2.0-15.0 inches of water gage. Many applications exist for a variety of contaminants.	Freezing and excess water accumulation are a problem for wet collectors; corrosive solutions can be a problem if equipment is not designed for them.
Electrostatic precipitators Air passes between charged electrodes; particles pick up charge and are drawn to collecting electrode.	Very efficient collection of fine particles, 0.01-50 μm diameters.	Negligible pressure drops are associated with these types of collectors.	High operating costs in terms of power; requires large area for installation in emission stream; collected particles must be removed periodically.
Settling chambers Velocity of air drops as it enters the chamber, causing larger particles to fall from air stream.	Good for large, coarse particles (e.g., chips).	Often used as precleaners for other air cleaning devices, or as traps for bulky particles.	Excessive space requirements; large pressure drop
Filters Use mechanical filtration to remove particles from air stream; made of cotton or synthetic fibers, fiberglass, wool. Examples include bag houses, filter banks.	Recommended for dry dusts only.	Efficiency increases with loading of filter; may remove 99 percent of incoming dust.	High pressure drops. Periodic cleaning or replacement is required. Can be expensive to maintain.

Table 7-1: Comparison of some common air cleaners. Most air cleaners are effective in removal of a specific type or size range of contaminants from the air.

Fans

Fans are placed near the end of the LEV system but their location by no means indicates that they are less important. They are a critical element in the system as they create the pressure difference that makes the entire system work. A fan must be selected based on its volume capabilities and on the system's design requirements. There are two types of fan design: centrifugal and axial (see Figures 7-3 and 7-4).

Centrifugal fans are wheel-type fans; the blades are arranged like the spokes of a wheel, around the outer edge. These fans are also called squirrel-cage fans and are used in many HVAC systems. Air enters these fans from the side and leaves at a right angle. Centrifugal fans with straight blades are ca-

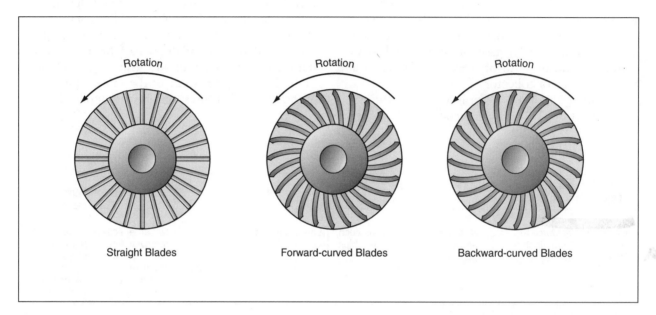

Figure 7-3: Centrifugal fans resemble the exercise wheels sometimes used for gerbils and hamsters, which has led to their being called squirrel-cage fans. The blades may be straight or curved/tilted forward in the direction of rotation, or backward in the opposite direction of rotation. Backward-curved blades are preferred for applications involving particulate since these blades tend to become less laden with material from the airstream.

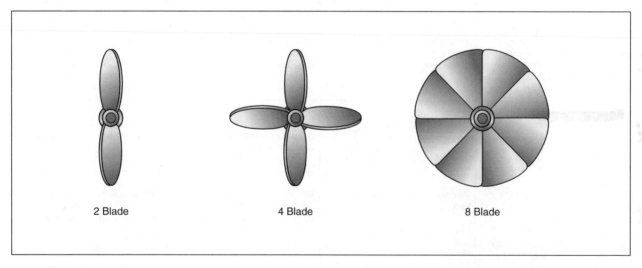

Figure 7-4: Axial fans, also called propeller fans, resemble the familiar box fans that get heavy use in hot weather; they have limited applications in LEV systems.

pable of moving air that contains particulate without the particles accumulating on the blades. These fans have the honor of being the workhorse of industrial ventilation systems, especially for applications involving lint, wood chips, and dusts. Centrifugal fans may have curved blades either in the direction of rotation – called **forward blades** – or in the opposite direction – called **backward blades**. Material tends to accumulate on both types of curved blades – although to a lesser degree on backward blades; this limits their applications to HVAC systems or to other situations where the air is precleaned before reaching the fan. Backward blade centrifugal fans are somewhat more efficient than straight or forward blades, which require more power to overcome the increase in air resistance that accompanies increased flow. Consequently, motors driving straight and forward blade fans can become overloaded and fail. By comparison, backward blades reach a maximum power requirement regardless of airflow and their drive motors cannot be overloaded (see Figure 7-3).

Axial-flow fans have their blades arranged like the propellers on a small aircraft. They can move large volumes of air but are incapable of operating against the usual pressures necessary to move air through a duct. Their applications are generally as stand-mounted cooling fans for personnel comfort. Some are used in LEV systems, but these applications are limited to systems with low pressure and clean (no particulate contaminants) air.

Exhaust

The final piece of the LEV system is the system outlet or exhaust. The best design will allow release of air with minimal added pressure. A straight path of outlet is the most efficient design. The common practice of placing a rain cover over the top of an outlet stack is not recommended as it impedes airflow. The many configurations of exhaust stacks on the roofs of commercial and office buildings are more likely to be the result of aesthetic concerns than operational requirements of the LEV system.

Measurements for Evaluating LEV System Performance

The effectiveness of any LEV system will be directly related to its design, installation, and operation. Assuming that the design is adequate and that the system is installed according to the design specifications, the last remaining aspect is the operation. A necessary part of assuring continued effective operation is the routine maintenance of the system, usually performed by the crew or company that performs the maintenance function for the whole facility. The industrial hygiene professional has a role in assuring proper operations through periodic evaluation of the system. This evaluation usually includes periodic basic measurements to assure that the LEV system is operating properly.

One measurement that might be performed is the checking of the **face velocity**, which is the average velocity of the air as it enters the hood. Face velocity is measured at the plane defined by the opening of the hood; several measurements can be taken over this area to determine the average value. Once the face velocity is known, the volumetric flow rate at which air is being drawn into the system can be determined using the equation $Q = VA$, as discussed earlier in this section. If the face velocity or volumetric flow rate differ significantly from the design specifications, there may be a problem with the system.

Another measurement that might be performed is called a **Pitot** (pronounced "pea-toe") **traverse** (Figure 7-5). This is actually the result of a number of measurements taken across the duct area. The location of each measurement depends on the diameter of the duct; smaller ducts require fewer measurements. Tables are available that contain the distances for taking each of these readings, depending on the duct diameter and its shape. The ACGIH "Vent Manual" is one source for these tables. Numerous readings are taken because air inside the LEV system flows at different velocities, depending on the location in the duct. This is due to the friction between air and the inside surfaces of the duct as well as turbulence induced by hood entries, elbows, connections, and other features of the system. The location for performing a Pitot traverse should be at a good distance from such disturbances in airflow. As a rule of thumb, the location should be at a distance that corresponds to approximately 7.5 times the duct diameter from such disturbances. For example, in a 6-inch duct the ideal location for a Pitot traverse would be one that is at least 45 inches from any elbows, hoods, or other connections that disturb airflow.

The instrument used to perform a Pitot traverse is called a **Pitot tube** (see Figure 7-6), named after

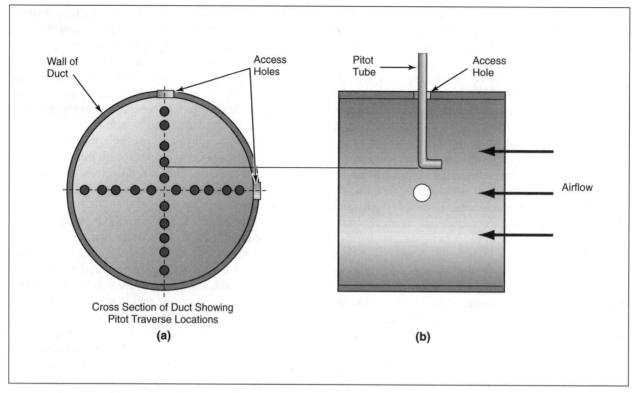

Figure 7-5: A Pitot traverse (a) consists of a number of measurements taken across the diameter of a duct in order to determine the average values for total, velocity, and static pressures inside the duct. The Pitot tube is inserted into the duct through small holes drilled in the side and top and positioned at the desired location (b). Distances for individual readings can be marked on the side of the Pitot tube, making it easy to determine when the end of the tube is in the desired location. For convenience, locations used regularly for performing Pitot traverses can be fitted with removable covers over the holes in the duct.

Box 7-1 ■ Case Study – Velocity Measurement and Fan Rotation

Workers were assigned to assemble small plastic parts using an adhesive that contained methyl ethyl ketone and some other equally odorous solvents. Each workstation was provided with its own small hood, part of an LEV system that served the entire operation. After a few months, it was necessary to move the process to another location in the plant. This was achieved over a weekend. The following week, persons working at the process called the safety department to complain that the solvent odors were not being effectively removed by the LEV system.

The safety professional took some readings of face velocity and found them to be on the low side, barely within the design specifications. The maintenance crew responsible for moving and reassembling the system was contacted and asked to check the current system installation against the design. During this check, it was determined that the fan had been wired backwards (the leads had been reversed). Because the fan rotated and was moving air, the maintenance crew thought that the system was operating properly. Once the wiring was changed to the proper orientation, the fan blades rotated forward, restoring the system's original design flow.

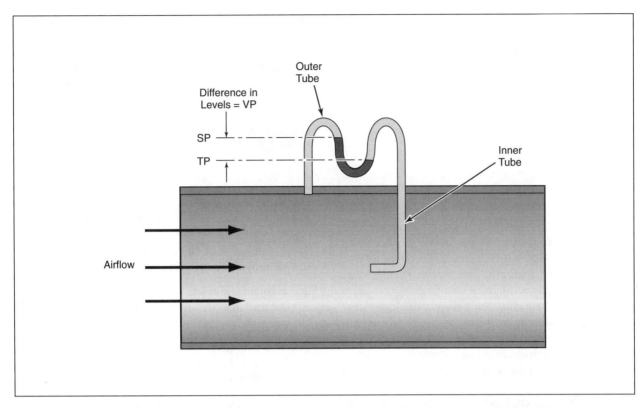

Figure 7-6: This diagram illustrates the principle of operation of the Pitot tube, which is a standard tool for measuring pressures inside of ventilation systems. These pressures can be used to determine the velocity of airflow in the LEV system, useful for evaluating system performance.

its inventor, Henri Pitot. Pitot tubes function based on the principle of conservation of energy for fluid flow, described by Bernoulli's theorem. A Pitot tube is really two tubes, one inside the other. The inner tube's end is open and faces the oncoming airflow in the duct. The outer tube is sealed except for openings that are perpendicular to the flow of air. The inner tube measures the total pressure (velocity pressure plus static pressure), while the outer tube measures static pressure. Since TP = SP + VP, the Pitot tube can be used to determine the velocity pressure (VP).

The two tubes that comprise a Pitot tube may be connected forming a U-shaped liquid-filled tube called a **manometer**. When this is the case, the liquid will be at different levels in each side of the manometer's U-shaped loop. The liquid level on the side connected to the inner tube reflects the total pressure, while on the other side of the loop it reflects the static pressure being exerted on the surface of the outer tube. The difference in the level of

the liquids connected to each side of the loop represents the VP. Most Pitot tubes available today provide an electronic or analog readout of the VP or V. Once the VP is known, the velocity of airflow can be computed using Bernoulli's equation, $V = 4,005 \times \sqrt{VP}$. Data obtained during a Pitot traverse can be used to compare system performance against design specifications to aid in identifying problem areas within the system.

Here is an example of a calculation of airflow velocity with resulting conclusions. A process involving the use of a solvent-based adhesive was relocated to another part of the plant. Since the move, workers have been reporting that the LEV system used for controlling the vapors does not seem to be as effective as it was before. The industrial hygienist knows from the design drawings that the system should be operating at a minimum velocity of 1,200 fpm. A Pitot traverse is performed and the average VP is found to be 0.06 inches of water gage. What is the velocity of airflow in the system?

We can use the equation:

$$V = 4{,}005 \times \sqrt{VP}$$

Substituting:

$$V = 4{,}005 \times VP^{1/2}$$

$$V = 4{,}005 \times 0.06^{1/2}$$

$$V = 4{,}005 \times 0.245$$

$$V = 981 \text{ fpm}$$

The system is not moving air at the required velocity. Further investigation as to the cause is necessary; it is possible that in the moving process, an unintended change has occurred. This could be due to placement of a length duct in the wrong location, or the backward wiring of a fan.

Other signs or symptoms of possible problems with an LEV system are observation of worker-installed alterations to the system, such as blocking of a hood or inlet, installation of a diversion, or addition of ducts or hoods to the system. Discolorations at seams and turns may indicate leakage or the accumulation of material. Relocation of processes and their associated LEV systems should always be an occasion for rechecking the system. Fans that are wired incorrectly may turn, but in the reverse direction; sometimes air is still moving, but the necessary volumetric flow rate is not achieved, and the system will therefore not function properly. Slipping drive belts and burned-out motors are other possible causes for poor airflow.

OSHA standards contain many references to minimum ventilation rates for a number of processes; examples include abrasive blasting; welding in confined spaces; grinding, polishing, and buffing; open surface tanks; sawmills; and spray finishing. For additional information see specific OSHA standards in 29 CFR 1910 or 1926.

Checking Your Understanding

1. Name each of the five components of an LEV system and briefly describe the function of each.

2. Explain the relationship between capture velocity and the volume of air drawn into an LEV system.

3. What features or aspects of ducts will result in losses of kinetic energy of air moving through the LEV system?

4. What types of measurements and observations might indicate the LEV system needs repairs or maintenance?

5. What is the function of a Pitot tube?

6. Define static pressure, velocity pressure, and total pressure.

7-4 Dilution Ventilation for Contaminant Control

Situations appropriate for use of dilution ventilation are those involving non-toxic and nuisance contaminants. Dilution ventilation is also sometimes used to prevent the accumulation of volatile vapors or dusts in concentrations that will support combustion. Depending on the location of the facility, the air supply used for dilution may need to be heated or cooled as the season dictates. Appropriate conditions for the use of dilution ventilation include:

—Low toxicity contaminants;

—Well-characterized, steady emissions into the air;

—The availability of a clean, adequate supply of air;

—Other methods for controlling contaminant levels are infeasible.

As with LEV systems, the placement of the source of dilution ventilation and air movements throughout the area is important for achieving the desired outcome. It is generally desirable to locate the air supply so that incoming air moves contaminants away from worker's breathing zones. Some method for exhausting air is often desirable, especially for situations where the dilution ventilation is being used to control worker exposures. The exhaust is often provided by a fan placed behind or otherwise near the point of emission, with the inlet for the supply air located behind or at some distance from the worker. Figure 7-7 illustrates some good and bad locations for fans and air inlets, relative to worker location.

Some good general practices for dilution ventilation are:

—Use the appropriate volume of air to achieve the desired dilution.

—Place exhausts as near as possible to the source of contaminant and inlets so that the source of contamination is between the worker and the exhaust.

—Avoid placements of inlets that result in drawing exhausted air back into the facility.

—If there are adjoining spaces in the facility, operate the system with slightly more exhaust than supply. If diluting the air in one room only, provide slightly more supply than exhaust.

The use of dilution ventilation for controlling exposures to hazardous materials is not recommended for materials with high relative toxicity. However, for lower toxicity substances that are being generated or released in low concentrations at a steady rate, dilution ventilation may be an effective method. The calculation for determining the required volume flow rate uses the following equation:

$$Q_{dil} = \frac{387 \times pounds \times 10^6 \times K_{mix}}{MW \times t \times C_a \times d}$$

Where

Q_{dil}	= the required volumetric flow rate of dilution air, in cubic feet per minute (cfm);
t	= time, in minutes, for a known amount of contaminant to evaporate;
MW	= the molecular weight of the contaminant;
pounds	= the amount of contaminant evaporated, in pounds;
C_a	= the acceptable concentration, in ppm;
K_{mix}	= the mixing factor, to account for air movement and mixing in the space;
d	= air density correction factor.

The value 387 is the volume, in cubic feet, that is occupied by one pound of air at standard temperature and pressure. The air density correction factor is 1 at the standard temperature of 70°F and at the standard atmospheric pressure at sea level. The correction factor drops to as low as 0.66 as altitude and temperature increase. This correction factor becomes more significant for calculations that involve large volumes of air; if a factor of 1 is always used, it will bias the result toward a more conservative estimate.

The **mixing factor**, symbolized K_{mix}, is a number used to express the degree to which mixing occurs in the air that occupies a specific space. In most instances, complete mixing does not occur. Factors that affect mixing include the presence of walls, par-

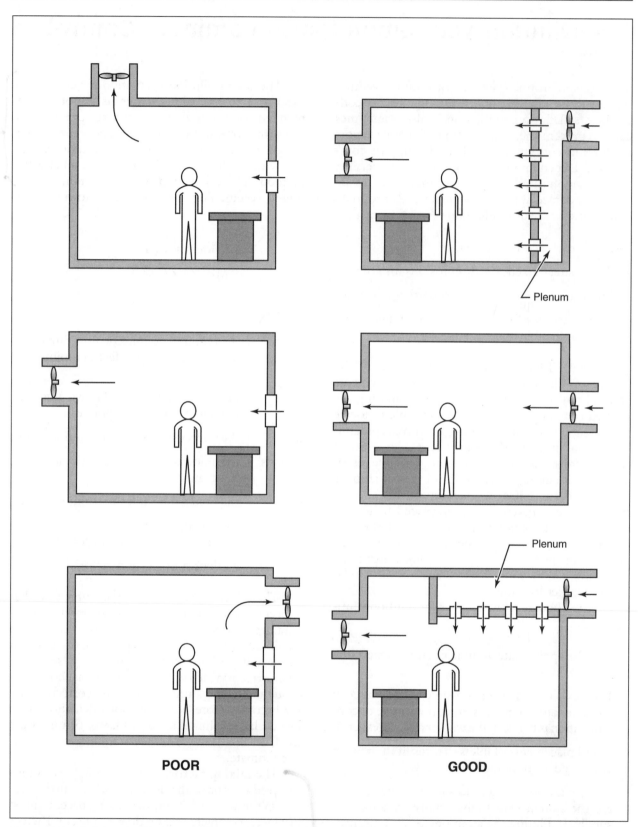

POOR

GOOD

Figure 7-7: If relying on dilution ventilation to control contaminants, placement of the inlet and patterns of air movement relative to the location of the worker are critical so that contaminated air is drawn away from the breathing zone of the worker.

titions, and poor locations for the inlet and exhaust used for dilution ventilation. K varies from 1.5 – representing optimum conditions such as wide open spaces with good supply and return locations – to 3.0 in locations with many walls, partitions, and poor supply and return locations. The use of a value of 2.5 or 3.0 is adequate for conservative estimates; however, if the situation meets all the good general practices for dilution ventilation listed above, a value of 1.5 or 2.0 might be appropriate.

Using metric units, the equation becomes:

$$Q_{dil} = \frac{0.0244 \times grams \times 10^6 \times K_{mix}}{MW \times t \times C_a \times d}$$

Where

Q_{dil} = the required volumetric flow rate of dilution air, in cubic meters per second (m^3/s);

t = time, in seconds, for a known amount of contaminant to evaporate;

MW = the molecular weight of the contaminant;

grams = the amount of contaminant evaporated, in grams;

C_a = the acceptable concentration, in ppm;

K_{mix} = the mixing factor, to account for air movement and mixing in the space;

d = air density correction factor.

We will apply this formula to an example. What dilution flow rate should be used if 50 pounds of toluene are evaporated from a process during an 8-hour work shift? Assume that C_a = 10 ppm, K_{mix} = 1.5, d = 1, and MW = 92.1.

$$Q_{dil} = \frac{387 \times pounds \times 10^6 \times K_{mix}}{MW \times t \times C_a \times d}$$

$$Q_{dil} = \frac{387 \times 50 \times 10^6 \times 1.5}{92.1 \times 480 \times 10 \times 1}$$

$$Q_{dil} = \frac{2.9025 \times 10^{10}}{4.4208 \times 10^5}$$

Q_{dil} = 66,000 cfm (rounded up)

As a rule, acceptable concentrations can be estimated as 50 percent or less of the OSHA PEL or ACGIH TLV for the contaminants of concern. Again, for carcinogens, sensitizers, and strong irritants, dilution ventilation is not an appropriate method for controlling exposures. Check the basis for the exposure limit in the ACGIH publication *Documentation of TLVs* and other literature, before proceeding with dilution ventilation as the control method of choice. Keep in mind that some personnel may still report odors or physical symptoms even when contaminant levels are well below regulatory or other limits.

Dilution ventilation may be used to prevent buildup of atmospheres that support combustion, but is never appropriate for situations where the workers will be exposed to the contaminated atmosphere. This is because for most volatile liquids that support combustion, the lower explosive limit, or LEL, is several orders of magnitude above the allowable level of exposure. To illustrate, compare the PEL and LEL of toluene. The OSHA allowable level of exposure is 200 ppm. The LEL is approximately one percent, or 10,000 ppm. Most dilution ventilation for control of flammable atmospheres is used to maintain a dilution of no more than 25 percent of the LEL. In the case of toluene, 25 percent of the LEL would be a concentration of 2,500 ppm, more than ten times the allowable limit and well in excess of the 500 ppm IDLH level. Dilution ventilation is appropriate for situations such as maintaining concentrations of flammable vapors below the LEL in ovens and driers of processes that involve flammable solvents.

Checking Your Understanding

1. What conditions are appropriate for the use of dilution ventilation to control exposure to airborne contaminants?

2. List three good practices to follow for use of dilution ventilation.

3. What is a mixing factor?

4. Under what conditions is the use of dilution ventilation acceptable for controlling worker exposures to airborne hazards?

Summary

Methods for controlling airborne hazards include engineered and administrative controls as well as the use of personal protective equipment. These controls may be used in combination to achieve the desired level of worker protection. Ventilation is one of the most widely used methods for controlling airborne hazards. There are two approaches: local exhaust and dilution ventilation. Local exhaust ventilation systems are best for use where the airborne contaminants:

— Pose a health or fire/explosion hazard;

— Are irritating, or create an unacceptable nuisance such as impaired visibility;

— Create significant housekeeping problems;

— Emissions are irregular in timing and volume, or occur in or near the breathing zones of workers; or

— State, local, or federal regulations require that an LEV system be used.

The components of an LEV system are 1) the inlet or hood, 2) the ducts, 3) the air cleaner, 4) the fan, and 5) the exhaust. Inlets should be placed as close as possible to the source of emissions. Ducts transport the entrained contaminants toward the air cleaner, where contaminants are removed prior to the exhausting of the air. The kinetic energy of the air moving through the ducts creates pressure; some energy is lost to friction on the inside of the ducts and as the air moves through the cleaner. The fan must be capable of moving a large enough volume of air to meet the design requirements for velocity and pressure in the LEV system. Methods for balancing an LEV system include balance by design or the use of slide gates to block and adjust flow. Usually the use of a balanced design is the best option. Some general principles of LEV system design are the following:

— A balanced system should not be altered in any way once it is installed.

— Proper maintenance of components is essential to the effective performance of the LEV system.

— Periodic checks of air volume or velocity should be conducted to identify problems, verify effectiveness, and/or meet regulatory requirements.

Air cleaners are recommended for removal of materials that pose a nuisance, a human health, or and environmental hazard. The choice of air cleaner will depend upon the characteristics of the contaminant and the process which generates it, as well as the desired degree of cleanliness of air. There are several types of air cleaners, each having different performance capabilities, power requirements, and space and maintenance needs.

Dilution ventilation can be used for controlling airborne hazards when the material is of relatively low toxicity. Good general practices for use of dilution ventilation include:

— Use the appropriate volume of air to achieve the desired dilution.

— Place exhausts as near as possible to the source of contaminant and inlets so that the source of contamination is between the worker and the exhaust.

— Avoid placements of inlets that result in drawing exhausted air back into the facility.

— If there are adjoining spaces in the facility, operate the system with slightly more exhaust than supply. If diluting the air in one room only, provide slightly more supply than exhaust.

It is possible to calculate the volume of air necessary to achieve a target concentration if the emission rate is known. The use of dilution ventilation for controlling flammability hazards and worker exposures at the same time is not recommended, as even a small percentage of an LEL concentration exceeds occupational limits and often exceeds IDLH levels.

Critical Thinking Questions

1. Compare and contrast the two methods for achieving balance in an LEV system.

2. Compare settling chambers with centrifugal collectors in terms of applications, power consumption, and advantages/disadvantages.

3. What type of fan is appropriate for the following applications, and why?
   a. Use in an LEV system which removes wood dust.

b. Use in an LEV system which operates under very low pressure to remove traces of a gas.
c. Provision of mixing and additional worker comfort in a large open shop area.

4. What is a mixing factor, and why is it important?

5. What volumetric flow rate of dilution ventilation is needed where 12 pounds of toluene are evaporated from a dip tank over an 8-hour shift? Assume that $C_a = 20$ ppm, $K_{mix} = 2.0$, $d = 1$, and MW = 92.1.

6. The production manager wants to use dilution ventilation to control flammable atmospheres in the plant and thinks this is a good idea since he assumed that it will control worker exposures to levels below the PELs, too. What is your response?

8

Occupational Skin Disorders

Chapter Objectives

1. **Identify** the major anatomical components/tissues of the skin on a cross-sectional diagram.

2. **Describe** the function/role and limitations of the skin in providing protection against hazardous chemical and physical agents.

3. **Describe** different types of skin responses to hazardous chemical and physical agents and illustrate each with an example.

4. **List** occupations or activities that pose skin injury hazards and name the hazards.

5. **List** several techniques that may be used to prevent occupational skin injuries.

8-1 Introduction

Despite more than 20 years of worker protection regulations being in place in the United States, skin disorders continue to be among the top occupational injuries reported each year. According to information reported to the Bureau of Labor Statistics, there were more than 60,000 cases of occupational dermatitis in the United States in 1993; this is believed to be an underestimation of the actual number of cases that occurred.

Any injury or illness that compromises the integrity of the skin is potentially serious since damaged skin is more susceptible to all forms of skin hazards, from chemicals to bacterial and fungal infections. Dermal injuries may therefore contribute to or increase a worker's risk of absorption of hazardous materials.

This chapter will describe 1) some of the materials and operations that have a potential for skin damage, 2) types of skin reactions to hazardous chemicals and agents, and 3) preventive and protective measures that industrial hygienists might recommend to prevent skin damage. The discussion will begin with a review of the function and anatomy of the skin.

8-2 Function and Anatomy of the Skin

Our skin is one of the most overlooked but most important organs. It is our first line of defense against assaults from the outside world. Heat or cold, wetness or dryness, corrosive materials, irritating chemicals, abrasives, sharp objects, insect bites, bacteria, sunlight – all of these we may ignore until skin damage has occurred. Many of us endure lifelong exposure to these agents without suffering permanent skin damage. If we look closely, all of us probably have some trace of past exposure to one or more skin-damaging events or materials.

The average human has nearly 2 m^2 (about 21 ft^2) of skin surface, which provides a physical barrier between our internal organs and outside agents, including chemicals and biological invaders. It also retains moisture, shields us from damaging radiation (sunlight), and serves as an organ of sensory perception. There are more than 70 feet of nerves in one square inch of skin, including receptors for heat and cold, pain, and pressure. The skin helps regulate internal body temperature through sweating. In addition, blood vessels, hairs, as well as sweat and oil glands are present in skin. The thickness of human skin ranges from about 0.5 to 4 mm, depending on the different locations on the body; the thickness on the palms of our hands, for example, is greater than around the eyes.

Three main layers of tissue comprise the skin. The outer layer is the **epidermis**, which, in turn, is composed of the basal layer and the **keratin layer**. The keratin layer – also called the **horny layer** or **stratum corneum** – is composed of dead, keratin-filled cells. This layer provides an effective barrier to most substances, except for alkaline materials and fat-soluble materials (such as organic solvents). The keratin layer wears off gradually and is continually replaced by new cells that are produced by the underlying basal layer. Also present in the epidermis are specialized cells that are part of the immune system, called **Langerhans cells**. These are sites for chemically binding haptens that enter the dermis and cause an allergic response (see Chapter 2 for more information on allergic reactions). The basal layer contains the pigment-producing cells, or **melanocytes**. The **melanin** (pigment) is what determines the color of a person's skin and hair. Exposure to the ultraviolet wavelengths of light stimulates production of melanin, resulting in freckles or a sun-tan; it can also stimulate thickening of the skin. Low melanin production results in a condition called **albinism**; individuals with this condition are called albinos and have very pale skin coloration. Exposure to some chemicals may cause loss of pigment from the exposed area of the skin; phenolic compounds, germicidal cleaners, metalworking fluids, paints, and plastic resins are examples of chemicals that have been associated with localized loss of pigment. The mechanisms by which pigment is lost are interference with melanin production, destruction of melanin-containing cells, or a combination of the two.

The next layer of skin is the **dermis**. This layer, also called the **corium**, contains connective tissue that supports the skin. Leather goods are made of this portion of an animal's hide after tanning. This layer contains the blood vessels, nerves, hair follicles, some muscles, as well as oil and sweat glands. The oils and fats that are present in the skin provide some water repellence to the outer surface, while main-

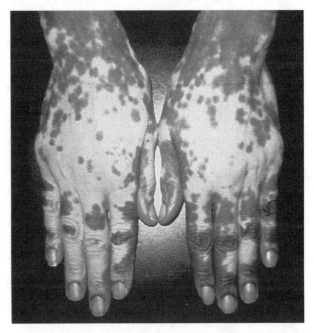

Figure 8-1: This image shows the depigmented skin of a hospital worker exposed to a phenolic germicidal cleaner. Depigmentation from damage to the melanocytes can be permanent.

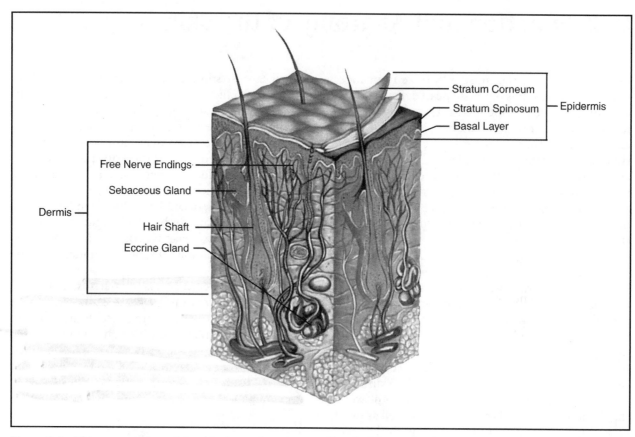

Figure 8-2: This cross-section of the skin shows the structures found within each of the layers of skin. Many materials used throughout industry damage specific layers or structures found in the skin.

taining moisture and flexibility in the skin. Some materials can irritate and cause inflammation of the hair follicles. This condition is called **folliculitis**.

> The suffix -itis, from a Greek word that means inflammation, is added to many words to indicate an inflamed condition; in this case folliculitis means the inflammation of the hair follicle.

Beneath the epidermis and the dermis lies the **subcutaneous** layer, which is the third layer. It contains fatty and connective tissue; it provides a cush-ion and insulative base for the skin; and it also binds the skin to underlying tissues that cover organs, muscles, and bones.

Checking Your Understanding

1. Name three main layers in the skin and describe the types of structures and features of each layer.

2. What are the primary functions of each layer?

3. The skin is not a good barrier for two types of substances; what are they?

8-3 Causes of Occupational Skin Injuries

The response of the skin to a hazardous material is dependent on a number of variables, such as:

— The physical condition of the skin;

— The environmental conditions under which the exposure occurs;

— The amount of moisture present or relative dryness of the skin;

— The amount of pigmentation present in the skin;

— The location on the body where the material contacts the skin;

— The age and gender of the worker;

— Pre-existing damage or allergies;

— The personal hygiene habits of the worker.

The physical condition of the worker's skin is of critical importance when considering the protective abilities of the skin. Dryness and cracking, irritation, sunburn, cuts, and other damage will lower the skin's ability to provide an effective protective barrier. The environmental conditions under which the exposure occurs can also influence the skin's ability to function as an effective barrier. High temperatures and humidity can create a layer of moisture on the skin, where hazardous materials may dissolve. In a hot environment, enlarged pores and increased moisture from sweating enhance the permeability of the skin; this facilitates absorption of some materials. Highly pigmented skin may be able to withstand increased exposure to ultraviolet radiation without suffering permanent damage; however, even dark skin is susceptible to sunburn. Also, thin skin will generally be more susceptible to damage, and absorption of materials may occur at an increased rate at locations with thin skin. The age and gender of the worker may play a role, since both factors may be linked to the overall health and physical condition of the skin tissue. Injuries, allergies or sensitization, as well as other conditions affecting the health of the skin will certainly play an important part in determining an individual's susceptibil-

ity to further skin damage. Finally, good personal hygiene habits – such as washing, moisturizing, and wearing clean gloves and clothing – that help maintain clean, intact skin can be among the most effective methods for preventing injuries to the skin.

Damage to the skin can occur in an occupational setting through several mechanisms. Mechanical injuries include those due to physical damage (cuts and bruises), temperature extremes, and ionizing and nonionizing radiation. Chemical injury may result from exposure to solvents, acids, and other materials. For outdoor activities and occupations, ultraviolet radiation presents a hazard. Biological agents – such as bacteria, fungi, and parasitic organisms – as well as botanical agents – such as poison oak, ivy, and sumac – may also cause skin injury.

Mechanical Damage

Mechanical skin damage may be in the form of friction resulting in blisters; abrasion of the skin due to contact with rough surfaces; or physical damage from sharp objects. An example of the latter is the irritation that occurs when small bits of glass fibers become lodged in the surface of the skin, resulting in redness, itching, and swelling of skin tissues. Extreme heat may result in burns, ranging in severity from the superficial reddening of a first-degree burn to more severe third-degree burns, which can be life threatening if enough surface area is involved. Exposure to extreme cold results in decreased circulation, possibly causing frostbite or actual freezing of the tissues. This causes tissue damage and infection, which may progress to more serious conditions, such as gangrene. Treatment of gangrenous extremities often involves amputation; many mountaineers have had several toes and fingers amputated due to this type of injury. Individuals whose occupations require spending extended time in environments of extreme temperatures, or the handling of materials or tools that are very hot or cold, are at risk for thermal injuries to the skin.

Chemical Damage

Chemicals are the most common cause of occupational skin damage and disease. They may affect the skin through simple contact irritation, or they may act as sensitizers. As mentioned in Chapter 2, sensitizing agents may produce a response – sometimes severe – only after repeated exposure and only in some individuals. Responses of this type are called **allergic contact dermatitis** (see Figure 8-3). This condition is characterized by redness, swelling, and cracking of the exposed skin, and sometimes more severe reactions involving the entire immune system. Systemic responses include difficulty breathing, inflammation of the airways, and pulmonary edema. Common chemicals associated with this type of response include phenols, epoxies, rubbers, acrylics, nickel compounds, and plant resins. Chemicals that are not sensitizers may also cause damage to the exposed skin, with localized symptoms similar to those produced by sensitizing agents. In fact, it is often difficult to distinguish between allergic contact dermatitis, and **irritant contact dermatitis** caused by exposure to chemicals such as solvents, acids, and bases. Figure 8-4 shows an irritant contact dermatitis.

Degreasing solvents dissolve and remove grease, but in the case of repeated contact with the skin, they also remove the oily secretions and proteins that provide water resistance in the outer layer of skin. Long-term exposure to degreasing solvents causes permanent damage to the skin. **Halogenated** compounds – which are compounds that contain chlorine, bromine, or iodine – are used in many industrial chemicals, including solvents, insecticides, and refrigerants (freons); they tend to cause skin irritation that resembles adolescent acne. This type

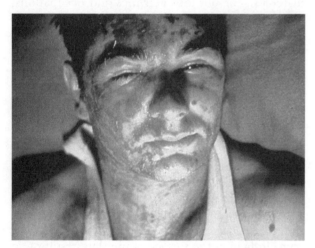

Figure 8-3: This allergic contact dermatitis is the result of exposure to phenol-formaldehyde resins. These types of reactions are also called delayed hypersensitivity reactions.

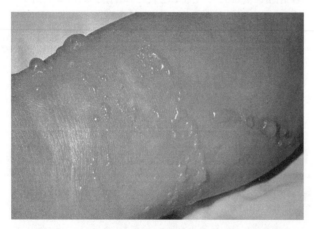

Figure 8-4: This acute contact dermatitis resulted from exposure to ethylene oxide, which is a powerful skin irritant. These reactions usually occur fairly soon following exposure to the offending agent.

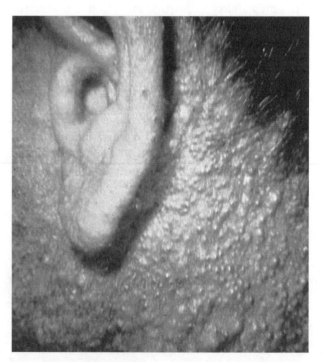

Figure 8-5: The occurrence of folliculitis or occupational acne, is usually associated with exposure to insoluble oils, lubricants, greases, and chlorinated compounds. A case of chloracne is pictured here.

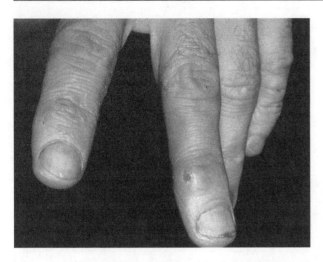

Figure 8-6: Chrome holes are ulcerative lesions associated with exposure to Cr^{6+}. The nasal septum can be similarly affected.

of condition, when associated with chlorinated compounds, is called **chloracne**. Oily or greasy compounds may also cause acne-like eruptions on exposed skin; water-insoluble cutting oils, creosote, and tar are among the more common causes of this **occupational acne**. Some chemicals, such as tars, arsenic compounds, and plant sensitizers (poison oak or ivy) may stimulate the production of melanin, resulting in darkening of the skin. However, darkening of the skin may also result from chronic physical irritation, like itching. Regardless of the cause, the condition is called **hyperpigmentation**.

Corrosive materials, including acids and bases, cause localized areas of damage ranging from **erythema** (reddening of the skin) to open sores or ulcers at the point of contact. Some materials have such a strong link with these types of injury that they have been given names associated with them, such as the chrome holes caused by exposure to chrome-containing compounds (see Figure 8-6).

The use of industrial solvents as skin cleansers not only causes chemical damage to the skin, but also provides a means for exposure to other hazardous materials. This is especially true if the workers are using solvents that have been mixed with other materials, or that have been used to remove or dissolve other substances – grease, dirt, surface coatings – as part of an industrial process. Depending on the solvent in use, this may increase the likelihood of one of these hazardous materials being absorbed by the skin.

Damage from Ultraviolet Radiation

Ultraviolet light may also cause skin damage, a point that has been widely publicized in recent years as the population of Europe and the United States is experiencing an increase in the incidence of skin cancer. These cancers may be occupationally related, especially among persons whose work requires spending long hours or days outdoors. Examples of these occupations include construction work, road building, roofing, landscaping, forestry, and others. In addition to sunburn and skin cancer, two other types of reactions – **phototoxic** and **photoallergic** – may occur that involve a chemical along with the presence of the UV radiation. Phototoxins may cause localized areas of tenderness or pain at the exposed location. An example of a phototoxic response is one where a roofing worker becomes more easily sunburned on parts of the skin that have been exposed to hot roofing tar or the fumes that this tar generates. A photoallergic response involves the immune system and – since it is an allergic response – does not occur in all exposed individuals. Figure 8-7 shows the blistery eruptions that formed on the hands of a bartender who squeezed limes while working outdoors in the sun. The eruptions were an allergic response of the skin to the furocoumarins

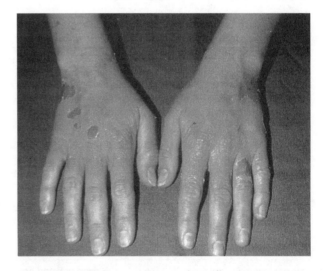

Figure 8-7: This image shows a photoallergic reaction involving sunlight and exposure to lime juice. Hyperpigmentation can also occur as the result of a photoallergic reaction.

present in the limejuice, triggered by the UV rays in the sunlight.

Table 8-1 contains examples of some common skin-damaging chemicals. The list is by no means exhaustive, and the number and variety of chemicals used by workers is increasing daily.

Biological Agents

Biological agents can affect the skin, sometimes producing a condition that progresses to cause a systemic reaction, if left untreated. Infectious agents that affect the skin include bacteria, fungi, insects

Chemical Class/Name	Layer Affected	Response
Inorganic and organic acids	Epidermis (keratin layer)	Contact irritation; chemical burns; (picric and chromic acids are also sensitizers)
Alkaline solutions	Epidermis (keratin layer)	Contact irritation; blisters, drying and cracking of skin
Petroleum products; coal tar materials	Epidermis (keratin layer) Dermis (hair follicles)	Contact irritation (coal tar materials are associated with skin cancer)
Organic solvents (degreasing chemicals)	Epidermis (keratin layer)	Drying and cracking of skin (some halogenated solvents are also sensitizers)
Metals or metal salts of mercury, arsenic, chrome, nickel, zinc	Epidermis (keratin layer) Dermis (hair follicles)	Skin ulcers; acne; spotty pigmentation; sensitizers; skin cancer
Pesticides	Epidermis (keratin layer) Dermis (hair follicles)	Dermatitis; skin cancer; folliculitis; chemical burns; hyperpigmentation; (many are also sensitizers)
Metalworking fluids (cutting oils and coolants)	Dermis (hair follicles) Epidermis (melanocytes)	Oil acne; loss of pigment (some coolants also produce dermatitis)
Resins/coatings and adhesives (epoxies, shellacs, vinyls, acrylates, isocyanates)	Epidermis (keratin layer)	Dermatitis; loss of pigment (most are sensitizers; some produce severe systemic reactions)

Table 8-1: Occupational skin disorders have a high incidence rate among workers. This table summarizes major categories of materials associated with skin damage.

Population at Risk of Exposure	Biological Agent	Condition
Animal care workers, breeders, veterinarians, farm workers	a. *Tinia unguium* and *Tinia corporis* b. *Candida candidiasis*	a. Ringworm b. Yeast infections
Dishwashers, bartenders, others performing wet work with their hands	*Candida candidiasis*	Yeast infections primarily around fingernails
Tannery workers	*Bacillus anthracis*	Cutaneous anthrax
Granary workers, grocers, personnel who handle/transport foods	a. Parasitic mites b. Ticks	a. Itching, dermatitis b. Itching, swelling, secondary infections
Construction workers, masonry workers, earth handlers in southeastern United States	Hookworm larvae	Possible subsequent infestation of the intestinal tract
Hikers, campers, outdoor occupations	a. Poison oak and poison ivy b. Snake bites c. Ticks	a. Contact dermatitis b. Systemic reactions c. Lyme disease, Rocky Mountain spotted fever

Table 8-2: Biological-agent effects of the skin are most likely to occur among workers who have agricultural or outdoor occupations. Recreational use of the out-of-doors also puts the general public at risk.

and parasitic organisms, as well as toxins produced by plants (poison ivy) and animals such as snakes and jellyfish. Examples of some of these biological agents and the at-risk populations are listed in Table 8-2.

Some compounds produced by plants and animals cause an almost immediate reaction on the skin, with swelling, reddening, and sometimes small fluid-containing eruptions. This type of reaction, called **urticaria**, is usually the result of an allergic reaction to the proteins in the secretion of the plant or animal. Natural latex rubber can elicit this type of response among some individuals (see Figure 8-8).

Some organisms, such as ticks, are **vectors**, or carriers of infectious agents. Rocky Mountain spotted fever is a tick-borne febrile (fever-causing) illness caused by *Rickettsia rickettsii*. The ticks themselves are not affected by the virus, which they pass on by biting the unlucky victim. Infection with this virus can cause a range of symptoms from mild fever and aches to multiple organ failure and even death. Symptoms include fever, headache, muscle aches, and a rash on the palms of the hands and soles of the feet.

Lyme disease is another febrile illness associated with a tick-borne agent, in this case a bacterium called *Borrelia burgdorferi*. Lyme disease is the most commonly reported vector-borne disease in the United States. The first sign that appears after the bite by an infected tick is a small red circle resembling a bull's eye that develops at the bite site and gradually expands to a large circle (2-4 inches). Symptoms include fever, aches, headache, nausea, and vomiting; in some cases the heart can also be affected. Death is rare, but it does occur. Figure 8-9 shows a typical tick bite reaction.

Some infectious agents do not need a live vector. *Bacillus anthracis*, for example – the bacterium that causes anthrax – lives in the hair of sheep and other animals and can also live in soil for a number of years without losing its ability to cause infection. Workers at risk of exposure in the United States include persons who handle imported wool and animal hides. The bacterium enters the body through a cut on the skin or at a mucous membrane – say a worker rubs his eyes – resulting in localized infection and a sore, which is usually noticed. This type of infection is called cutaneous anthrax (see Figure 8-10). Inhalation is another route of entry for the bacterium, resulting in pulmonary anthrax, which advances quickly. Pulmonary anthrax may be misdiagnosed and therefore go untreated, possibly resulting in death. Both forms of anthrax can be successfully treated with antibiotics.

Tetanus is another potential infectious agent encountered by anyone engaged in outdoor activities. The bacterium – *Clostridium tetani* – that causes the infection is more or less everywhere and commonly enters the body through a wound. People who work outdoors should be encouraged to obtain a vaccination against tetanus at least once every 5-10 years.

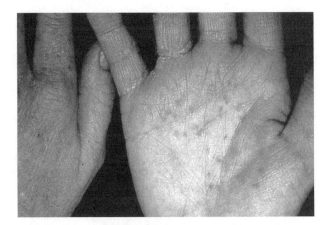

Figure 8-8: This image shows contact urticaria (hives) from exposure to natural latex rubber. The recent rise in the incidence of these types of reactions in the health care industry may be linked to the increased use of latex-containing products such as gloves, respirators, and goggles used for biological substance isolation.

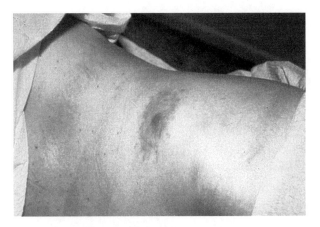

Figure 8-9: This is a classic bull's-eye pattern often seen at the site of a tick bite. It is not uncommon for these reactions to take two weeks to go away.

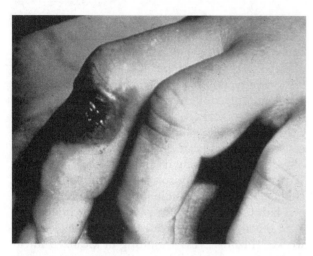

Figure 8-10: Cutaneous anthrax is difficult to miss, and the worker will usually obtain prompt treatment for this condition. Pulmonary anthrax is less obvious and may go unnoticed, or be misdiagnosed for some time before treatment is administered.

Checking Your Understanding

1. Name five factors that can affect the skin's ability to withstand exposure to a potentially damaging material.

2. List the mechanisms or ways in which the skin may be injured.

3. Explain the difference between allergic and contact irritation dermatitis.

4. What effect do degreasing solvents have on the skin?

5. Plant or animal toxins often cause a reaction called urticaria. Describe this condition.

6. Name at least four infectious agents that enter the body through the skin and describe their effects.

8-4 Preventing Occupational Skin Damage

Preventing injury to workers' skin is among the most challenging problems faced by the industrial hygiene professional. There are several approaches to preventing skin injuries: as for other protective measures employed in the occupational environment, they fall into the three basic categories of engineered controls, administrative controls, and the use of personal protective equipment.

Engineered controls to protect skin include the same options used to protect the worker against inhaled contaminants. Among these choices is the substitution of the offending ingredient or material with another that does not pose a skin injury hazard. However, the substitution process should guard against replacing one hazard with another. For example, replacing a skin irritant would not be a solution if the replacement material posed serious eye and respiratory irritation. The potential hazards of fire and explosion should also be considered when evaluating possible substitute materials.

Another engineered control involves process design, or redesign, so that the worker is not at risk of contacting the material. These controls might include either mechanizing the process, or installing a closed system to contain the hazardous material, or relocating workers to areas where they will not be exposed it. A fact that is sometimes overlooked in the design is that service and maintenance activities usually cannot be protected through these controls. Workers assigned to do the repairs cannot avoid opening process feed lines or entering vessels where the hazardous material is contained. Leaks and spills, which can pose both skin and airborne hazards, must also be considered and planned for in the design process.

Administrative controls for protection against skin hazards are very common and usually accompany other controls in use. Educating workers about the need to avoid contact with chemicals and about the importance of personal hygiene should be part of any hazard communication program. Added emphasis should be placed on these types of controls where workers encounter skin hazards on a regular basis. To encourage good hygiene, wash facilities must be provided in locations that are accessible to workers, with supplies of warm water, soap or other cleansers, towels, and a moisturizing skin cream. Workers should be encouraged and reminded to wash their hands and all skin exposed to materials that pose a skin hazard. Washing hands and exposed skin should always be the first step when taking breaks, going to lunch, and at the end of the shift. Depending on the materials in use, there may be specific OSHA requirements for the provision of hygiene facilities; for example, the OSHA standard for lead requires employers to provide shower facilities and clean changing rooms. In 29 CFR 1910.141, Sanitation, OSHA requires that all places of employment contain a lavatory – a basin or similar place for washing of the hands, arms, face, and head. These facilities must be provided with hot and cold (or tepid) running water, soap or similar cleansers, and clean cloth or paper hand towels or warm air blowers.

Barrier creams are often mentioned in discussions of skin hazards. These creams or lotions are applied to the skin and allowed to dry, which creates a thin film that provides protection with little tactile or dexterity interference. While these creams may have some specific applications, their use in general as an effective method of skin protection is not as widespread as it once was. There are several reasons for this. One is that **topical** (applied to the surface of the skin) applications will vary in thickness, depending on the amount used and the particular way in which the employee distributes it on the skin. Another issue is that of effective coverage of all exposed skin: unless the creams contain a colorant, it may not be possible to determine that all skin has been adequately covered. Barrier creams may also wear off, be wiped off, or be removed from the hands by the donning and doffing of gloves as well as by the use and handling of cloth or paper wipes. Also of concern is that hazardous materials may dissolve in the layer of cream, and be held against the skin for an extended period. From this, it is clear that the use of a physical barrier, such as the one offered by gloves, is always preferable to the use of barrier creams.

Personal protective equipment (PPE) with regard to skin hazards comes in the form of gloves, aprons, boots, full-body coverings, face shields, and goggles. These items are used to provide a physical barrier between the material and the worker's skin, often covering much or all of the body. When selecting the appropriate protective gear, a number of issues must be considered:

—The chemical properties of the hazardous material;

—The conditions under which the protective gear must be used;

—The constraints or requirements for the personnel performing work while wearing the equipment.

An entire industry has arisen out of the demand for more and better protective equipment; many different types of clothing and equipment are produced, some quite specialized in their application. Gloves, for example, are made of natural rubber as well different kinds of polymers, such as butyl rubber, polyvinyl alcohol, polyvinyl chloride, Viton, and others. Items of protective clothing may provide a simple barrier to dust and dirt, or they may be composed of a woven material with a rubber or polymer coating that also provides some chemical protection. Some protective suits are constructed of thick rubber or polymer material and provide an airtight environment for workers exposed to extremely hazardous materials or infectious agents.

Personal protective equipment for use against chemical hazards must be carefully selected, since some provide effective barriers against one or more classes of compounds, while being quite permeable to others. Permeability refers to the degree to which a material – in this case the glove, coverall, apron,

or boot material being worn to protect the skin – lets different kinds of substances pass through. Manufacturers of personal protective gear test their materials against many of the more common industrial chemicals; the resulting data are provided to the customer in the form of tables or charts. A number of manufacturers also furnish complimentary software to aid in the selection of protective gear against specific chemicals.

To perform a permeation test, the material to be tested is placed in a bracket or holder that separates two chambers. In one chamber, there is a sensor capable of detecting the solvent or chemical to be tested. The other chamber is filled with the solvent or chemical. The chemical remains in contact with the barrier material until the sensor detects the chemical, indicating that it has passed through the barrier material. The test is repeated several times, and the detection time is noted at each test. Some manufacturers present these data in terms of the average time for permeation to occur, usually minutes or sometimes hours. Others assign a rating to the barrier material, such as excellent, good, fair, or poor. It is important to read all of the information provided on manufacturer's charts and data sheets, since manufacturers' rating parameters may vary. In addition, the performance characteristics of the barrier material may vary depending on manufacturers. This is probably attributable to differences in

Chemical	Manufacturer A Glove Type and Ratings	Manufacturer B Glove Type and Ratings
Acetone	Natural rubber Degradation: Excellent Permeation: 10 min. Overall: Fair	Natural rubber Degradation: Excellent Permeation: 9 min. Overall: Not recommended
Butyl Cellosolve	PVC Degradation: Poor Permeation: Not rated Overall: Not recommended	PVC Degradation: Excellent Permeation: None detected Overall: Excellent
Nitric acid, 70%	Neoprene Degradation: Good Permeation: None detected Overall: Not recommended	Neoprene Degradation: Excellent Permeation: None detected Overall: Excellent
Phenol	Neoprene Degradation: Excellent Permeation: 3.5 hours Overall: Good	Neoprene Degradation: Good Permeation: 2.45 hours Overall: Poor

Table 8-3: This table compares some glove ratings from two manufacturers. Although the gloves are listed as being made of the same material, their performance in degradation and permeation tests is quite different.

manufacturing processes, such as the thickness of the materials, the coatings used as well as the construction methods. A comparison of glove material test data from two different manufacturers is presented in Table 8-3.

The use of permeation data is not as straightforward as it seems. Manufacturers' test data are obtained under controlled laboratory conditions, and each chemical is tested individually. The actual use conditions probably bear little resemblance to the conditions under which the protective gear was evaluated! For example, field and process conditions usually involve some variations in temperature, sometimes to extremes. In addition, it is more likely to encounter mixtures of chemicals rather than a single one. The forces of abrasion from use of tools and equipment, and the fact that personnel clad in impermeable clothing will be perspiring, causing the inside of the protective gear to be wet, will also affect permeability of the protective gear. Another consideration is that the use conditions may change through the course of a project, sometimes in the span of a single day or work shift. Industrial hygiene professionals must take these factors into account and decide what changes or additional measures are needed to assure worker protection. We will examine these issues in more detail in the chapter on personal protective equipment.

The needs of the worker must also be addressed. Protective gear that is too uncomfortable or incompatible with other necessary gear or equipment will be used or worn incorrectly, or not at all. Input from those who must do the work while wearing the gear is invaluable in settling these issues. They may have a suggestion for different gear, a different sequence for doing the work, or perhaps a method for doing the job that eliminates the need for use of protective gear altogether. It is also important – and an OSHA requirement – that the workers understand the limitations of the protective gear they are assigned to wear, as well as how to use it to provide the highest level of protection.

Finally, consideration should be given to the fact that use of personal protective gear may introduce or exacerbate hazards to the workers. Clothing that prevents evaporation of perspiration will contribute to heat stress. Moisture condensing on a face shield, on goggles, or on safety glasses will interfere with vision. Layers of protective clothing may interfere with mobility, while the use of gloves may limit tactile senses. The primary objective is to provide adequate worker protection and comfort, while allowing the work to be done.

Checking Your Understanding

1. Explain the significance of personal hygiene habits as an effective preventive measure for occupational dermatitis.

2. Name three reasons why a topically applied barrier cream may not provide adequate skin protection.

3. What are some reasons workers might refuse to use gloves and other skin-protecting gear?

4. Do you think that gloves from either manufacturer in Table 8-3 can be used interchangeably for protection against the same chemical hazard?

Summary

The skin is one of our most important but often overlooked organs. It provides a protective barrier between our bodies and the outside world; it retains moisture; it provides protection from ultraviolet radiation; it allows perception of heat, cold, pressure, pain, and pleasure; and it helps regulate internal body temperature. The skin is composed of three main layers: the epidermis, the dermis, and the subcutaneous layers. Each layer provides a slightly different function; together they complement each other to provide all of the functions listed above.

The response of the skin to a hazardous material is dependent on:

—The physical condition of the skin;

—The environmental conditions under which the exposure occurs;

—The amount of moisture present/relative dryness of the skin;

—The amount of pigmentation present in the skin;

—The location on the body where the material contacts the skin;

—The age and gender of the worker;

—Pre-existing damage or allergies;

—The personal hygiene habits of the worker.

Several potential skin disorders can affect workers. Chemical agents may damage the skin by removing the protective oils and secretions that maintain the skin's moisture. Physical damage may occur to the skin through friction, cuts, and exposure to temperature extremes. Ultraviolet radiation may cause skin cancer. Biological agents may cause local infections or serious diseases. Skin disorders persist among the most commonly occurring occupational diseases and disorders, despite advances in process technology and the availability of more – and better – personal protective gear.

Once the skin is damaged, it becomes more permeable, making it easier for chemicals to enter the body. Exposure of the skin to chemicals can result in the development of a number of conditions or injuries, including irritant contact dermatitis, allergic contact dermatitis, erythema, loss of pigment, and occupational acne. Some materials are associated with conditions named for the offending agent, as with chrome holes. Many materials are capable of sensitizing the exposed worker, so that subsequent exposures result in more severe reactions. This type of reaction has been noted, for example, among workers whose skin is exposed to nickel compounds. A true allergic reaction is another possible response of the skin. Latex rubber is capable of causing an allergic reaction among some individuals; this reaction, urticaria, does not require ultraviolet light to facilitate the reaction. For some sensitizing compounds, the allergic reaction occurs only in the presence of ultraviolet radiation and is then defined as a photoallergic response. A phototoxic response to irritating chemicals also occurs in the presence of UV radiation, but does not involve the immune system.

Some biological agents capable of causing damage or disease include pathologic organisms that are passed to workers through insect bites, contact with soil containing the organism, or through broken or damaged skin contacting a contaminated surface. Rocky Mountain spotted fever, tetanus, fungal and yeast infections, and cutaneous anthrax are examples.

Preventing damage to the skin may be accomplished through engineered controls – which eliminate the need for workers to contact the materials that pose a dermal hazard – or through administrative controls that provide similar protections. The use of personal protective gear such as gloves, aprons, and full-body coverings may be used when other measures are not feasible or effective. In addition to PPE, personal hygiene habits should be encouraged. These include regular washing of skin that has been in contact with hazardous materials and wearing of clean clothing. Barrier creams are not generally recognized as an effective protective measure against a hazardous material.

The continued incidence of occupational skin disorders might be linked to issues such as reluctance to wear ill-fitting PPE, or inability to perform job functions while wearing the recommended PPE. The industrial hygiene professional should consider the needs of the worker to perform the job when selecting PPE, and seek input from those who will be required to use the equipment.

Critical Thinking Questions

1. Give some possible reasons for the continued high incidence of occupational skin disorders.

2. Name and describe several factors that will influence the skin's susceptibility to damage.

3. Name the three main layers of skin tissue, describe the primary function of each, and name at least one way or mechanism by which damage to that layer might occur.

4. Isolation of a process, waste stream, or operator's station may provide protection against skin hazards for persons whose duties are at those locations, but will most likely not protect all workers. Name the two or three groups of workers who may not be protected, and describe the conditions under which their exposures might occur.

5. Your assignment is to select gloves for painters who use a variety of coatings and solvents. What issues must you consider in selecting the gloves? Will it be possible to select a single glove for all-purpose use? Why or why not?

9

Occupational Noise Exposure

Chapter Objectives

Upon completing this chapter, the student will be able to:

1. **Describe** the physics of sound as energy in the form of waves.

2. **Identify** the anatomical parts of the human ear and the basic mechanics of how we hear.

3. **Apply** equations that describe how sound interacts with other sound and with the surrounding environment.

4. **Define** terms and expressions used in the discussion of noise and hearing loss.

5. **Know** how OSHA and ACGIH exposure limits for noise compare.

6. **List** and explain the importance of each of the required elements of a hearing conservation program.

7. **Describe** some of the instruments, requirements, and measurement techniques for evaluating noise levels and worker exposures.

8. **Describe** basic engineering methods for reducing or controlling worker exposure to noise.

9-1 Introduction

The link between exposure to high levels of noise and hearing loss is well documented. Most of us know that exposure to loud noises can damage our hearing – witness the popularity of earmuffs at shooting ranges and gun clubs. The continued occurrence of noise-induced hearing loss among workers is unfortunate because it is irreversible and usually preventable. In addition to causing hearing loss, noise exposure has been linked to increased levels of stress, decreased productivity, as well as speech and communication problems.

The terms noise and sound are often used interchangeably, although usually the term noise is used to refer to sound that is unwanted or inappropriate. This chapter addresses noise and hearing loss in the occupational setting. We will begin with a discussion of the physics of noise and move on to the anatomy of the ear and to how we hear. We will then progress to other topics including evaluating noise exposure, limits for exposure to noise, OSHA regulatory requirements, and methods for controlling noise exposure.

9-2 The Physics of Sound

Noise, or unwanted sound, is similar to light in that it is also energy in the form of waves. However, sounds are the result of cyclic pressure changes that are produced by a source. These cyclic pressure changes, or oscillations, create **sound pressure** that can be measured using portable field instruments. As other energy in the form of waves, sounds can be described by a frequency (f), a speed (c), and a wavelength, (symbolized by the Greek symbol lambda, λ). These three are related as shown by the following equation:

$$c = f \lambda$$

The distance that a sound wave travels through a medium (such as air or water) during one pressure cycle is the wavelength. Wavelengths are usually given in feet or meters, or fractions thereof. The speed of the sound, c, is usually expressed in units of meters per second (m/sec) or feet per second (ft/sec).

Sound waves travel at different speeds through different materials, depending on the material's density. Generally, the denser the material, the faster it will transmit sound. For example, sound moves at about 344 m/sec in air, 1,500 m/sec in water, and 6,100 m/sec in steel. The speed of sound in air, 344 m/sec, can be considered a constant, although temperature, pressure, and humidity can have slight effects. In some materials, sound moves so slowly that it is absorbed. Other materials are capable of amplifying or reflecting sound. These characteristics can be used to the advantage of noise control design, as we shall see later in the chapter.

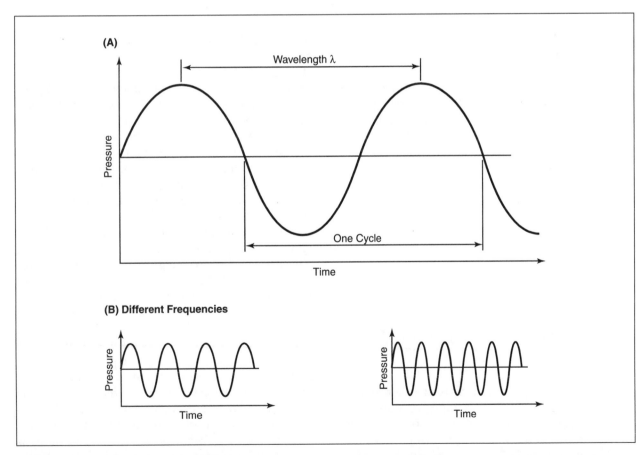

Figure 9-1: (A) Sound, like other forms of energy that travels in waves, can be described in terms of frequency and wavelength. One complete pressure oscillation is one cycle. The distance traveled during one cycle is the wavelength, λ. (B) The number of cycles per second is the frequency and is measured in Hertz, Hz.

The frequency, f, is the rate at which the pressure oscillations occur, and is expressed in **Hertz (Hz)**. One Hz equals one cycle per second; a frequency of one Hz means that one pressure oscillation, or cycle, occurs each second. Some sounds – such as the ones generated by tuning forks – produce a single tone, which is a sound with only one frequency. Other sounds are comprised of mixtures of frequencies – most industrial noises are made up of multiple frequencies. The **pitch** of a sound is related to its frequency; the human ear perceives increasingly higher frequencies as increasingly higher-pitched tones. A healthy, young person with normal hearing can detect sounds ranging from 20 to about 20,000 Hz. Age and exposure to noise, especially loud noise, contributes to the decrease in a person's audible range throughout their lifetime.

Loud noises pack a lot of energy, called **sound power**. The unit for expressing this energy is the **acoustic watt** (stereo speakers are rated in terms of how many watts of power they are capable of expressing). We cannot measure sound power directly; however, we do know that sound power is proportional to the square of the sound pressure, which occupational health professionals are able to measure relatively easily.

Because we measure sound pressure, we express sound levels in units that express pressure, which is a force per unit area. The SI (International System of Units) unit for pressure is the pascal, symbol Pa. All sound pressures are related to a reference sound pressure, which for occupational measurements is 20 μPa (this is the approximate lower threshold for human hearing at 1,000 Hz). The equation for expressing sound pressure is:

$$L_p = 20 \log_{10} \frac{P}{20\ \mu Pa}$$

Where

L_p = the sound pressure level, in decibels (dB)

P = the measured sound pressure, in Pa, and

20 μPa = the reference sound pressure

It is relatively rare to hear or read about sounds being expressed as Pa. For example, one does not expect to hear about a "25 Pa rock concert." The more common and convenient unit for expressing sound pressure levels (SPLs) is the **decibel (dB)**.

Noise source	Sound pressure level (SPL) in dB (ref. 20 μPa)
Jackhammer	110-120
Rock band	110-115
Gasoline lawn mower (as heard by the operator)	95-98
Garbage disposal	80
Vacuum cleaner	70
Normal conversation	55
Quiet room	40

Table 9-1: Some common sources of noise and the sound pressure level of each. Because decibels describe a logarithmic relationship, a 10 dB increase represents a 100-fold increase in sound power.

The decibel is a dimensionless quantity based on the logarithm of a ratio or comparison, in this case the ratio of the measured sound pressure to a reference sound pressure (see the above equation). The decibel is also a more convenient measure for working with sound pressure level values than the Pa or μPa. For example, the sound pressure level of a jackhammer is 20 Pa, while the sound pressure level in a quiet room might be 0.002 Pa. The same sound pressure levels expressed in dB would be 120 and 40 dB, respectively. Table 9-1 lists some noises and their sound pressure levels for comparison.

It is common to see decibels with an added designation of A, B, or C in parentheses, as in 85 dB(A). These letters refer to **weighting scales** for measuring sound. Each scale approximates the response of the human ear at different ranges of sound pressure levels. These scales were derived from an experiment in which test subjects were presented with a reference tone and a test tone, then asked to adjust the volume of the test tone until it equaled that of the reference tone. The experiment demonstrated that different sound pressure levels are required for some frequencies to sound equally as loud as others to the human ear. In other words, a noise with a frequency of 1,000 Hz and an SPL of 20 dB sounds as loud as a noise with frequency of 500 Hz at SPL

25. The human ear is more sensitive to higher frequencies; therefore, it takes less sound pressure for the noise to be perceived as loud as a noise with a lower frequency. The A weighting scale approximates the response of the human ear and provides a good estimate of the potential for noise to cause hearing loss in humans. This scale is the one that is specified by OSHA for use in evaluating occupational noise exposure. A reading of 85 dB(A) means that the sound pressure level was measured at 85 dB on the A scale. The B scale is most appropriate for medium sound pressure levels and does not get much use. The C scale is good for evaluating high sound pressure levels, such as those produced in explosions and other impact-type noise sources; this scale is commonly used to evaluate industrial and other noises with high sound pressure levels.

Checking Your Understanding

1. The terms noise and sound are often used interchangeably. How would you distinguish between the two?

2. Define the following terms: frequency, pitch, wavelength.

3. Explain why the decibel is a convenient unit for expressing sound pressure level.

4. Explain why it is preferable to express noise/sound levels in units of sound pressure?

5. What are the A, B, and C weighting scales and what are they used for?

9-3 The Anatomy of the Ear

Our discussion of noise exposure continues with an understanding of how the human ear translates sound waves into the many different sounds that comprise a conversation, the roar of a jet engine, or a piece of music.

Figure 9-2 illustrates the human ear. The outer cartilaginous and most visible part of the ear is called the **pinna**. The pinna amplifies and channels sound waves into the **ear canal**. The ear canal, also called the auditory canal, further amplifies sounds. In fact, the pinna and ear canal work together to amplify sounds in the 2-4 kHz range, increasing them by 10 or 15 dB. Once in the ear canal, sound pressure waves move toward the middle ear and impact on the **ear drum**, a thin piece of skin that covers the opening to the middle ear. Vibration of the eardrum causes the three tiny bones in the middle ear to vibrate. The bones are named after objects they re-

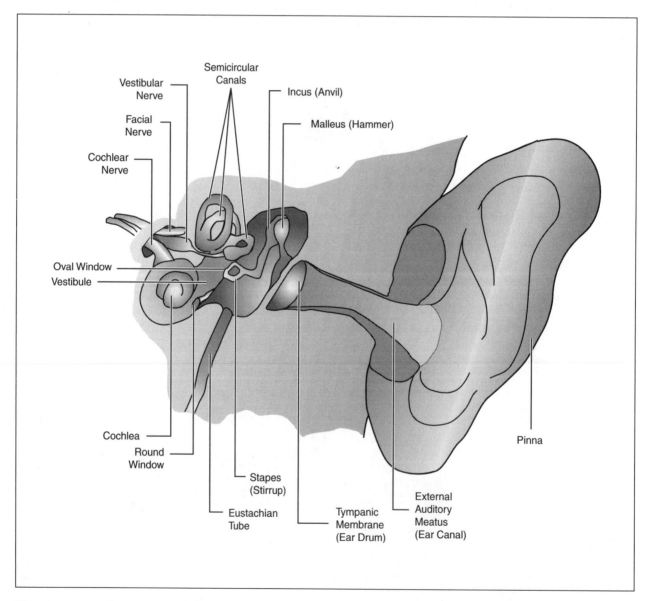

Figure 9-2: In order to understand the mechanics of how we hear, a familiarity with the anatomy of the internal structures of the ear is necessary.

semble. The hammer is positioned against the eardrum; the anvil is in the middle, and the stirrup is located against another membrane, called the **oval window**, which covers the entrance into the inner ear. The snail-shaped organ of the inner ear is called the **cochlea**. Vibrations from sound pressure impacting the eardrum are conducted down through the bones of the middle ear to the oval window, which also vibrates. These pressure-induced vibrations are transmitted from the oval window into the fluid that fills the cochlea.

> The Latin names for the three bones of the inner ear are the *malleus*, which means hammer; the *incus*, or anvil, and the *stapes*, or stirrup. You may see the bones identified with these names on some diagrams of ear anatomy.

The interior of the cochlea is lined with a **basilar membrane** that supports around 25,000 specialized cells, called **hair cells**; these are attached to nerves. There is another membrane, called the **tectorial membrane**, which lies above the hair cells. The pressure exerted on the fluid of the inner ear causes the tectorial membrane to respond, which in turn causes the hair cells to bend. The bending of the hair cells generates a nerve impulse, which travels from the hair cells to the brain where it is interpreted as sound. The shape of the cochlear canals changes along their length, being straighter and stiffer at the terminal or inner end than at the base. The higher frequencies tend to stimulate the hair cells near the wider base of the cochlea, while lower frequencies seem to act on hair cells located further along the canals.

Attached to the cochlea are three **semicircular canals**, each in a different plane. Movement of the fluid in these canals provides the sense of balance and perception of our body's position relative to our surroundings. Amusement park rides that involve spinning can create a temporary dysfunction of these canals, resulting in a dizzy, uncoordinated feeling. Traumatic injuries to the head can also damage these canals and create orientation problems in the affected individual.

The **eustachian tube** connects the middle ear and the throat. This tiny tube provides a mechanism for equalization of pressure in the middle ear with the surrounding atmospheric pressure. When traveling by air or driving over a mountain pass, you may have noticed your ears "popping," or you may

have experienced a slight loss in hearing. This can usually be corrected by taking a deep breath and forcefully pushing the air from the lungs into the upper respiratory tract without exhaling. The result is an equalization of the pressure in the middle ear, and your hearing returns to normal.

Hearing Loss

The ability to hear sounds requires that the components of the outer, middle, and inner ear function properly. Interruptions along the conductive pathway of sound energy to the inner ear causes **conductive hearing loss**. The net result of any such interruption is that the hair cells of the inner ear will not receive as much stimulation, and the perception of noise will consequently be lowered. Some of the more common causes of conductive hearing loss include excessive earwax, **otitis media** (fluid in the middle ear), or a ruptured or damaged eardrum. Other possible causes include damage or misalignment of the tiny bones of the middle ear, breakage of the links between the bones, and abnormal growth or fusion of the bones. Damage to the middle ear structures can occur as the result of a hard blow to the head, such as might occur in a fall or automobile crash. Exposure to very loud noises such as explosions can damage the eardrum due to the rapid change in pressure that occurs during such events. Some drugs as well as some infectious diseases can also damage the structures of the middle ear.

It is important to note that causes of conductive hearing loss are often correctable through surgery or medication. This distinguishes conductive hearing loss from **sensory hearing loss**, which is not reversible. Sensory hearing loss is caused by damage to tissues of the inner ear, such as the hair cells, which translate sound pressure into the nerve impulses that our brains interpret as sound. Exposure to noise will produce sensory hearing loss. Under a microscope, noise-damaged cochleae show changes such as twisting and swelling of the hair cells, tangles in the hairs, and breaks in the connection between the tectorial membrane and the hair cells. All of these changes result in less efficient transfer of sound energy to the hair cells, and it takes increasingly stronger stimuli to cause the hair cells to react. After repeated exposure to noise, the hair cells are permanently damaged (See Figure 9-3). The hairs may disappear, stick together, or die. The membrane that formerly supported the hair cells may atrophy, as do the nerves that were connected to the hair cells.

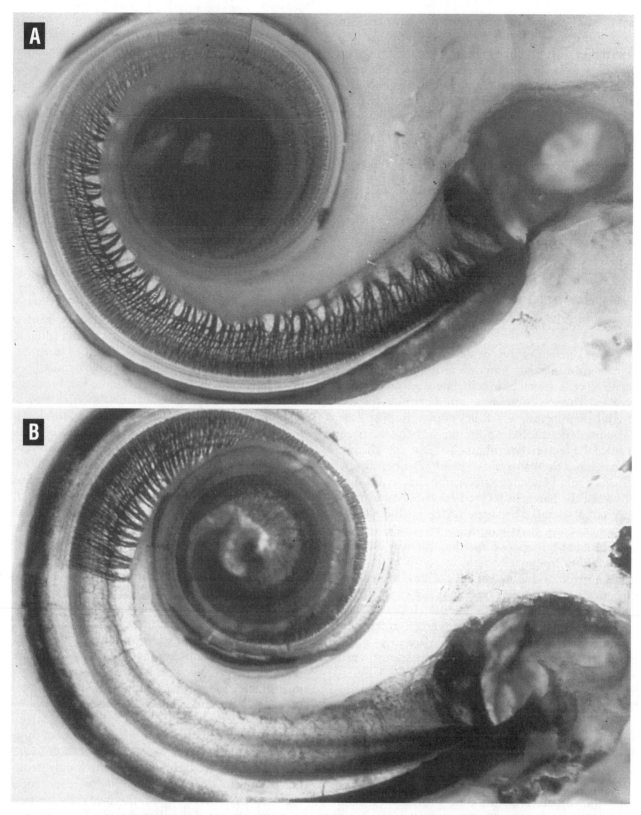

Figure 9-3: Photomicrograph of cochlea showing healthy (A) and damaged (B) hair cells. The cochlea in B displays significant noise-induced hearing loss.

These changes, and the associated loss of hearing, are permanent.

Several terms are used to describe different types of sensory hearing loss. For example, most of us know that as people grow older, their ability to hear seems to decrease. This hearing loss attributed to aging is called **presbycusis**. **Noise-induced hearing loss** is the term used to refer to hearing loss caused by exposure to noise. The noise exposure can take place in an occupational setting, but can also occur as the result of participating in activities such as attending rock concerts or shooting firearms without the use of hearing protection. The term for hearing loss as a the result of being exposed to the noises of every day life is **sociacusis**. The term **nosacusis** is used to refer to hearing loss that results from other causes, including diseases, heredity, drugs, exposure to sudden and severe pressure changes, or traumatic head injuries. Most individuals have hearing loss from a combination of exposures; this makes it impossible to distinguish between the amount of sensory hearing loss attributable to each of the different causes.

Manifestations of Hearing Loss

Significant hearing loss can have a profound effect on a person's life. It impairs one's ability to understand speech, to hear warning signals and alarms, to listen to music, to hear the wind blowing in the trees. We think of hearing loss as not being able to hear sounds, but mostly it is frequency-specific and therefore only partial. Profound and total deafness is not a common manifestation of noise-induced hearing loss.

Among some individuals, sounds may be heard, but incorrectly. An example is a musical note or tone. The affected person hears a tone, but it is not the one being produced. Music can be heard as containing notes of the wrong pitch. This incorrect perception of sounds is called **paracusis**. Another manifestation that often accompanies hearing loss from noise exposure is **tinnitus**, which is the perception of noises that are not there, like ringing, roaring, or hissing sounds. Sometimes tinnitus follows a traumatic exposure to loud noise; it may be permanent, or it may go away with time. Usually, tinnitus resulting from noise exposure is associated with some amount of permanent hearing loss.

A more common and perhaps distressing manifestation of hearing loss involves speech misperception. Similar to paracusis, speech sounds may be heard, but heard incorrectly. Normal conversational speech contains many frequencies, with the consonant sounds being mostly high frequency sounds and vowels being relatively low frequency sounds. The sounds of "s," "f," "sh," and "th" as well as "p," "t," and "k" contain subtle differences that may not be detected among some hearing-damaged individuals. Some individuals can compensate for these losses by lip reading if they can see the face of the person talking.

Checking Your Understanding

1. Describe the path that sound travels as it makes its way from the pinna to the brain. Name each major anatomical element in the path, and describe its function in the mechanics of hearing.

2. What is the eustachian tube and what is its function?

3. Explain the difference between sensory and conductive hearing loss.

4. Define the following terms: presbycusis, sociacusis, nosacusis, paracusis, tinnitus.

5. Describe some of the ways that hearing loss can affect a person's life.

9-4 Evaluating Hearing Ability and Hearing Loss

Audiograms

Since it is not possible to examine the internal surfaces of a human cochlea to evaluate the extent of damage that noise exposure has caused, we estimate the damage by measuring a person's hearing acuity. The test process is called **audiometry**, and the resulting report is called an **audiogram**. OSHA regulations require that all workers exposed to occupational noise at or above 85 dB(A) receive an initial audiogram, called a **baseline audiogram**, which provides a record of their hearing acuity before they are exposed to noise at work. Baseline audiograms are generally obtained as part of the new-hire physical evaluation and provide a basis for comparison with each subsequent audiogram for the worker. The individual being tested sits in a special booth, which has been soundproofed to reach a specific level of quietness relative to the surroundings. While sitting in the booth, the person receives instruction on how the process will work, puts on a set of headphones, and the door is closed. The measurement process involves a series of tones that are produced in the headphones at several frequencies. Each time the person hears a tone, he or she indicates this by pressing a button on a small hand-held device. The audiometric machine records the responses automatically. The sound pressure level of the tone starts out at about 40 dB and then decreases until the person cannot hear the tone, which is indicated by the individual not pressing the button. This continues across each frequency, through a range of dB levels. At the completion of the evaluation, the audiogram will show the lowest level at which the person is able to detect the test tone in each frequency. This low threshold is called a **hearing threshold level (HTL)**. The tested frequencies are 500, 1,000, 2,000, 3,000, 4,000, and 6,000 Hz.

The significance of testing across the specified range of frequencies, from 500 to 6,000 Hz, is that this range includes most of the frequencies that are detectable by the human ear. In addition, most speech falls in the range of 1,000-4,000 Hz. Significant hearing loss in the frequencies above 1,000 Hz will result in difficulty understanding spoken communications. Noise-induced hearing loss often first shows up on an audiogram in the 3,000-4,000 Hz frequencies. The corresponding dip in the graph on an audiogram is characteristic of noise-induced

high frequency hearing loss. See Figure 9-4.

Like other industrial hygiene measurements, the audiogram is subject to some uncertainty. There are several reasons for variations in test results, including ear wax buildup, a head cold or congestion, confusion about how to respond to the test tones, incorrect placement of the headphones, hair under the headphones, or problems with the audiometric equipment. The worker may not understand or may not desire to follow instructions. When the audiogram shows erratic responses, a repeat of the test is necessary. The OSHA standard for occupational noise exposure – 29 CFR 1910.95 – requires audiometric testing equipment to be maintained and calibrated regularly. In addition, the person conducting the audiogram must have demonstrated competence in operation of the equipment and in conducting the test. Many states require licensure of audiometric test technicians; OSHA considers licensure or certification as optional for some audiometric testing.

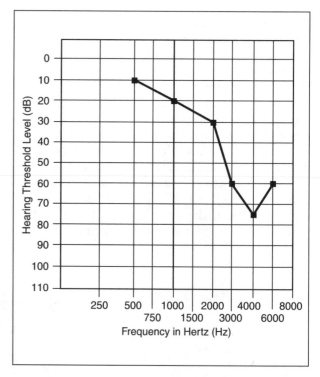

Figure 9-4: An audiogram is a graphic representation of a person's hearing acuity across six different frequencies. This audiogram indicates a significant hearing loss, most likely due to noise exposure, at 4,000 Hz.

Quantifying Hearing Loss

Audiograms provide an indication of a person's hearing threshold levels across a range of frequencies, as described above. Subsequent audiograms, which usually are performed annually, may show changes in the HTLs. Special terms are used to describe significant changes in HTLs.

A **standard threshold shift (STS)** is said to have occurred if the person's HTL increases by 10 dB or more at 2,000, 3,000, or 4,000 Hz, in either ear. Detection of an STS is important, since it represents a permanent loss of hearing of significant proportions. When an audiogram indicates a possible STS, retesting after at least 14 hours away from noise is required to make sure the shift is not a **temporary threshold shift (TTS)**. A TTS is a shift in HTL that goes away after the person has been in a quiet environment for a few hours.

> A simple self-test for detecting a temporary threshold shift can be done before and after your workday. At the beginning of the shift, when you park your automobile, before turning off the key dial the volume setting on the radio to the lowest level at which you can still hear the music or announcer. After work, start the car and sit for a few moments before moving. If you cannot hear the radio without turning up the volume, chances are good that you are suffering from a temporary shift in your hearing threshold due to the noise exposure you experienced during the day.

According to the OSHA regulation, the detection of an STS in an audiogram requires that the employee be notified of their hearing loss in writing. In addition, OSHA requires that these employees receive additional training on the effects of noise on hearing as well as instruction on the use of hearing protection. Hearing protection for workers whose audiograms show an STS is essential to prevent further hearing loss.

The issue of hearing impairment is an important one. The requirements that apply to worker's compensation claims for hearing loss as an occupational disability vary widely from state to state. Audiograms are used as a basis for determining a worker's ability to earn wages based on the amount of hearing loss. Many states require that the average hearing loss in both ears be greater than 25 dB across a range of frequencies, from 500 to 3,000 Hz. Other states allow a physician to evaluate the hearing loss and make a determination. The applied criteria are aimed at determining whether the hearing loss affects the person's ability to hear and understand speech during everyday activities. Since there is wide variation in defining "everyday activities," some difficulty exists in use of the current criteria. In addition, the OSHA PEL for noise exposure will not prevent all workers from suffering some hearing loss.

Checking Your Understanding

1. Why are baseline audiograms important for persons who work in noisy jobs?

2. Hearing loss in the higher frequencies (1,000 Hz and above) is potentially a serious problem for workers. Explain why.

3. Define the following: HTL, STS, TTS.

4. List some causes for inaccuracies in audiograms.

5. Why is retesting done to verify an STS?

9-5 Standards for Occupational Noise Exposure

OSHA 29 CFR 1910.95, Occupational Noise Exposure and the Hearing Conservation Amendment

The OSHA standard for occupational noise exposure can be found at 29 CFR 1910.95. This standard was passed in 1971 and consisted at the time only of paragraphs (a) and (b) of the current standard. The contents of these two paragraphs address requirements for use of engineering or administrative controls, or provision of hearing protective equipment to ensure that worker exposures not exceed prescribed levels of noise. The 8-hour permissible noise exposure was set at 90 dB(A), which remains the OSHA PEL for noise. In 1983, the regulation was amended. These added sections, collectively called the Hearing Conservation Amendment, contain additional requirements for ensuring that the hearing of workers is protected against noise-induced hearing loss from occupational exposure.

The first provision of the Hearing Conservation Amendment is the requirement for a hearing conservation program. Such a program is required whenever employee exposures exceed 85 dB(A) as an 8-hour time-weighted average. An exposure of 85 dB(A) is half of the allowable noise exposure for an 8-hour workday, sometimes called the **daily noise dose (DND)**. OSHA's action level for noise is 85 dB(A), or 50 percent DND. Persons exposed at or above the action level are subject to inclusion in the hearing conservation program. The various elements of a hearing conservation program, as required by OSHA, include:

1. Exposure monitoring to evaluate employee noise exposure. The monitoring must be done in a manner that will identify those employees whose noise exposure equals or exceeds the action level. Monitoring must be repeated 1) whenever changes in production or process equipment increases noise to levels that might result in the inclusion of more employees in the hearing conservation program; or 2) if the methods used to control noise are no longer effective in reducing the noise to acceptable levels. Employees must be notified of the monitoring results. The equipment used for the monitoring must meet the ANSI Specification for Sound Level Meters, S1.4.

2. Audiometric testing programs are required for all employees exposed to noise at or above the action level. The program must be administered by a licensed physician, at no cost to employees. Testing requirements include a baseline audiogram, which must be preceded by 14 hours without exposure to noise; annual tests must be repeated thereafter. Detection of an STS requires written notification of the employee and additional training as well as mandatory use of hearing protective devices. Specific requirements for conducting the audiometric tests, including equipment calibration and operation, are included.

3. Hearing protective devices must be made available for use by employees. The employer is required to provide a variety of protectors that ensure adequate protection for the anticipated levels of noise exposure. Training in use and care of the hearing protection is also required. Specific methods for determining the amount of **attenuation**, or protection, the hearing protectors are capable of providing are included in the mandatory Appendix B to the regulation.

4. A training program must be implemented that addresses 1) the effects of noise on hearing, 2) the purpose and proper use of hearing protectors and their attenuation characteristics, and 3) the purpose for, and methods used in, the audiometric testing program.

5. The amendment contains a provision for providing employees with access to a copy of the OSHA standard for occupational noise exposure as well as a requirement for posting a copy in the workplace.

6. All records including noise exposure measurements as well as audiograms and audiometric

test calibration data must be retained. Noise exposure measurements must be retained for two years, audiograms for the length of the affected worker's employment. Employees must be allowed access to these records. There are also provisions for transfer of records to any successive employer if the business is sold.

Mandatory appendices to the regulation include:

— Appendix A, which addresses methods for computing employee noise exposure. This section contains two tables that are useful to the occupational health professional when evaluating worker noise exposure. Table G-16a (see Appendix 5 of this book) relates the A-weighted SPL and the allowable duration for exposure. The table is useful for situations where the average noise level over the entire day is known. It provides a sort of sliding scale for determining the allowable time for exposure to noises that exceed the 8-hour PEL of 90 dB(A). Table A-1 (see Appendix 6 of this book) provides a simple way to convert from 8-hour TWA values to the percent DND. We will try some of these calculations later in the chapter.

— Appendix B contains a description of some methods for determining the amount of attenuation provided by hearing protection devices.

— Appendix C contains the requirements for performance of the audiometric testing equipment.

— Appendix D specifies the maximum levels of noise that may be present in rooms (or booths) where audiometric tests are conducted.

— Appendix E describes the calibration requirements for audiometric testing equipment.

Appendices F and G are nonmandatory. Appendix F contains procedures for age-correcting audiograms. This allows the employer to deduct a portion of hearing loss and attribute it to presbycusis. The values for age correction are based on research done by NIOSH. Appendix G contains some information on noise monitoring. Two additional appendices, H and I, contain information on reference documents and definitions, respectively.

ACGIH Recommendations for Noise Exposure

The ACGIH has a set of recommendations for occupational exposure to noise as it does for other corresponding OSHA standards, discussed in other chapters. While there are significant similarities between the two, there are also some important differences. We will explore both in this section.

Perhaps the most obvious difference between OSHA and the ACGIH is the limit for allowable exposure to noise. The ACGIH TLV for an 8-hour TWA is 85 dB(A), compared to the OSHA PEL of 90 dB(A). Although the difference between the two values is only 5 dB, the ACGIH TLV represents a significantly lesser amount of noise exposure than the OSHA PEL of 90 dB(A). The reason for this is related to the fact that the decibel is logarithmic. This means that for every three dB increase in sound pressure, the energy of the sound is doubled. Conversely, if the sound pressure level decreases by three dB, the energy of the sound is reduced by one half. This doubling of sound energy for each increase of three dB is called the **exchange rate**. However, it takes an increase of about five dB for the human ear to perceive noises as being twice as loud, and OSHA exposure measurements stipulate the use of a five dB exchange rate for measurement of noise exposures. (A four dB exchange rate is used in some countries.) Sound measurement instruments must allow for this, so that sound pressure may be measured according to the prescribed exchange rate. Some instruments have manual switches for setting the exchange rate, while others allow the settings to be changed through software. The ACGIH recommends that a three dB exchange rate, which represents actual exposure conditions, be used for evaluating worker noise exposures.

Both OSHA and the ACGIH allow for a maximum exposure of 140 dB(C) as a peak level of unprotected exposure to **impulse noise**, or short-duration noise (less than three seconds). Exposures in excess of this amount are prohibited. A significant difference between OSHA and ACGIH recommendations is that for noises above 100 dB, the ACGIH recommends controlling exposure at the source, as opposed to using administrative controls such as worker rotation. OSHA allows employers to control noise exposure through limiting the amount of time that may be spent in high noise areas if engineering controls are not feasible.

As is the case with other exposure limits, dropping below a PEL or TLV value does not mean that the hazard to the worker diminishes abruptly or disappears. A significant reminder included near the beginning of the ACGIH recommendations for noise exposure reads in part: "...the application of the TLV for noise will not protect all workers from the adverse effects of noise exposure..." This is an additional similarity between the two: neither the OSHA nor the ACGIH exposure limits for noise are adequate to prevent hearing loss in all workers.

Checking Your Understanding

1. List the OSHA-required elements of a hearing conservation program.

2. What topics must be addressed in the employee training portion of a hearing conservation program?

3. What are the OSHA and ACGIH 8-hour TWA exposure levels for noise?

4. What is an exchange rate and what is the significance of this value? What exchange rate is most protective of the worker?

5. Why is the ACGIH TLV for noise exposure not a guarantee of protection for all workers?

9-6 Measuring Noise in the Occupational Setting

Earlier in the chapter the A, B, and C weighting scales were explained as being derived from experiments to determine the equivalent loudness of noises at different frequencies. Two of these scales, the A and the C scale, are used extensively in noise measurements.

The A scale simulates the human perception of sound levels and provides a good prediction of the harmful potential of noise on human hearing. Therefore, all worker noise exposure evaluations are based on measurements taken using this weighting scale. The instruments used for such evaluations are of two types: sound level meters and dosimeters. Sound level meters (SLMs) are generally used for performing area surveys to determine which areas around a piece of process machinery are likely to fall into the OSHA action level. Sound level meters can also be used to measure the peak sound pressure level (remember the maximum allowable exposure is 140 dB). Usually there is a peak setting or mode on the instrument for obtaining such a reading. Some SLMs are capable of functioning as dosimeters.

Noise dosimeters differ from SLMs primarily in that they are smaller – so that they can be worn like a sampling pump – with a microphone attached as near as possible to the ear of the worker. Most dosimeters are battery powered and can compute and store a large amount of data relative to the worker's shift-long exposure to noise. These data can often be downloaded into a computer for producing reports and preparing records of exposure. Dosimeters must be set with the appropriate exchange rate (3, 4, or 5 dB), as well as with the appropriate **criterion level** for computing the noise dose. The criterion level is the 8-hour TWA limit for noise exposure. For instance, a dosimeter set to OSHA sampling requirements would have a 5 dB exchange rate and a 90 dB criterion level.

Most meters have both fast and slow response settings; the use of the slow setting is commonly specified for noise exposure measurements. The setting is related to the speed at which the instrument reacts to a sudden increase in sound pressure. A fast response will show the extremes of variation that are occurring, but sometimes the readings come and go too fast – within 0.125 seconds – for the person reading the instrument to make a record (another good reason for recording and then downloading data). The slow response provides a value that represents the average of the observed sound pressure, and the time to reach the highest reading is more gradual (about one second).

Noise measurement equipment – like other industrial hygiene equipment – must be calibrated before and after use to ensure accurate measurements are obtained. Noise monitoring records resemble those for air sampling activities and include a detailed description of the work process being evaluated, identification of the personnel whose exposures are being measured, as well as calibration data and monitoring results. Sometimes the monitoring involves only area measurements; in these situations the records may not contain as much detail on personnel identification but should always include the calibration data. Area samples may sometimes yield noise maps, which are floor layouts that show the location of the machinery with the surrounding noise levels. These maps resemble topographic maps as contour lines are drawn on them to show the boundaries between each different noise level. These maps are useful for determining which workers' exposure to noise should be further evaluated for possible inclusion in the company hearing conservation program.

Adding Decibels

We will now demonstrate some situations associated with evaluating worker noise exposure. This should help the reader prepare for similar calculations involving actual situations in his or her work.

Our first task will be to determine the combined effects of two or more noise sources. Because decibels are logarithms, they cannot be added directly (85 dB + 92 dB, for example, does not equal 177 dB). Where two or more noise sources are present, the sum of their sound pressure levels is expressed by the following equation:

$$SPL_{total} = 10 \log \left(\sum_{i=1,2,\ldots}^{N} 10^{L_i/10} \right)$$

Where

N = the total number of different sound pressure levels and

L_i = the measured sound pressure level(s).

To better understand this equation, we will present a sample problem. Suppose there are three industrial sewing machines operating in a room. A sound pressure level is measured for each machine as follows:

Machine 1: 86 dB

Machine 2: 89 dB

Machine 3: 87 dB

What is the combined SPL of the three machines?

Starting with the equation:

$$SPL_{total} = 10 \log \left(\sum_{i=1,2,\ldots}^{N} 10^{L_i/10} \right)$$

Substitute the known SPL values into the equation:

$$SPL_{total} = 10 \log \left(10^{86/10} + 10^{89/10} + 10^{87/10} \right)$$

And simplify:

$$SPL_{total} = 10 \log \left(10^{8.6} + 10^{8.9} + 10^{8.7} \right)$$

$$SPL_{total} = 10 \log \left(3.98 \times 10^8 + 7.94 \times 10^8 + 5.01 \times 10^8 \right)$$

$$SPL_{total} = 10 \log \left(1.693 \times 10^9 \right)$$

$$SPL_{total} = 10 \, (9.23)$$

$$SPL_{total} = 92.3 \text{ dB}$$

The combined sound pressure level of the three machines is 92.3 decibels.

There is another method for adding sound pressure level readings in decibels; it is much simpler than the method above and nearly as accurate. It works like this:

Difference in decibel values:	Add to the higher value:
0 or 1 dB	3 dB
2 or 3 dB	2 dB
4 to 9 dB	1 dB
10 dB or more	0 dB

We will now demonstrate the same problem using the simple method.

With the same SPLs of 86 dB, 89 dB, and 87 dB we calculate the difference between the first two measurements:

89 dB – 86 dB = 3 dB

From the table above, we now add 2 dB to the higher measurement:

2 dB + 89 = 91 dB

Then we take the difference between the result of this calculation and the remaining measurement:

91 dB – 87 dB = 4 dB

From the table above, we now add 1 dB to the previous calculated value:

1 dB + 91 dB = 92 dB.

(If additional measurements are present, the last two steps are repeated until each of the measurements is used.)

We can see that this result is only slightly different from the 92.3 dB result calculated with the previous method. Of course, if the most accurate result is desired, it is best to use the long equation we reviewed first.

Computing the Daily Noise Dose and Calculating 8-hour TWAs

Under OSHA requirements, a worker's exposure to noise is limited to 100 percent of the daily noise dose, or an 8-hour TWA of 90 dB(A). We will look at some equations for computing these values in this section.

The percent of the daily noise dose is obtained by a simple calculation. The relationship is described by the following equation:

$$D = 100 \, (C_1/T_1 + C_2/T_2 + C_3/T_3 + \ldots C_n/T_n)$$

Where

D = the daily noise dose, in percent,

C = the total time of exposure at the measured noise level, and

T = the referenced allowed duration for that noise level taken from Table G-16a of Appendix A to OSHA 29 CFR 1910.95 (see Appendix 5 of this book).

The referenced allowed duration values, T, listed in the table are derived using the following equation:

$$T = \frac{8}{2^{(L-90)/5}}$$

Where L is the measured sound pressure level in dB(A), and the values are based on an 8-hour work shift.

If the noise exposure has been the same level throughout the day, the equation becomes:

$$D = 100 \, C/T$$

Where

C = the total length of the work day, in hours, and

T = the reference duration corresponding to the measured sound level.

We will try a sample problem and will need to refer to Table G-16a from the OSHA noise standard.

A worker's noise exposure was monitored with the following results:

4 hours at 92 dB(A)

4 hours at 87 dB(A)

What percentage of the allowable DND is indicated by the worker's exposure?

Start with the equation:

$$D = 100 \, (C_1/T_1 + C_2/T_2)$$

We know our values for C_1 and C_2, but need to look at Table G-16a in the OSHA regulation to determine the reference durations (T_1 and T_2) allowed for exposure at 92 and 87 dB(A). From the table, we can see that 6.1 hours and 12.1 hours are the appropriate durations.

Substituting the data using values from the table:

$D = 100 \, [(4/6.1) + (4/12.1)]$
$D = 100 \, [0.656 + 0.331]$
$D = 98.7\%$

The worker has received 98.7 percent of the allowable daily noise dose.

We will now try a problem going in the other direction, from knowing the DND to determining the worker's 8-hour TWA exposure as it relates to the OSHA PEL.

Use the DND we calculated above. The DND value of 98.7 percent must be rounded to 99 percent since the OSHA Table A-1 lists only whole numbers. From the table, a 99 percent dose exposure is equivalent to an 8-hour TWA of 89.9 dB(A), which is not sufficiently lower than the 90 dB PEL for noise to

DND = 203 = 95 dB(A)

debate the issue. This worker would be required to participate in the employer's hearing conservation program since the exposure exceeds the OSHA action level of 85 dB(A).

Checking Your Understanding

1. Two planing machines in a sawmill operate at the same time. Their individual sound pressure levels are measured and found to be 89 dB(A) for Planer #1 and 92 dB(A) for Planer #2. What is the combined sound pressure level for the planers? (Compute the answer using both the long and short methods for adding decibels).

2. What is the DND for a worker exposed to 93 dB(A) for six hours and 91 dB(A) for two hours? What is this worker's 8-hour TWA?

9-7 Controlling Noise

The engineering of noise controls is an industry in itself. There are manufacturers of insulative curtains, coverings for noise-reflective floors, ceilings, and walls, vibration isolation devices, and others. One of the basic principles applied in these industries stems from the wave-like nature of noise. As such, noise emanates in a straight line from the source. It cannot turn corners or go around objects. This **directivity** of noise makes the use of simple barriers and enclosures effective methods for controlling noise. A good illustration of this principle of noise control is the marked quiet that follows closing a window when a lawnmower is operating in the yard outside. The simple barrier of the window glass blocks much of the sound pressure waves from entering the house.

The same lawnmower example illustrates another phenomenon involving **reflection**. The person operating the mower may notice that it seems to get louder when mowing areas next to the house. This is because the sound waves impact the walls of the house and change direction, moving back toward the source and producing a slight additive effect in sound pressure. The degree to which the reflected sound pressure increases is related to the surface reflecting the waves, both in terms of type of material and configuration. Surfaces of dense materials tend to be more efficient reflectors of sound energy: the side of a building, for example, would be a better reflector of sound energy than a tall evergreen hedge. The reflection of a sound greater than the amount that originally impacted the surface is called **amplification**. Glass, metal, and glossy concrete surfaces sometimes exhibit this phenomenon, depending on their shape. Another possible interaction between sound waves and surfaces is **resonance**. This is where the material vibrates at the same frequency as the emitted sound. The sound energy is not necessarily amplified, but it resonates into areas where in some cases it might be bothersome. An example is the hum or vibration of a heating duct that results from the vibration of the fan in the HVAC unit. This can be avoided by placing a flexible cloth or plastic connection sleeve, called a **vibration isolator**, between the source of the vibration – in this case, the fan – and the material that is resonating – here, the air distribution duct. Other ways of isolating a source of vibration include the use of rubber feet or mountings under a vibrating motor, or the use of a thick rubber mat under a vibrating machine.

Some surfaces exhibit **absorbance**, which means they do not reflect sound energy very well. Some good examples of noise-absorbing surfaces are the fiber-filled cloth-covered partitions used for office cubicles. Noise control curtains that can be used in a factory setting are often filled with fiberglass or some other sound-absorbing material.

We should not neglect the effect that regular, preventive maintenance of noise sources can have on maintaining acceptable noise levels. The parts of a machine that tend to produce noise include items such as motors, bearings, drive belts, pumps, and others. Often the operator of the machine will be the first to notice something; it might be as subtle as a slight change in the pitch of the whine of a drive motor, or a gravelly grinding sound warning of bad bearings. Attention to the parts that wear often prevents the development of a noisy problem. The installed barriers, vibration isolators, and other engineered noise controls also must be maintained to ensure their continued effectiveness. Enclosures and access doors must be returned to their original positions following maintenance, adjustments, and repairs.

Another possible method for controlling exposure to noise is to move either the workers or the noise source to a location that is far enough away that the noise level drops below 85 dB(A). The relationship between noise and distance follows the inverse square law; for example, doubling the distance reduces the sound pressure level by 1/4. Tripling the distance would reduce the sound pressure by 1/9.

Administrative Controls and Hearing Protective Devices

The OSHA standard allows employers to utilize administrative controls and personal protective equipment in situations where the engineered controls are infeasible, or while engineered controls are being designed and installed. Administrative controls consist of actions taken by management such as limiting the amount of time spent in a high noise area; rotation of workers to limit the noise exposure of each; and preventing unnecessary exposures to personnel by limiting access to high noise areas to as few individuals as possible. These measures can be effective when the restrictions are adhered to by all affected workers.

The use of hearing protective devices is a convenient method for controlling noise exposure when engineered controls are not available, or when they are not effective in reducing noise to acceptable levels. The OSHA standard specifies conditions for use of hearing protection as follows:

1. All workers exposed at or above the action level of 85 dB(A) as an 8-hour TWA must be provided with hearing protective devices, at no cost to the employees.

2. Employers must ensure that workers who need hearing protection actually wear the devices. The mandatory use of hearing protection, according to OSHA, applies to all workers exposed at or above the action level who have not yet had a baseline audiogram. Use of hearing protection is also mandatory for any employee whose audiogram shows a standard threshold shift.

3. Employees must have a variety of hearing protectors available to them, in order to select a device that is comfortable to wear (see Figure 9-5). This should help assure that workers will actually use the hearing protectors. Hearing protective devices available today include earmuffs, various shapes and sizes of foam ear plugs, and other devices that are not inserted into the ear canal but that cover it, called canal caps. Canal caps usually are mounted on a flexible headband that allows them to be placed and removed easily and quickly. They can be worn around the neck and are particularly useful for people who are in and out of high noise areas many times a day.

Earmuffs are comfortable in all but hot and humid environments, where they have a tendency to slide around on moist skin. Muffs are available in many styles; some muffs are attached directly to hard hats and can be flipped up off the ears when not in use. Some muffs, originally manufactured for use in the shooting sports, allow the wearer to hear at normal levels but do not allow sudden impact noise – such as

a gunshot – to pass through. Earmuffs do wear out on the soft foam surface around the edge of the muff, which is the part that forms a seal against the head of the user. As the foam ages and hardens, it is less able to form a seal around the ear. Even a small opening will allow sound to enter. Wearing safety glasses or corrective lenses can interfere with the seal formed by earmuffs, allowing potentially damaging sound waves to enter the ear.

Earplugs made of foam are probably the most common hearing protectors in use today. Most manufacturers now provide foam plugs in at least two different diameters, increasing their comfort and wearability. Proper placement of earplugs is vital to their performance. Hands should be clean and dry to avoid introducing dirt and potential infectious organisms into the ear canal. The plug must then be rolled into a tight cylinder, without creases in its surface, and inserted into the ear canal. The plug is then held in place with light finger pressure while it expands to fill the canal. Plugs that are not placed far enough into the canal will not provide protection. Sometimes, grasping the top of the pinna and gently pulling back and up will straighten the canal and allow better placement of the plugs.

Custom earplugs are another possible option. These are molded to the actual contours of a person's ear canal. These plugs are the most comfortable to wear, and are likely to be worn and protect the worker's hearing because of the increased comfort level.

4. OSHA requires employers to train employees to use the hearing protectors and care for them so that the effectiveness of the devices is maintained.

5. Hearing protector attenuation must be evaluated by the employer in accordance with Appendix B to 29 CFR 1910.95. This appendix describes how to determine the amount of protection afforded by a hearing protector. We will review the process here.

All hearing protective devices are issued a **noise reduction rating (NRR)** in accordance with EPA regulations. The NRR is a numerical value – given in decibels – that indicates the amount

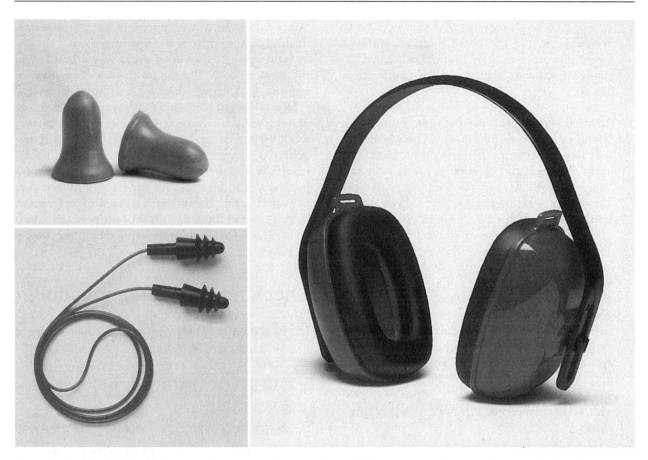

Figure 9-5: The OSHA standard requires that employers make available a variety of hearing protection devices for employees who are required to use them. The ability to select the most comfortable device that provides adequate protection will hopefully help motivate workers to use their hearing protection.

of reduction in noise exposure provided by a properly fitted and used device. NRR values are determined in a laboratory, often using mannequins or artificial/model ear canals. Typical NRR values range from 22 to 30 dB, with a few plugs having slightly higher ratings.

In order to evaluate the attenuation afforded by a particular hearing protective device, it is necessary to first obtain a sound pressure level TWA value for the anticipated exposure. This can be determined using a meter with either an A or a C weighting scale.

If a C-weighting scale is used, the NRR value is subtracted from the measured SPL for the resulting A-weighted TWA exposure. For example, say our meter, using the C-weighted scale, indicated an 8-hour TWA of 95 dB(C), and our

hearing protectors are foam plugs with NRR of 27. To determine the A-weighted TWA while wearing the plugs, subtract the NRR from the measured SPL: 95 – 27 = 68 dB(A). These plugs, if worn correctly, will reduce the employee's noise exposure to about 68 dB(A), well below the OSHA action level.

If an A-weighting scale is used to determine the 8-hour TWA noise exposure, OSHA requires that an additional seven dB be deducted from the NRR. Assume we used a meter with an A-weighting scale to measure an 8-hour TWA of 95 dB(A). Our earmuffs have an NRR of 22. We must subtract 7 dB from the NRR, 22 – 7 = 15. Now we subtract the corrected NRR value from the 8-hour TWA: 95 – 15 = 80 dB(A). Not quite as much attenuation as the foam plugs above, but still enough to get below the action level. OSHA also requires that employees who have experienced a standard threshold shift use hearing protection with attenuation that is capable of reducing their exposure to 85 dB or lower.

In some situations – for example, noises with TWAs in excess of 100 dB(A) – it may be necessary to use two forms of hearing protection. The most common combination is muffs with earplugs beneath. Use of two hearing protectors does not provide attenuation that is the sum total of the NRRs of the two devices! Studies have shown that at frequencies above 2,000 Hz, the combined effects lessen. The plugs and muffs tested showed an effective combined NRR of only an additional 3-10 dB. One reason for this is that at higher frequencies, (2,000 Hz and above) sound can be transmitted through the bones of the skull. The reduction in transmittance via air is reduced by the hearing protective devices, but there is a point of diminishing return. Covering the entire head (as with an insulated helmet) in such situations may provide additional protection but is not very often done.

The NRR is a laboratory-determined value. Studies have shown that tests used to determine the NRR do not take into account several factors affecting the real-world performance of hearing protectors. These include issues of comfort; improper fit; periods during the day when the protectors are not worn, shift in position, protectors worn incorrectly; worker modifications to hearing protectors; and removal of the protectors to allow the ears to "take a break" – usually related to discomfort caused by the devices. The fact that NRR ratings do not allow for these lapses in protection has led some occupational health professionals to advocate reducing NRRs by at least 10 dB as more representative and realistic for the level of protection that can be expected.

A few more words about worker's compensation and noise-induced hearing loss. The worker's compensation awards for occupational hearing loss vary from state to state, with the maximum award in the early 1990s being approximately $30,000.

Some states do not compensate for occupational hearing loss; in others, the loss must be a total loss in both ears. At least four states levy a penalty on the worker for willful failure to wear hearing protection, and many allow for a credit against the measured loss to account for presbycusis. Some compensate only for hearing loss due to a traumatic, work-related injury. Clearly, workers rarely are compensated adequately for their loss. The current status of occupational hearing loss compensation procedures and statutes for the state should be part of every hearing compensation training program so that the worker understands fully the potential long-term effects that this injury can have on their life.

Checking Your Understanding

1. Explain the following terms as they relate to noise: directivity; absorbance; resonance; amplification.

2. Name and describe six ways to reduce or control noise.

3. Name three types of hearing protective devices.

4. Why is proper placement of hearing protectors important?

5. What is an NRR? How is it used?

Summary

Noise, or unwanted sound, can cause irreversible hearing loss. This is usually preventable through the use of engineered or administrative controls or through the use of personal protective equipment. Because it is energy in the form of waves, sound can be described using frequency and wavelength; higher-frequency sounds are perceived by humans as higher-pitched.

The oscillations of a sound wave create changes in pressure that can be easily measured. The sound pressure level is converted into decibels, a dimensionless unit that is more convenient to work with than the standard unit for pressure, the pascal. The decibel, or dB, is based on the logarithm of the ratio of the measured sound pressure to a reference sound pressure. Because the dB represents a logarithmic relationship, increases in decibel levels are not linear. This means that a 10 dB increase in sound pressure represents an approximate 100-fold increase in sound power.

There are three weighting scales for measuring sound: the A, B, and C scales. The A scale is most like the response of the human ear, and is used for evaluating worker exposures to noise. The B scale is not used much, while the C scale is used for impact-type noise and evaluating high pressure level noise in industrial settings.

Sound waves enter the human ear canal and strike the eardrum, which vibrates; as a consequence, three bones –the hammer, the anvil, and the stirrup – vibrate and conduct the sound wave energy to the oval window, which covers the entrance to the cochlea. Interruptions in the transfer of sound waves along this path result in conductive hearing loss. These interruptions may be due to a ruptured eardrum, ear wax buildup, or damage to the small bones of the middle ear. Many conductive hearing losses can be corrected with surgery or medication.

From the oval window, sound waves are transferred to the fluid in the cochlea, and continue to move along the length of this snail-shaped organ. Small hair-like structures inside the cochlea detect the sound waves and conduct electric impulses to the brain. These impulses are interpreted as sound. Damage to the internal structures of the cochlea results in sensory hearing loss, which is not reversible. Loss of the ability to hear may affect frequencies that comprise most speech sounds, resulting in difficulty communicating with others. The potential negative consequences of this for workers include the inability to understand conversation, verbal instructions and warnings, and a decline in the quality of life outside of work.

OSHA requires that employees who are exposed to noise at levels of 85 dB(A) or higher participate in the employer's hearing conservation program. Such a program must include employee exposure assessments, training, audiometric testing, and provision of hearing protective devices. The OSHA PEL for an 8-hour TWA for noise exposure is 90 dB(A). The ACGIH 8-hour TLV for noise is 85 dB(A). These limits are significantly different since the ACGIH uses a 3 dB exchange rate compared to the 5 dB exchange rate used by OSHA. Instrumentation used for measuring noise includes sound level meters for area measurements and noise dosimeters used for worker exposure measurements. These instruments must be able to accommodate settings for exchange rate and criterion levels in order to provide data that are relevant to the appropriate standard.

Methods for controlling noise include engineered controls, administrative controls, and the use of PPE. Most engineered controls take advantage of the wave-like nature of sound as well as the phenomena of directivity and absorbance. Examples of engineered controls include use of barriers, vibration isolators, and sound-absorbing curtains or enclosures. Administrative controls include rotation of workers or limiting the amount of time workers spend in high-noise areas. Personal protective equipment consists of hearing protection devices, also called hearing protectors. These come in a variety of types: muffs, plugs, and canal caps are the most common. Each hearing protector is tested by EPA and assigned a noise reduction rating to indicate the amount of attenuation provide by the device. Both OSHA and the ACGIH advocate reducing the NRR value by seven dB to obtain a more realistic value for actual use conditions. Some studies have shown that derating the NRR by 10 dB is more realistic.

Hearing loss is compensated at widely varying levels across the United States. It is not uncommon that for compensation to take place the loss must be total in one or both ears, accompanied by an inability to work at any job. Some states penalize workers who failed to wear their hearing protectors; many allow the employer to reduce the amount of the worker's hearing loss to account for presbycusis. Others allow compensation only in the case of traumatic injury resulting in hearing loss.

Critical Thinking Questions

1. Compare and contrast the OSHA noise standard and the ACGIH noise exposure recommendations.

2. Explain how to determine the actual NRR provided by foam plugs with an NRR of 30 in an exposure situation where the 8-hour TWA is 102 dB(A). Do the same for a measured exposure of 102 dB(C).

3. Two paper banding machines operate at the same time. Their individual sound pressure levels are measured and found to be 99 dB(A) for Banding machine #1 and 94 dB(A) for Banding machine #2. Compute their combined SPL using both the long and short methods for adding decibels.

4. What is the DND for a worker exposed to 95 dB(A) for three hours and 96 dB(A) for five hours? What is this worker's 8-hour TWA?

5. Locate a noise source and recommend some simple ways to reduce the noise level, taking advantage of the phenomena associated with the wave-like nature of energy.

Long: $SPL_{total} = 10 \log \left(\sum_{i=1,2...}^{N} 10^{L_i/10} \right)$

$SPL_{total} = 10 \log(10^{99/10} + 10^{94/10})$

$= 10 \log(10^{9.9} + 10^{9.4})$

$= 10 \log(7.94 \times 10^9 + 2.51 \times 10^9)$

$= 10 \log(1.05 \times 10^{10})$

$= 10 (10.02)$

$= 100.2 \, dB$

A.) $D = 100\left(\frac{C_1}{T_1} + \frac{C_2}{T_2}\right)$

$= 100\left(\frac{4}{3} + \frac{35}{5}\right)$

$= 100(1.333 + .7)$

$= 100(2.033$

$= 203.3 \%$

B.) $DND = 203$

$200 = 95 \, dB(A)$

$95 \, dB(A) = 8 \, hour \, TBA$

Short: $99 - 94 = 5$, $2 + 99 = \boxed{101 \, dB}$

10

Ionizing and Nonionizing Radiation

Chapter Objectives

Upon completing this chapter, the student will be able to:

1. **Locate** the relative position of each type of radiation on the electromagnetic spectrum.

2. **Describe** the major types of ionizing radiation hazards.

3. **Describe** the health effects associated with exposure to the different types of ionizing and nonionizing radiation.

4. **Name** some ways to prevent tissue damage and other health effects resulting from exposure to ionizing and nonionizing radiation.

5. **Locate** exposure limits and worker protection information in OSHA, ANSI, and other standards.

10-1 Introduction

The electromagnetic spectrum (Figure 10-1) encompasses both ionizing and nonionizing radiation, from cosmic rays with very short wavelengths to radio frequencies with very long wavelengths. All electromagnetic waves are composed of photons. They differ only in their wavelength, which determines their frequency, and their intensity, which determines their amplitude. Ionizing radiation is that part of the spectrum that is associated with nuclear energy and x-rays and has wavelengths of 10^{-6} cm or shorter. Nonionizing radiation includes electricity, radio waves, microwaves, infrared light (IR), visible light, and ultraviolet (UV) light with wavelengths between 10^9 and

10^{-6} cm. Within this range are practical applications such as radar, television, radios, microwave ovens, cellular telephones, welding, and laser printers.

Many occupations involve exposures to radiation hazards. Radiology and other health care workers, workers at hazardous waste sites, welders, and glass workers are some of the groups at risk. This chapter will introduce some of the basic occupational health problems presented by both ionizing and nonionizing radiation. The different types of radiation will be discussed as well as the health hazards posed by each and the possible ways to evaluate and control exposures.

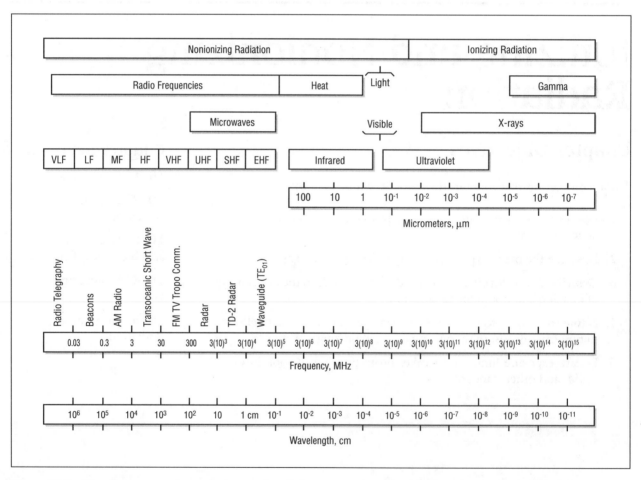

Figure 10-1: The electromagnetic spectrum contains all the wave energies that allow modern conveniences such as cellular phones, medical x-rays, pagers, television sets, and radios to operate.

10-2 Ionizing Radiation

We learned in basic chemistry that all matter is composed of atoms; each atom consists of a central nucleus made up of positively charged particles, called **protons** and particles with no charge, called **neutrons. Electrons** are negatively charged particles that surround the nucleus.

Ionizing radiation is defined as radiation capable of causing ionization, either directly or secondarily, through its interaction with matter. **Ionization** is an energy transfer that changes the electrical balance of the atom. Atoms are electrically neutral, which means they have an equal number of protons and electrons. If an atom absorbs enough energy to cause one of its electrons to be lost, the resulting particle would then have a positive one (+) charge. If the lost electron were to become attached to another atom, the second atom would then have more electrons than protons and would become negatively (–) charged. Ionization always results in atoms either losing or gaining electrons and forming charged particles called **ions**, e.g., sodium atoms (Na) become sodium ions (Na^+).

Ionizing radiation is also capable of breaking molecules into very reactive **free radicals**. Free radicals are short-lived molecular fragments that have one or more unpaired electrons – typically denoted by a dot, e.g., $Cl\cdot$ or $(C_2H_5)\cdot$. These unpaired electrons are capable of causing other substances to readily gain and lose electrons. This is called **secondary ionization**. If secondary ionization occurs within cells it may – depending on the ions that are formed – inactivate enzymes, break molecular bonds, affect cell membrane permeability, and be responsible for the mutation of genetic material.

Radioactive materials in an unwanted place are referred to as **contamination**. Radioactive materials are composed of atoms with an unstable proton/neutron ratio within their nucleus. The process whereby the nucleus of one of these atoms undergoes a spontaneous release of matter and energy – bringing the ratio closer to the most stable one-to-one configuration – is referred to as **radiation**. Every radioactive substance is composed of a mixture of **isotopes**. These are atoms of the same chemical element with differing numbers of neutrons. The **half-life** of a radioactive isotope is the amount of time it takes for one half of it to decay. This decay is loss of energy and results in the emission of radioactive particles and/or waveform energy. This decay process may be a single or multiple step process before the stable form of the isotope is reached. The most common types of radiation emitted from the decay process are alpha particles, beta particles, x-rays, gamma rays, and neutrons. We will discuss each of them in turn.

Elements with relatively high atomic numbers, such as plutonium, uranium, and radium, tend to emit alpha particles to improve the stability of their nuclei. **Alpha particles** are positively charged particles that originate in the nucleus of a disintegrating atom. Consisting of two protons and two neutrons, alpha particles are relatively large and slow moving, with a charge of positive two (2+). They are not capable of traveling very far through air – maximum range two to four inches – and can be stopped by simple barriers such as paper, clothing, or the outer layer of skin. However, because of the amount of energy they can transmit to tissue, they are very dangerous when released within the body. Target organs affected by alpha emitters include the bones, kidneys, liver, lungs, and spleen.

Beta particles are fast moving electrons that also originate in the nucleus of a disintegrating atom. Once emitted, beta particles can travel through the air from 6 to 60 feet, depending on their energy. They pose a limited external hazard, but at high exposures, can cause skin burns and damage to the cornea and lens of the eye. Being smaller in size and traveling faster, their penetrating ability makes them somewhat more hazardous than alpha particles from an external exposure standpoint. Once inside the body, however, they produce a limited amount of tissue damage due to their small mass. Nuclear disintegration sometimes produces a similar damaging particle called **positron**. The positron is identical to an electron or a beta particle in every respect, except that it carries a charge of positive one (+).

Beta particles and positrons can be shielded against by a layer of paper, plastic, or aluminum. When a beta particle is slowed by a dense shielding material, a type of x-radiation called **bremsstrahlung** (from a German term meaning braking radiation) is produced. Bremsstrahlung is the electromagnetic radiation that is released when a fast moving, charged particle quickly loses its energy upon being deflected by a large positively charged nucleus. This principle is the basis for the modern x-ray machine. Shielding materials with a lower density, such as Plexiglas and aluminum, result in less bremsstrahlung being produced.

X-rays are high energy, short wavelength electromagnetic radiation. They originate outside of the nucleus as electrons drop from higher energy to lower energy levels closer to the nucleus. This form of energy poses an external exposure hazard since it is capable of penetrating well into tissue. This penetrating ability, however, becomes a useful tool in the medical field through the widespread use of x-rays as well as in industrial applications such as metering the flow of materials through pipes and examining welds. **Gamma radiation** originates in the nucleus and is often among the products of nuclear disintegration. Its characteristics are similar to x-radiation, except that it has an even shorter wavelength and higher energy. Both x- and gamma radiation can be shielded against by several inches of lead or several feet of concrete, depending on the energy of the radiation.

Neutrons are not commonly encountered in industrial settings, but potential exposures may occur at laboratories where there is a neutron source, at particle accelerators, or nuclear reactors. Neutrons have no charge and are released from the nucleus when an atom disintegrates. Their range of travel in air varies, depending on the originating source. Their penetrating ability varies as well, but it is usually quite high compared to other subatomic particles. Once inside the body, neutrons can collide with the nuclei of hydrogen, nitrogen, or other atoms, imparting energy to these atoms. Under certain conditions,

neutrons can be absorbed by the nucleus of an atom, causing the atom to become a radioactive isotope of that element. During this process – called **activation** – the activated atom loses the energy it gained as a result of the absorption of the neutron. This energy may be emitted as alpha, beta, or positron particles, but is usually emitted as gamma rays when the activated species decays to a more stable configuration. These emissions are what damages tissues, and it is difficult to evaluate the extent of internal damage resulting from neutron exposure. Massive exposure to neutrons – which takes place only in criticality accidents – can be measured by determining how much of the sodium in the bloodstream has been activated. Effective shielding against neutrons can be obtained with materials such as water and polyethylene, which contain hydrogen atoms and are very good neutron absorbers.

Checking Your Understanding

1. Define each of the following types of radiation: alpha, beta, gamma, x-ray, and neutron.

2. Compare the above radioactive particles in terms of their ability to travel through air and penetrate human tissue. What shielding is effective for each type of particle?

3. Define the following terms: activation, contamination, bremsstrahlung, secondary ionization.

10-3 Units of Exposure and Biological Effects

Several terms have been used to describe and/or quantify exposure to ionizing radiation. The **roentgen**, abbreviated **R**, is an indicator of the degree of ionization that a particle is capable of producing in air. One roentgen is the amount of x- or gamma radiation that produces one unit of charge in one cubic centimeter of dry air. While not a dose unit, the roentgen has been used as an indicator of exposure rate, typically seen as mR/hr, less often as R/hr. The Chernobyl incident in Russia resulted in many people who worked on the initial response effort being exposed at rates of several R/hour. The **roentgen absorbed dose (rad)** is a unit for measuring the absorbed dose of ionizing radiation. The **roentgen equivalent man (rem)** is a measure of the dose of any radiation to body tissue in terms of its estimated biological effect, relative to the dose received from an exposure to an ionizing unit of x-rays. The rem, often expressed in **millirem (mrem)** – equaling 0.001 rem – is useful since it allows us to compare the biological effects of the different types of radiation.

Biological Effects of Ionizing Radiation

There are many naturally occurring sources of radiation. Cosmic radiation from the sun, stars, and other bodies in outer space penetrates the earth's atmosphere. Gamma, alpha, and beta radiation originating in ores and minerals are found in the earth's surface. Uranium, thorium, and radiocarbon are radioactive isotopes found in many organic materials, in drinking water, and in the air. In the United States, the average annual dosage is approximately 100-300 mrem per person per year. This comes from the food we eat, (e.g., potassium-40 in bananas); breathing, taking airplane trips, and undergoing an occasional x-ray at the doctor's office. The usual dose from medical procedures is 50-70 mrem/year. Depending upon the type of particle, dose, and type of cells or tissues exposed, ionizing radiation can affect cell division, damage or alter cell structure, or change the way materials are metabolized, stored, or otherwise used in the body.

Students often confuse the terms radiation and contamination. You can think about this using the following analogy with cow manure. Radiation is when it's in the pasture and contamination is when it's on your boots. Contamination is radioactive materials in an undesired place; it can be either **fixed**, meaning it is nonremovable, or **removable** (sometimes called smearable), which means it can be removed with light pressure. From an exposure standpoint, removable contamination is more dangerous because it might become airborne or be spread to other locations. It can usually be removed from the skin with soap and water, lifted off with adhesive tape or duct tape, or vacuumed from surfaces using a high-efficiency particulate air filter-equipped vacuum cleaner. Sealants – such as paint or epoxy – can be applied to surfaces with removable contamination to fix it in place, thus preventing it from becoming airborne or spreading to other areas. Fixed contamination can only be displaced if the integrity of the surface itself is disturbed or damaged by processes such as cutting, grinding or welding.

Internal exposure results from radioactive materials entering the body. Like other hazardous materials, radioactive particles can enter the body through inhalation, ingestion, or absorption. Unlike external contamination, internal radioactive sources cannot easily be washed or lifted out; therefore, it is more desirable – and effective – to prevent the material from entering the body to begin with. This can be accomplished through use of protective clothing and other PPE providing specific shielding for the particles of concern. Radiation workers commonly refer to the coverall worn for shielding as **anti-C** (anti-contamination clothing) or **PC** (protective clothing). Respiratory protection may also be used, especially if airborne contamination is a concern.

The human body is capable of tolerating some ionizing radiation exposure without permanent, damaging effects. However, at high levels of exposure, the biological effects of ionizing radiation are potentially quite serious (see Table 10-1). As stated above, the effects resulting from exposure depend upon the type of radiation involved, the level of exposure, and the parts of the body that are exposed.

Dose	Effect
5-25 rad	Minimal dose detectable by chromosome analysis.
50-75 rad	Minimal acute dose readily detectable in a specific individual.
75-125 rad	Possible nausea and vomiting; diarrhea; anxiety; irregular heartbeat; fatal to five percent of those exposed if prompt medical attention is not provided.
150-200 rad	Nausea with vomiting; diarrhea; weakness; fatigue.
300 rad	Median lethal dose for single short exposure; lethal to approximately 50 percent of those exposed if no medical treatment is given.
600-1,000 rad	Severe nausea and vomiting; diarrhea; fatal within two weeks.
>1,000 rad	Burning sensation within minutes; nausea, vomiting within 10 minutes; confusion, prostration within one hour; fatal in 1-2 weeks.

Table 10-1: Effects of acute whole body exposure (equivalent to x- or gamma-absorbed dose). The physical effects of exposure to ionizing radiation range from effects on chromosomes – which are not detectable unless specialized diagnostic tests are used – to the more visible effects of nausea, weakness, and fatigue.

Most of the data we have on human exposure was obtained through exposures that occurred during early experiments with radioactive materials, exposures to health care workers, the survivors of the atomic bombs dropped on Hiroshima and Nagasaki, and from accidents that resulted in large releases, such as the Chernobyl reactor incident. The potential hazards of many radioactive metals are due to the toxic effects of the metal itself as well as to its ionizing radiation. Both should be considered when assessing the hazard.

Exposure of tissues to ionizing radiation can result in somatic or genetic effects. **Somatic effects** take place in the cells and tissues of the individual who was exposed to the ionizing radiation. **Genetic effects** can be passed on to children of the exposed individual. Cells that undergo rapid division seem to be more susceptible to the effects of ionizing radiation. Examples include decreased production of red blood cells and the potential reproductive effects of damage to eggs and sperm as well as developmental effects on the fetus if the mother is exposed to high levels during pregnancy. Somatic effects include cataracts, disruptions of the gastrointestinal tract; and cancers or tissue damage to the thyroid, kidneys, spleen, pancreas, and prostate.

Checking Your Understanding

1. Define three terms used to describe or quantify exposure to ionizing radiation.

2. Explain the difference between somatic and genetic effects of ionizing radiation.

3. Why are the cells of a developing fetus susceptible to the effects of ionizing radiation? What other tissues or cells in the human body are similarly susceptible?

4. What is a half-life?

10-4 Protecting Workers

A significant and widely practiced concept in controlling exposures to radioactive materials is **ALARA**, an acronym that stands for **as low as reasonably achievable**. The ALARA approach involves taking steps throughout the process that will reduce exposures. The three primary components of ALARA are: 1) time, 2) distance, and 3) shielding.

1. To protect workers, exposure time is minimized. Jobs that involve exposure to radioactive materials can be planned, practiced, or divided into smaller jobs to reduce the amount of exposure time.

2. Distance is used to protect workers by increasing the physical distance separating the source of radiation and the worker. The inverse square law that describes how distance from a noise source reduces the sound pressure also applies to ionizing radiation. Therefore, doubling the distance decreases the intensity of the radiation by 1/4. This is one of the easier ways to reduce exposure.

3. Shielding is an effective method for protecting workers from external sources of radiation; it involves placement of a physical barrier between the worker and the source of radiation. Selection of the best shielding method requires knowledge about the type of hazardous radiation. Generally, the denser a material, the better shielding it will provide. However, because neutrons interact only with the nuclei of atoms, effective neutron shielding is composed of materials that are good neutron absorbers. Examples include water, liquid sodium, and plastics. Table 10-2 summarizes the types of shielding that are most effective for each type of radiation.

Exposure Limits

Exposure limits for occupational exposures to ionizing radiation, both internal and external, are published by the National Council on Radiation Protection and Measurements (NCRP). The Nuclear Regulatory Commission (NRC) also publishes permissible doses, levels, and concentrations in the Federal Regulations at 10 CFR 20. Additional guidelines for maximum permissible external exposures are published by the International Commission of Radiological Protection and Measurements (ICRP). The existence of these sources of exposure limits does not supersede the ALARA principle for keeping exposures as low as possible; these exposure limits should always be considered upper limits. There may be some situations that need closer evaluation: a job assignment performed by a pregnant worker, for example, requires that the recommended levels of 50 mrem/month not be exceeded in order to protect the growing fetus.

Radiation Type	Shielding Material
Alpha	Several inches of paper or wood
Beta	Several inches of plastic or aluminum
X-ray	Several inches of lead; several feet of concrete
Neutron	Several feet of water, or another hydrogen-rich material

Table 10-2: Shielding materials for different radioactive particles. The type and thickness of shielding that is necessary depends upon the type of radiation and its energy. Sometimes a combination of shielding materials is appropriate.

Measurements and Monitoring

As with other occupational hazards, ionizing radiation measurement requires specialized instrumentation designed to detect the type of radiation of concern. Failure to select the proper measurement technique will result in inaccurate estimation of exposures. Monitoring techniques may be aimed at measurement of individual exposures with dosimeters, or evaluation of an area using a direct-reading instrument. We will take a brief look at some of the more common methods used.

Exposures	Recommended Limits
Occupational Exposures	
Annual dose limit	5 rem
Cumulative dose	1 rem multiplied by age in years
Lens of the eye	15 rem
Skin, hands and feet	50 rem
Public Exposures	
Continuous/frequent exposure	100 mrem
Infrequent exposures	500 mrem
Lens of eye	1.5 rem
Skin, hands and feet	5 rem
Embryo/fetal exposure (monthly)	50 mrem
Note: medical exposures not included in listed limits.	

Table 10-3: Some NCRP recommended limits for exposure to ionizing radiation. The extremities (hands and feet) are able to withstand higher levels of exposure without permanent adverse effects, as compared to the internal organs.

Thermoluminescent detectors (TLD) are used for monitoring beta, gamma, and x-radiation. TLDs are worn by the worker, either as a badge-like device or a finger ring, depending on the part of the body where the highest exposure is likely or expected. They contain a small chip of lithium fluoride that absorbs radiation energy; this energy displaces electrons in the chip. The chip is heated in order to evaluate the amount of energy absorbed. The heat causes the excited electrons to return to their ground state, which prompts them to release light energy. The amount of light released is proportional to the amount of energy absorbed by the dosimeter. Some TLDs contain two chips for increased accuracy; each chip can be read for comparison of the results.

Film badges are worn on the outside of the worker's clothing; they can also be worn as rings. Like TLDs, they are used to measure beta, gamma, and x-radiation. They contain a small piece of photographic film that is exposed or darkened by the ionizing radiation. The film is compared to an unexposed control film in order to evaluate the radiation exposure of the badge.

Pocket dosimeters are tube-like devices that resemble pens. They can be clipped to the worker's clothing to measure gamma and x-radiation. Inside of the dosimeter are an electrically charged quartz fiber and a scale. When the dosimeter is exposed to ionizing radiation, some of the atoms of the gas inside the tube become ionized, and the electric charge on the quartz fiber changes in direct proportion to the amount of radiation. When the charge on the fiber changes, it deflects or bends to a new position that can be read on the scale. These dosimeters are convenient for worker use as their scale can be checked visually as often as necessary. Some recent models have internal lights that make them easier to read, with the added benefit of decreasing the likelihood of contamination around the eye.

Other dosimeters contain an electronic alarm that produces a beep or chirp to alert the worker to significant exposures as they occur. These dosimeters usually contain a film or chip for recording integrated exposures; some can produce readouts or printouts of data when connected to a computer with the appropriate software.

Direct-reading instruments can be used for measuring alpha, beta, gamma, and x-radiation. One of the most common is the Geiger-Mueller counter. These instruments contain an **ionization chamber**, which is a gas-filled container with a positive and a negative electrode inside. When the chamber is exposed to ionizing radiation, the electrons that are displaced from the gas atoms are attracted to the positive electrode. This allows for the direct measurement of the radiation. Different ionization chamber designs allow for many applications. Calibration of these instruments is an essential part of their operation and maintenance.

As with other hazardous materials, radioactive materials should be treated with respect and care. A radiation safety program should be established for purposes of controlling worker exposures through training, control of sources, tracking source locations, and ensuring that shielding and containers maintain their effectiveness and integrity. Depending upon the radiation source, a license from the Nuclear Regulatory Commission may be required. Occupational health professionals with responsibility for radioactive materials as part of their company's safety program will want to seek additional specialized training.

Checking Your Understanding

1. Explain what the acronym ALARA means.

2. Explain how time, distance, and shielding can be used to reduce worker exposure to ionizing radiation.

3. What are the NCRP recommended limits for occupational and public exposure? Fetal exposure?

4. Describe how a TLD works.

5. Why do you think some dosimeters are designed to be worn as finger rings?

10-5 Nonionizing Radiation

Nonionizing radiation is associated with radiation that does not have enough energy to cause ionization. This portion of the electromagnetic spectrum includes electric fields, radio and television frequencies, microwaves, infrared light, visible light, and ultraviolet light. This portion of the electromagnetic spectrum is also referred to as optical radiation; the wavelengths are generally expressed in nanometers. The ranges of most concern from an occupational standpoint include ultraviolet, infrared, and microwave radiation. Our discussion will be limited to occupational hazards posed by ultraviolet (UV) and infrared (IR) radiation. The bands for IR and UV radiation start at the edges of visible light and extend outward; hence the terms **near** (referring to the wavelengths closest to visible light) and **far** (wavelengths farthest from the visible light region). Table 10-4 depicts the bands of optical radiation.

Biological Effects on the Eyes

The target organs of ultraviolet and infrared radia-

tion are the skin and eyes. The eye is particularly susceptible since most of it does not have pain receptors; burns to eye tissues and damage from energy absorbance are therefore not readily felt. In addition, the focusing function of the eye concentrates light energy on the retina. Figure 10-2 shows the anatomy of the human eye.

Optical radiation enters the eye through the transparent outer cornea, through which visible light and near IR can pass; behind the cornea is a fluid called the vitreous humor. The iris is the pigmented part of the eye; its muscles constrict to restrict the amount of light passing through. Next, light passes through the lens, which is made thicker or thinner by the surrounding ciliary muscles. Altering the thickness of the lens is what allows the image to be focused on the retina at the back of the eye. Visible light and visible IR radiation can pass through the lens while the other bands of IR and UV are absorbed, if they were not already absorbed by the cornea. Visible light strikes the retina, and its photoreceptors – called rods and cones – transfer this stimulation to the brain, which interprets it into the images we see. The rods are sensitive to light inten-

Wavelength	Band Name
> 1 mm	Microwaves
3 µm - 1 mm	Far infrared (IR-C)
1.4 µm - 3 µm	Intermediate IR (IR-B)
760 - 1400 nm	Near IR (IR-A)
400 - 760 nm	Visible light
315 - 400 nm	Near UV (UV-A); black light
280 - 315 nm	Intermediate UV (UV-B)
100 - 280 nm	Far UV (UV-C)

Table 10-4: Optical radiation bands. The letters A, B, and C are used to designate bandwidths in the IR and UV ranges away from visible light; A corresponds to near, B to intermediate, and C to far in either direction.

Band	Site of Damage and Injury Type
Far IR	Corneal burns
Intermediate IR	Corneal burns
Near IR	Retinal burns; cataracts of the lens (glassblower's cataracts)
Visible IR	Retinal burns; degraded night and/or color vision
Near UV	Lens cataracts
Intermediate UV	Lens cataracts, corneal damage (welder's flash)
Far UV	Corneal damage (welder's flash)

Table 10-5: Tissues of the eyes may be damaged by exposure to IR and UV radiation. The site of damage depends upon the wavelength of the radiation involved. Effects – some of which are permanent – range from thermal burns of the cornea and retina to cataracts.

sity, while the cones are responsible for perception of colors.

As you might expect, the portions of the eye most susceptible to damage by each band of radiation correlate to their absorbance of that band. The cornea is susceptible to burns from intermediate and far IR as well as intermediate and far UV. The retina, on which visible and near IR light is focused, is susceptible to overheating and burns. The lens absorbs near UV, and cataracts may result from overexposure to this band of radiation. Table 10-5 summarizes the portions of the eye susceptible to damage from each band and the type of resulting injury.

Infrared radiation is emitted by very hot materials, such as molten glass or metal. Cataracts from exposure to near IR emitted by hot glass was so common among glassblowers that it resulted in the assignment of the name **glassblower's cataracts**, which may take 10 years or more to form. This cloudiness of the lens is usually not reversible. Damage to the cornea from exposure to the intermediate and far UV bands produced by welding processes resulted in the name **welder's flash** being associated with that type of damage. Welder's flash is rarely experienced by welders, since they are usually viewing the work through a tinted shield; it is the helper

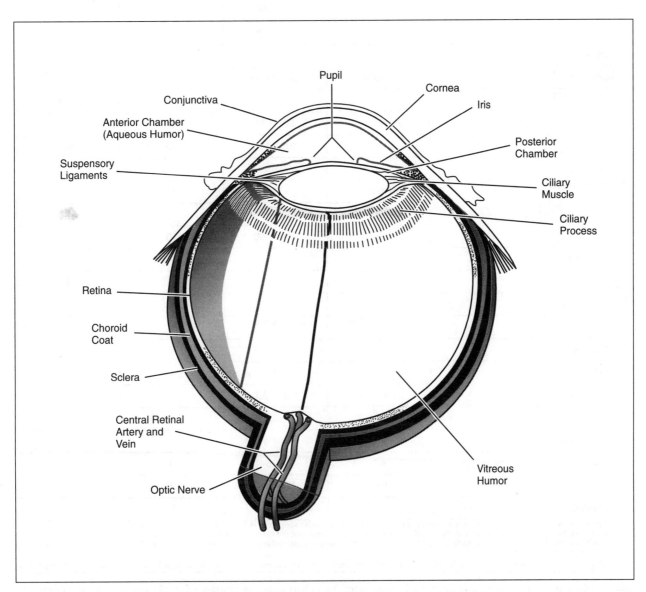

Figure 10-2: The human eye is susceptible to damage from ionizing and nonionizing radiation. Many of these injuries are the result of absorption of energy by the tissues. Some eye tissues preferentially absorb a specific wavelength.

Operation	Electrode Size, Inches	Arc Current, Volts	Minimum Shade[a]	Suggested Shade
Shielded metal arc welding	< 3	< 60	7	–
	3 - 5	60 - 160	8	10
	5 - 8	160 - 250	10	12
	> 8	250 - 550	11	14
Gas metal arc welding & Flux cored arc welding	–	< 60	7	–
	–	60 - 160	10	11
	–	160 - 250	10	12
	–	250 - 550	10	14
Gas tungsten arc welding	–	< 50	8	10
	–	50 - 150	6	12
	–	150 - 500	10	14
Air carbon	(Light)	< 500	10	12
Air cutting	(Heavy)	500-1,000	11	14
Plasma arc welding	–	< 20	6	6 - 8
	–	20 - 100	8	10
	–	100 - 400	10	12
	–	400 - 800	11	14
Plasma arc cutting	Light[b]	< 300	8	9
	Medium[b]	300 - 400	9	12
	Heavy[b]	400 - 800	10	14
Torch brazing	–	–	–	3 - 4
Torch soldering	–	–	–	2
Torch soldering	–	–	–	14
	Plate Thickness (inches)	Plate Thickness (mm)		
Gas Welding				
Light	< 1/8	< 3.2	4	4 - 5
Medium	1/8 - 1/2	3.2 - 12.7	5	5 - 6
Heavy	–	> 12.7	6	6 - 8
Oxygen Cutting				
Light	< 1	< 25	3	3 - 4
Medium	1 - 6	25 - 150	4	4 - 5
Heavy	> 6	> 150	5	5 - 6

a. This is the minimum shade contained in the OSHA standard for eye and face protection, 29 CFR 1910.133. Start with a filter lens that is too dark to see the weld zone, then go to a lighter shade that allows sufficient view of the zone. However, the selected shade must not be lighter than the minimum shade specified in the table. In oxyfuel gas welding or cutting where the torch produces a yellow flame, it is desirable to use a filter that absorbs light in the yellow or sodium range of the light spectrum.

b. These values apply where the arc is clearly seen. Experience has shown that lighter filters may be used when the arc is hidden by the item being worked on.

Table 10-6: Filter lenses for radiant energy selection chart. This chart contains guidance for selection of eye protection against IR and UV radiation. The recommendations in the chart should be considered minimums to be adjusted according to actual use conditions.

or someone who stops to watch the work that is affected. The symptoms of welder's flash are irritation and inflammation of the soft tissues surrounding the eyes, and a feeling of sand or similar grit in the eyes. Snow blindness, brought on by the reflection of the sun (UV) on white snow, is a similar condition.

The body possesses some natural means of protecting the eyes from damage. Under the retina is a vascularized membrane called the **choroid**. Blood circulating through the choroid acts as a coolant to remove excess heat from the retina. This mechanism can be overwhelmed by actions such as looking at the sun through binoculars or a telescope. The optics of such devices gather a large amount of light energy, which is then focused onto the retina, producing the damage. The blink reflex provides some protection against sudden bursts of light as well as flying objects.

Effective methods for protecting the eyes against damage from IR and UV light include various styles of eye and face protectors. Safety glasses with side shields prevent foreign objects from entering the eyes; some safety glasses, capable of providing UV protection, are now marketed for use outdoors. Welding curtains, tinted or shaded glasses, goggles, and face shields in different densities are available for protection against the damaging effects of IR and UV radiation from welding, torch cutting, and similar processes. The OSHA standard for eye protection, including radiant energy protection, is 29 CFR 1910.133. The ANSI standard that contains recommendations for eye protection is AWS/ANSI Z49.1-1988. Table 10-6 is an adaptation of tables found in these standards. It can be used as a guide in selecting the proper lens for protection against radiant energy. The chapter on selection and use of personal protective equipment contains additional information on eye protection.

Biological Effects on the Skin

The melanin found in the outer layer of the skin has a protective function in that it absorbs energy in the UV bandwidth (see Figure 8-2 for a cross section of the skin). UV-A is the band that stimulates production of melanin, which produces the bronze glow that sunbathers seek. UV-B is the band that causes leathery changes in the skin among people with long-term exposure. UV-B is also the band that most topical sunscreen lotions absorb, since this is the bandwidth thought to be most active in causing skin

cancer. Recent evidence shows that UV-A is also capable of causing changes that result in cancer. Aside from the risk of cancer associated with UV exposure, persons working at outdoor occupations such as road construction and roofing can be exposed to coal tar emissions, which have been shown to have a synergistic effect with UV exposure for causing skin cancer. Other reactions that occur following exposure to UV radiation and chemical agents are discussed in Chapter 8. The liberal use of a sunscreen, wearing a broad-rimmed hat and sunglasses, and covering exposed skin are the most effective methods of protecting workers' skin from damaging exposure to UV radiation.

The other bands of UV and IR radiation are capable of causing thermal burns to skin tissue. Methods for protecting workers from these hazards include process isolation or automation, use of insulated barriers or clothing, or employee rotation to limit exposures.

Checking Your Understanding

1. Where are the near, far, and intermediate IR and UV bands located relative to visible light?

2. Explain the terms glassblower's cataracts and welder's flash.

3. List the different bands of UV and IR radiation and the types and location of damage to the eye that can result from each.

4. What is the choroid and how does it protect the eye?

5. What types of skin damage can result from IR and UV exposure?

6. List at least three methods that can be used to protect workers exposed to IR radiation. Do the same for UV radiation.

Summary

The electromagnetic spectrum describes wave energy in the ionizing and nonionizing regions. Specific names given to each region are usually based on uses or physical characteristics of the energy of that region.

Ionizing radiation includes energy with wavelengths of 10^{-6} cm and shorter. The most common types of ionizing radiation are alpha, beta, gamma, x-rays, and neutrons. Beta, gamma, and x-radiation primarily present external hazards; alpha and neutron pose internal hazards. Radiation is the spontaneous release of energy/matter from an atom moving toward a more stable configuration of protons and neutrons. Radioactive materials are composed of atoms with an unstable proton/neutron ratio. Contamination is radioactive materials in an unwanted place; it may be fixed or removable. Removable contamination poses the greater hazard as it can become airborne or it may be spread to other areas.

Units for measurement of ionizing radiation include units for exposure (roentgen) and absorbed dose (rad, rem, millirem). Exposures can be measured using direct-reading instruments or user-readable dosimeters, or through use of film badges or TLDs. The latter ones must be evaluated in a laboratory to determine the amount of radiation the device (and wearer of the device) absorbed.

The effects of ionizing radiation may be somatic or genetic, depending on the type of radiation as well as the level and duration of exposure. Cells that undergo rapid division and/or growth – like bone marrow and germ cells – are most susceptible to the effects of ionizing radiation. Some tissues and organs may be target organs for radioactive isotopes that enter the body. For example, bones, kidneys, and spleen are targeted by radioactive heavy metals such as uranium and plutonium, which can exert their toxic effects both as a metal and as a source of ionizing radiation. Effects on the eyes, reproductive system, lungs, pancreas, and intestinal tract are also possible.

Methods for protecting workers include application of the ALARA principles of time, distance, and shielding. Shortening the amount of exposure and increasing the distance between workers and the source of radiation are relatively easy concepts to apply. Shielding is also straightforward but the types of radiation being shielded against must be known to select the most effective type of material. Paper and wood work well for alpha; plastic and aluminum are good shielding for beta; lead and concrete are good for x-rays and gamma rays; water and other hydrogen-rich materials make good neutron shielding.

There are several sources of exposure standards for occupational exposure to ionizing radiation. They include the Nuclear Regulatory Commission, National Council on Radiation Protection and Measurements, and the International Commission of Radiological Protection and Measurements. All of the agency limits should be considered upper limits, and the principle of keeping exposures at the lowest reasonable level should be applied first.

Nonionizing radiation includes infrared, ultraviolet, electric, radio, microwaves, and visible light. Our discussion focuses on two with occupational importance: ultraviolet and infrared. The eyes and skin are most susceptible to these two types of radiation. Eye tissues vary in their absorbance of these types of radiation, therefore different eye tissues are more susceptible to damage from specific wavelengths. Some eye injuries resulting from nonionizing radiation are named for the occupations in which they occur; glassblower's cataracts (from IR radiation) and welder's flash (from UV radiation) are examples. Bystanders are more likely to suffer from welder's flash than welders themselves, as they are protected by a tinted shield. The choroid is a vascular membrane that protects the retina from damage due to overheating; however, this mechanism can be overpowered by viewing through a telescope or binoculars. The blink reflex provides limited protection against flying objects. Tinted lenses and filters can provide effective protection for the eyes against UV and IR radiation. Both OSHA and ANSI have issued guidance for selecting eye protection against radiant energy (see Table 10-6).

Skin damage resulting from nonionizing radiation includes thermal burns from IR and UV radiation and skin cancer from UV radiation. Some chemicals, such as coal tar volatiles, have been shown to have a synergistic effect with UV radiation for causing skin damage. The use of sunscreen and sunglasses, keeping the skin covered, and avoiding unnecessary exposure can reduce or eliminate many of these hazards. IR radiation poses the added threat of heat-related illnesses. Insulated barriers, employee rotation, and gradual exposure to hot environments can lessen the chance of heat-induced illness occurring among workers.

Critical Thinking Questions

1. Apply the ALARA concept for worker protection during replacement of a pump in a system where removable contamination and radiation are expected to be hazards.

2. What was the outcome for workers and others who were part of the initial response team at the Chernobyl reactor incident?

3. What shade would you recommend for plasma arc cutting using a medium weight, 350-volt electrode?

4. In addition to ventilation for fume control and protective gear worn by the welder and helper, what other recommendations would you make for a welding station?

11

Ergonomics and Temperature Extremes

Chapter Objectives

Upon completing this chapter, the student will be able to:

1. **Describe** some of the common potential injuries associated with performance of tasks that present ergonomic hazards.

2. **Evaluate** a task to identify potential ergonomic hazards.

3. **Apply** ergonomic concepts to the prevention, control, or elimination of hazards.

4. **Describe** illnesses associated with exposure to hot environments.

5. **Describe** some of the approaches used to evaluate potential heat stress situations.

6. **Apply** strategies for prevention and control of heat stress.

7. **Describe** some common cold injuries and situations that pose cold exposure hazards.

8. **Apply** strategies for preventing cold injuries.

11-1 Introduction

This chapter addresses the topics of ergonomics and temperature extremes in separate sections. Ergonomics is a general term for the practice of fitting the task or job to the physical characteristics of the worker, a principle that seems logical but that isn't always applied. Like other disciplines, ergonomics has its own language and its own safety professionals called ergonomists. These professionals have been trained or are experienced in evaluating a task in terms of its impact on the human body and its capabilities and limitations, including variations in stature and physical strength.

Both very high and very low temperatures pose hazards to workers, although for most safety professionals, heat-related illnesses are the most likely to be encountered. The hazards associated with exposure to extremes of temperature can be quite severe, as evidenced by the loss of fingers and toes to frostbite. Heat stroke is a serious emergency that can result in death if the victim is not provided with prompt medical care.

11-2 Ergonomics

Ergonomics is a science dealing with the application of information on human physical and psychological characteristics to the design of the work environment. Application of ergonomic principles includes consideration of many factors that affect how a person interacts with their environment, including:

—Spatial relationships between the equipment and the worker's movement in the area;

—The physical dimensions and stature of the worker;

—The type and number of motions necessary for performing the task;

—The methods used for manipulating, securing, and otherwise using hand tools, including the force required for gripping or actuating the tool;

—Whether the task involves handling vibrating, cold, or hot tools;

—Whether the job involves sitting or standing as well as the duration of time involved;

—The surrounding environment, including temperature and lighting.

The term ergonomics was first used by a team of researchers in the United Kingdom sometime around 1950. The team was composed of professionals with different backgrounds whose goal was to apply physiological, physical, and psychological knowledge to equipment and work task designs. Similar teams in the United States took up the task of work environment design. Ergonomics is sometimes used synonymously with the term **human factors**, although some professionals use this term to refer to the psychological and sociological aspects of the issues involved. Persons whose area of expertise is ergonomics are referred to as **ergonomists**.

Injuries attributable to ergonomic hazards are sometimes called **repetitive trauma** or **repetitive use injuries**; an older but still common term is **cumulative trauma disorders**. These terms are apt descriptions of the probable cause of the injury: the repeated use of the body to perform a task or motion that is unnatural, or that the human body is poorly adapted to perform. Assembly line workers and keyboard-entry positions are two examples of jobs that may require the worker to spend extended periods performing similar motions over and over. After weeks, months, or years, the soft tissues that are involved – generally muscles, tendons, and/or nerves – begin to show evidence of this overuse. Not surprisingly, aging workers are more susceptible to these types of injuries. However, the tremendous amount of individual variability among the working population makes it difficult to single out any convenient method for accurately predicting who will be susceptible to ergonomic hazards. Without additional information, the occupational health professional should assume that all workers are potentially at risk.

A repetitive trauma injury that has received a lot of press is **carpal tunnel syndrome**. This condition results when the median nerve becomes compressed in a passage through bones in the wrist, called the **carpal tunnel** (See Figure 11-1). This opening provides a path for the median nerve and a number of tendons to reach the hand from the forearm. The compression results from inflammation of the tendons and is associated with continued or repeated flexing of the wrist. Tasks that can present risk of this type of injury include the use of straight-handled pliers, screwdrivers, or similar tools, to perform a job that requires repeated twisting motions of the wrist. Workers in poultry and meat processing factories are a good example of a population at risk because of the motions of their wrists while wielding knives. Attention to this situation was a factor in OSHA's proposing a standard for controlling ergonomic hazards.

The use of vibrating tools may cause a disorder called **white finger** – also called **dead finger** or **Raynaud's syndrome**. This condition is caused by loss or lessening of blood flow to the digits, due to spasms in blood vessels that are brought on by vibration. The skin on the fingers turns white, and the person may notice numbness and tingling. Sensation and control of movement may also be lost. White finger is made worse if the vibrating tool is cold, or if it is used in a cold environment. **Trigger finger** is an injury characterized by inability to bend or straighten a finger due to a locked or immobile tendon. Sometimes the tendon may still be able to move, but the movements may be jerky. Some air-

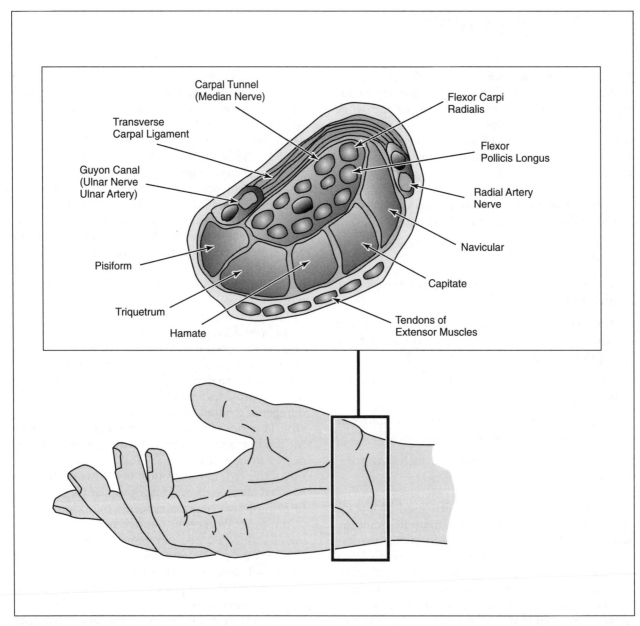

Figure 11-1: A cross-section of the human forearm showing the location of the carpal tunnel in the bones of the wrist and the median nerve that passes through it.

powered drivers used on assembly lines are trigger-operated; if resistance on the trigger switch requires too much force to actuate the tool, this type of injury can result. Not all of the ergonomic injuries have elaborate names; conditions such as muscle soreness, stiffness, back strain, and blisters on the palms or fingers are other examples of injuries that can be caused by repeated, forceful motions.

Anthropometry

When designing workstations and tasks so that the potential ergonomic hazards are minimized, the most important factor is adjustability. By providing some built-in flexibility for customizing the work to fit the worker, it is possible to engineer out many of the hazards. However, how much flexibility is

Dimension	Percentile		
	5th	50th	95th
Stature	152.8/164.7	162.94/175.58	173.7/186.6
Overhead reach	200.6/216.7	215.34/232.80	231.3/249.4
Sitting height	79.5/85.8	85.2/91.39	91.0/97.2
Knee height, sitting	47.4/51.4	51.54/55.88	56.0/60.6
Popliteal, sitting	35.1/39.5	38.94/43.41	42.9/47.6
Thigh clearance, sitting	14.0/14.9	15.89/16.82	18.0/19.0

Table 11-1: Some body dimensions (in cm) for U.S. adults representing the 5th, 50th, and 95th percentiles. Dimensions are given for women/men.

necessary to accommodate the majority of the workers? Ergonomists and engineers have made use of **anthropometry** – which is the measurement of humans in terms of heights, depths, breadths, and distances – by incorporating the extremes of these measurements into workstation design. Generally these are straight-line, point-to-point measures between landmarks on the body – the length of the forearm from elbow to wrist, or the length of the femur from hip to knee. Both the sitting and the standing posture are used for obtaining the measurements. Many of the data in use were obtained from measurements taken on military personnel, whose body dimensions seem consistent with the general population (although body weights tend to be higher among the civilian population). While few work situations allow the person to sit or stand in a stationary position, the measurements and relationships are still useful for work design applications. The data are often divided into groups that represent the extremes at each end (smaller and larger individuals) as well as the population in between. These groups are called percentiles, referring to the portion of the population that fits into that category. Each of the two extremes comprises about five percent of the total population; on a scale from 0 to 100, they are represented by the 5th and 95th percentiles. The 50th percentile represents an approximation of the population average. Table 11-1 shows some body dimensions for general populations of adults in the United States.

Applications of Ergonomic Principles

A workstation should accommodate people with above- or below-average body dimensions. These extremes are represented by a small percentage of the population, but by accommodating both, workers should be able to find some setting of the adjustments that allows them to perform the work comfortably. Thus, a standup workstation designed to accommodate both the 5th and 95th percentiles would be adjustable for persons with heights between 153 and 187 cm. A seat with adjustability for the same range would need to have the capability of being set as low as 35 cm and as high as 47.6 cm to allow the users to sit with their feet flat on the floor. For this purpose, an important measurement for determining seat height is the **popliteal** dimension, which is the distance from the floor to the back of the knee.

The evaluation of any work task should not overlook the dimension of body weight. Heavy workers may require additional accommodation; other body dimensions, such as circumference, may not be within the range that the workstation was designed to accommodate. Manual material handling jobs are an example. Proper handling and movement of materials requires that the person secure the load close to their body. If their body prevents them from achieving this position, the arms will remain straight, and additional stress will be placed on the shoulder and elbows. Reaching overhead or across a table or counter, and sitting with legs and feet under a work surface are other examples of potential ergonomic issues that may need to be addressed. Sometimes the process itself has to be modified to achieve the necessary changes. Completely automating a task, for example, may eliminate the problem in question.

Designing tasks and workstations involves more than incorporating adjustable seats and tables. Attention also needs to be paid to the tools or other devices that the worker will hold, manipulate, or otherwise use to complete the task. Twisting motions of the hand and wrist as well as forceful gripping or pinching of handles should be avoided. An automatic brake for pneumatic or hydraulic drivers will avoid the inevitable twist of the operator's wrist that occurs when the fastener is fully tightened and the driver kicks back. The gripping surface of tools should have a diameter that fits comfortably in the worker's hand, which may be achieved by slip-on or wraparound grips that add dimension and some cushioning. These wraps also add some insulation, which is important if the tools are made of metal and the work environment is cold. Insulation can also aid in absorbing vibration.

In some instances, the tools allow the worker to keep the wrist straight, but the positioning of the tool over the work requires that the wrist be bent.

Including ergonomic concerns in the planning stages can help eliminate these issues. An example of this would be the introduction of adjustable brackets or fixtures that allow the workpiece to be repositioned as each of the fasteners are installed. This type of design requires that the engineers consider the angle of each fastener as the item is assembled or installed.

Gloves are an important item of personal protection for workers handling hazardous materials and sharp objects. Ergonomic issues involving gloves include their interference with the tactile sense, in terms of loss of dexterity and of gripping ability. These may be compensated for by using snug-fitting gloves for tasks that require dexterity and tactile ability, or by using gloves with a slightly abrasive

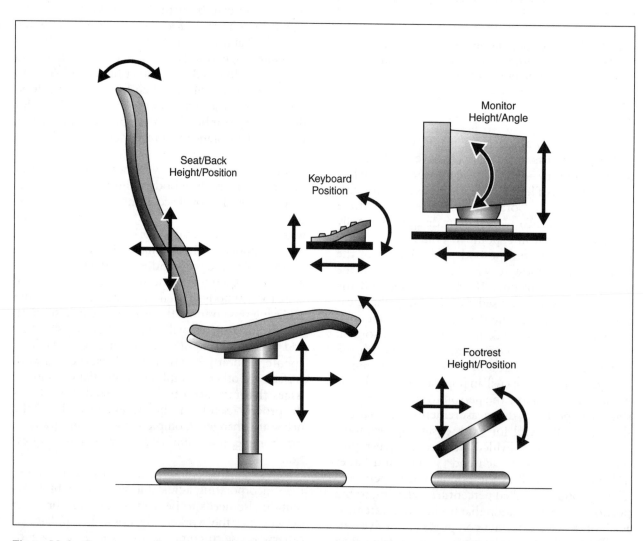

Figure 11-2: Computer workstations should incorporate adjustability into the position of all elements: monitor, keyboard, seating, and footrest. This allows each user to customize the workstation to fit individual needs for comfort and productivity.

surface to enhance grip. Of course, for some tasks – like cutting up chickens – the use of a protective glove to prevent cuts and lacerations is a logical and desirable choice. The use of gloves for a job that involves ergonomic hazards associated with repetitive or intricate hand and finger motions should be carefully evaluated.

Let's apply some of the ergonomic principles to a common workstation in many offices today: the computer workstation. Figure 11-2 shows the many features that should be adjustable to suit the person. A desirable addition to these would be monitor and keyboard adjustments that would allow the operator to alternate sitting with standing, as his or her own desires dictate. The seat should be adjustable in height from 38 to 51 cm (15 to 20 inches) to accommodate the 5th and 95th percentiles, and should be 41-51 cm (16-20 inches) deep. The padding or upholstery should be comfortable but not so thick that it puts pressure on the back of the knees. The back of the seat should have room at the bottom for the buttocks, but should have a slight protrusion in the area of the lower back to provide lumbar support. Ideally, the lumbar support should also be adjustable in height, from 15-23 cm (6 to 9 inches) above the seat surface. The remaining part of the backrest should be tall enough to support the entire back, as well as the neck. The neck or cervical support should also be adjustable, from 51 to 71 cm (20 to 28 inches) in height; the protrusion for this support should be adjustable to fit the concavity of the user's neck. As for the keyboard, it should be at about elbow height so that the wrists can be kept as straight as possible during use. The monitor should be positioned close enough to be clearly seen while the user is looking straight ahead. The relative positions of the work surface and user's body should allow for adequate clearances, so that body movements and occasional repositioning of feet or legs are possible and comfortable.

Manual materials handling – such as small parts packaging – also requires attention to ergonomic hazards in order to avoid back strain and other injuries. Some guidelines for designing tasks involving lifting and manipulating small packages or boxes (six cubic feet or less with weights of up to 20 pounds) are listed below.

— The width of the package should be 30 inches or less;

— Provide handles for gripping when possible;

— Design the workstation so that boxes can be lifted and held directly in front of the body;

— Avoid twisting motions during lifting or lowering.

Tasks that require the worker to move about an area from one station to another should be laid out with the sequence of the task in mind. Walkways should be kept clear of obstructions, and walking surfaces should be free of trip hazards and slippery spots. Lighting in the area should be adequate. Worker stations should have suitable clearance for feet and legs while standing, and the use of sit-or-stand posture might be offered. This is achieved with specially designed chairs –sometimes called stand-seats – that allow workers to operate from a seated or semi-seated position.

Many facilities have ergonomics committees, which may include the plant doctor or nurse, safety professionals, engineers, and other recruits from among the facility's work force. Ergonomics committees may be a separate group, or part of an in-plant safety committee that evaluates ergonomic hazards as one of several functions. Worker input is an invaluable source of information, which can often suggest practical changes to improve conditions. While it is possible to alter existing processes and workstations, the inclusion of ergonomic considerations in the initial design is often the most cost-effective method for eliminating these types of hazards. A formal, written ergonomics program is not a regulatory requirement, but it should be part of a comprehensive safety and health program. It should 1) identify its purpose, 2) assign specific responsibilities for its implementation, and 3) describe the procedures that will be used in the evaluation of ergonomic hazards. The ergonomic hazards evaluation should 1) include a description of the task, 2) list the identified hazards, 3) include recommendations for changes and improvements, and 4) be documented. Training employees to recognize and call attention to ergonomic hazards should also be included in the program. The training could be incorporated into the facility hazard communication program.

In 1995, OSHA proposed an ergonomics standard that never became a regulation and is now considered obsolete. This standard is no longer available from OSHA. Although there are no definitive regulations, most occupational health professionals and employers consider ergonomic hazards to be recognized hazards as defined in the original OSHA Act. Therefore, alleged violations that involve ergonomic issues can be cited by OSHA under the general duty clause.

Checking Your Understanding

1. Define the term ergonomics.

2. Describe four common repetitive use injuries.

3. What is anthropometry? How is it important in ergonomics?

4. Explain why designing a workstation or task to fit the average individual is not the optimum approach.

5. Describe some features or characteristics of hand tools and explain their ergonomic significance.

6. What single ergonomic factor has been shown to be important in improving worker comfort throughout the work shift?

11-3 Temperature Extremes

Heat-related Illnesses

Individuals working in hot environments are at an increased risk for development of heat-related illnesses. Work environment, relative effort involved, and clothing are all potential factors leading to heat stress condition. Occupational exposures to heat can occur among individuals with outdoor occupations, such as construction, landscaping, and utility workers. Indoor occupations with potential heat exposures include foundries, steel and glass manufacturing, hot processes involving shaping or molding of plastics and metals, and many others. PPE worn to protect workers from hazardous materials can pose added problems by impeding the natural mechanisms that cool the body. When the body's ability to cool itself is interfered with, body temperature, heart rate, and perspiration increase. If temperature does not fall to a near-normal or tolerable level, serious illness can result.

Perspiration is the mechanism by which the body cools itself. As perspiration evaporates from the skin, the cooling effect helps lower body temperature. Blood vessels in the outer surface of the skin dilate, allowing more blood flow to the skin surface where heat can be dissipated. Air movement from wind or fans as well as body movement help enhance the evaporation process. The body can also reduce its internal temperature through respiration, when cooler air is brought into the body (lungs) where it comes into contact with blood.

The following are descriptions of the forms of heat-related illness that can occur during work in hot weather, in hot ambient conditions, or when workers are wearing layers of protective gear that interfere with perspiration.

Heat rash, also called prickly heat, appears as little red bumps on the skin, which are in fact inflamed sweat glands. It usually appears on areas of the body that become and stay damp, as under sweat-soaked shirts, pants, and gloves. Heat rash is not usually serious, although it can become infected. Treatment includes allowing the skin to dry and keeping affected areas as dry as possible. Infections can be treated with a topical antibiotic ointment.

Heat cramps are caused by heavy perspiration with resultant loss of body fluid, causing an imbalance in the salts and minerals of the muscles, which in turn causes cramping. Heat cramps can be very painful but do not usually last very long and do not cause permanent disability. Treatment for heat cramps includes removing the individual from the hot environment and providing plenty of water to drink.

Heat syncope is a fainting or near-fainting condition that occurs among people who have been standing in one position for a period of time, usually in the sun, but it can occur in any warm environment. Standing still causes the blood to pool in the lower regions of the body, which leads to fainting after some time. In summer, this is not uncommon at outdoor receptions or weddings. It may also occur at construction sites, affecting the worker who stands on the street in the hot sun directing traffic. Individuals with heat syncope should lie down in a shady spot and drink water. Flexing leg muscles and moving around periodically during the work shift along with regular intake of water all help prevent the condition.

Heat exhaustion is a condition that usually develops among individuals who have experienced loss of body fluids due to heavy perspiration. Symptoms of heat exhaustion include nausea, dizziness, headaches, tiredness, and possibly fainting. An individual suffering from heat exhaustion is usually sweating profusely and may be confused or disoriented. Treatment includes removing the individual from the hot environment and providing cool water (45-50°F) to drink. The individual should be monitored by someone with first aid training and medical attention should be sought immediately if the condition deteriorates.

Heat stroke is the most serious form of heat-related illness. Individuals suffering from heat stroke may or may not be perspiring; they will have an elevated body temperature, at or above 104°F. Symptoms of heat stroke include a red, hot face and/or skin; lack of or reduced perspiration; erratic behavior, confusion, or dizziness; and collapse or unconsciousness. This condition is an extremely dangerous medical emergency. The person should be moved to a cool area and aggressively cooled, using wet blankets and fanning. Victims should be transported by a medical team to the nearest hospital. Outcomes include possible coma and death.

Several factors can affect the potential for workers to develop heat-induced conditions. They include:

—Acclimatization – Workers experience an **acclimatization** period during the first ten days to two weeks of work in a hot environment. During this time, their body gradually adjusts to operating in very warm conditions. After the body has acclimated, workers are less likely to experience heat-related problems. While individuals need 10 to 14 days to become heat-acclimated, they may lose this acclimation after only a few days away from the hot environment. For this reason, workers returning from long weekends or vacations should monitor themselves closely to detect early signs of heat stress.

—Physical Fitness – Workers who are in good physical condition are less likely to experience heat-related illnesses. Obesity also contributes to a worker's inability to handle heat stress, due to the added insulation that prevents the body from cooling efficiently.

—Age – Older workers may have more difficulty operating in a hot environment, and may take longer to become acclimated.

—Alcohol and Drug Usage – Alcohol consumption contributes to dehydration and makes workers much more likely to experience heat-related illness. Some prescription and over-the-counter drugs may also increase a worker's susceptibility to heat stress.

—Atmospheric Conditions – High humidity, direct sunlight, and radiant heat greatly increase heat stress conditions, which are likely with PPE usage at temperatures of 70° F or greater.

—Workload – Workers performing strenuous work are more likely to suffer from heat-induced illness since they are generally losing more body fluids through perspiration. In addition, the heat produced by the body's metabolism adds to the overall heat load of the body.

The best treatment for heat stress is prevention. In most situations, a combination of several preventive measures can provide an effective program for averting heat-related illness. If possible, heat and humidity produced by the process should be engineered out using local and general ventilation. Some possible measures are: 1) adjusting work schedules, 2) alternating work with breaks, and 3) monitoring workers for heat stress symptoms. Maintaining flexible and adequate work/rest schedules is relatively easy and an effective method for reducing heat stress. Working during early or late hours avoids the heat of midday. Workers should be able to take breaks to cool down and should be encouraged to drink water frequently. Coffee or caffeine-containing soft drinks are not advisable since their diuretic effect contributes to dehydration. Break and lunch areas should be cooler than the work area; they should be shaded and with good ventilation. Workers should also be monitoring themselves and each other for signs or symptoms of heat-related illness and should understand what to do if they detect them.

A number of approaches can be used for monitoring the work environment. The most common method is one published in the ACGIH TLV booklet. The ACGIH TLVs for heat stress are based on an index called the **wet bulb globe temperature (WBGT)** that provides information on the heat load of the environment. It measures temperatures with a dry bulb thermometer, a wet bulb thermometer, and a large, matte, black globe. The dry bulb thermometer measures the ambient temperature; this thermometer is exposed to the air just as any thermometer is used to measure air temperature. The wet bulb thermometer – in which the bulb is surrounded by a wet cloth sock or a sponge – provides a measurement that is the result of evaporative cooling. The dull, dark surface of the black globe effectively absorbs the solar heat and measures the radiant heat load, usually from the sun.

The temperature readings for the WBGT index are combined according to one of two equations. For indoor work situations or for work outdoors where there is shade or cloud cover – and thus no solar heat load – the equation is:

$$WBGT_{in} = 0.7\ T_{wb} + 0.3\ T_g$$

For outdoor work in direct sunlight, the equation is:

$$WBGT_{out} = 0.7\ T_{wb} + 0.2\ T_g + 0.1\ T_{db}$$

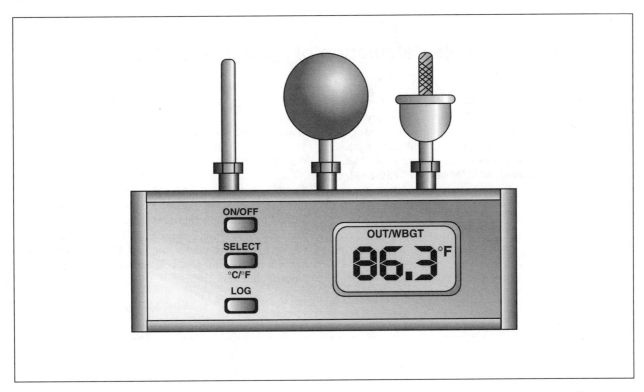

Figure 11-3: This heat stress monitor contains all the elements of a WBGT. It automatically computes the WBGT index value for indoor and/or outdoor environments and provides a readout in °C or °F, depending on the settings selected by the user.

In both equations, the symbols represent the following measurements:

—WBGT is the wet bulb globe temperature index. This number is cross-referenced to a chart in the TLV booklet that provides guidelines for work/rest schedules. It is given in standard units of temperature, and can be figured in degrees Celsius or Fahrenheit.

—T_{wb} is the temperature reading from the thermometer with the wet bulb.

—T_g is the temperature reading from the thermometer with the matte black globe.

—T_{db} is the temperature reading from the thermometer with the dry bulb.

If a device such as the one illustrated in Figure 11-3 is not available, the WBGT can be approximated using a homemade version. The WBGT method incorporates consideration of the level of work being performed and the type of clothing worn by the workers; additional guidance is provided for

Work Rest Regimen	Work Load		
	Light	Moderate	Heavy
Continuous work	30.0 (86)	26.7 (80)	25.0 (77)
75% Work – 25% Rest, each hour	30.6 (87)	28.0 (82)	25.9 (78)
50% Work – 50% Rest, each hour	31.4 (89)	29.4 (85)	27.9 (82)
25% Work – 75% Rest, each hour	32.2 (90)	31.1 (88)	30.0 (86)

As workload increases, the heat stress impact on an unacclimatized worker is exacerbated. For unacclimatized workers performing a moderate level of work, the permissible heat exposure TLV should be reduced by approximately 2.5°C.

Table 11-2: Examples of permissible heat exposure threshold limit values (values are given in °C and °F WBGT). Like other TLVs, these values are guidelines that will prevent heat-related illness among most, but not all, individuals. Acclimatization status, physical condition, PPE use, and other factors should also be considered by the IH.

Box 11-1 ■ Sample Calculation of Approximate WBGT

Tom and Mary are assigned to relocate some empty 10-gallon drums in an outdoor location on a sunny June day. The plant occupational safety professional has obtained the following measurements: Ambient air temperature is 77°F; the wet bulb temperature is 74°F, and the globe temperature is 80°F. What is the WBGT index value?

Because they are working outdoors in the sun, the appropriate equation to use is:

$$WBGT_{out} = 0.7\ T_{wb} + 0.2\ T_g + 0.1\ T_{db}$$

Substituting the temperatures in the equation,

$$WBGT_{out} = 0.7\ (74) + 0.2\ (80) + 0.1\ (77)$$

$$WBGT_{out} = 51.8 + 1.6 + 7.7$$

$$WBGT_{out} = 61.1°F$$

The resulting WBGT index value is 61.1°F. We can conservatively assume that this task is moderately strenuous; Table 11-2 indicates that heat stress would become an issue when the WBGT index is 80°F or higher for performing moderate work. Tom and Mary should be at low risk for developing a heat-related illness while performing their task.

Clothing Type	Clo Value*	WBGT Correction
Summer work uniform	0.6	0
Cotton coveralls	1.0	2
Winter work uniform	1.4	4
Water barrier, permeable	1.2	6

*Clo: Insulation value of clothing. One clo unit = 5.55 kcal/m²/hr of heat exchange by radiation and convection for each °C of temperature difference between the skin and adjusted dry-bulb temperature [the average of the ambient air dry bulb temperature and the mean radiant temperature, $t_{adb} = (t_a + t_r)/2$].

Table 11-3: TLV WBGT Correction factors in °C for clothing. These values are approximations. Additional reduction in the WBGT should be considered when PPE ensembles incorporate multiple layers of impermeable clothing.

workers who are not yet acclimated to the hot environment. Tables 11-2 and 11-3 summarize the WBGT indices and correction factors for clothing.

The use of a WBGT index to help set work/rest regimens should be coupled with monitoring of personnel for signs and symptoms of heat-related illness. Some situations may require more diligent or frequent monitoring of the physical state of the workers. The OSHA/EPA/NIOSH guidance document for hazardous waste work recommends a pulse rate monitoring method. This method is appropriate for situations where workers are performing heavy work in impermeable protective clothing. Workers can be trained to monitor their own pulse rate. The method for pulse monitoring is as follows:

1. Count the radial pulse during a 30-second period as early in the break period as possible.

2. If the rate exceeds 110 beats/minute at the beginning of the break period, shorten the next work period by one third, keeping the rest period the same.

3. If the rate still exceeds 110 beats/minute at the beginning of the next break period, shorten the following work period by an additional third.

The most important measure for preventing heat stress is adequate fluid replacement. Cool (45-50°F) water is the best replacement fluid for personnel working under heat stress conditions. Drinks like Gatorade, Powerade, and Squincher do not contribute excessive amounts of electrolytes and are generally well accepted by workers, especially if diluted with about half again as much water. Use of salt tablets is not recommended. Personal cooling devices – such as ice vests and cooling suits – are generally effective for only a short time, but may be useful for short-term tasks. Vests will add a heat-trapping layer of PPE once the ice melts, and full-body cooling suits will add a similar insulative layer if they should malfunction. Loose-fitting cotton clothing is the best alternative when added skin protection is not required.

Again, workers should be trained to recognize the symptoms of heat-related illness, be alert to monitoring the condition of themselves and their fellow workers, and know what actions they should take to prevent or treat heat-related illness.

Cold Stress

Cold stress is caused by the body's inability to keep its inner temperature within the normal range. Cold stress is more likely to occur during outdoor work in the fall, winter, and spring seasons when damp or wet conditions are common and wind may contribute an additional cooling effect (wind chill). Handling liquids, refueling vehicles, and working with cryogenic materials such as liquid nitrogen and dry ice are other examples of activities that can pose a cold stress hazard.

The primary goals in the case of cold hazards are to prevent deep body core temperature from falling below 36°C (95°F), and to prevent cold injury, such as frostbite, to the body's extremities. Normal body temperature is 98.6°F or 37°C.

The ultimate effect of cold stress is hypothermia, which is a decrease of the deep body core temperature. This can occur in temperatures as high as 30°-40°F, especially among people who have stripped off their insulative clothing while doing strenuous physical work. Predisposing factors for cold injury are similar to those for heat-related illnesses. They include age, alcohol and/or drug use, poor physical condition, and circulatory problems. Pain to the extremities is often the first sign of possible cold injury. Severe shivering should be a sign to workers that immediate warming is necessary. Signs and symptoms of more severe hypothermia include lack of shivering, slowed activity, slurred speech, and clumsiness. Table 11-4 shows what happens as the body temperature falls.

As body temperature decreases below 82.4°F, the victim's physical condition continues to deteriorate. Because of the potential risk to personnel, cold stress protection methods such as warm-up breaks and use of insulative, dry clothing are recommended for workers any time work is performed in air temperatures below 40°F (4°C). In windy conditions, the **wind chill cooling rate** must be considered as well. The wind chill cooling rate is an approximation of heat loss due to the combination of ambient temperatures and the velocity of the wind across the body. The higher the wind speed and the lower the ambient temperature, the higher the wind chill cooling rate. Table 11-5 shows the cooling effect that wind has on exposed skin. Similar effects can be created by ventilation systems or fans near the work area.

Working in cold environments can be made less hazardous by minimizing time spent standing or sitting in one position for an extended time. Manipulation of tools and controls in cold environments can be difficult if the hands become cold, affecting dexterity. Machine controls and tools should be designed to be used without having to remove insulative mittens. Gloves should not be used unless absolutely necessary. If bare hands are required for performing some tasks, a contact warming plate, a space heater, or other warming method should be

Temperature	Physical Reaction
96.8°F (36°C)	Metabolic rate increases to compensate for heat loss
95°F (35°C)	Severe shivering
91.4°F (33°C)	Severe hypothermia below this temperature
89.6°F (32°C)	Clouded consciousness; shivering ceases; decreasing blood pressure
86°F (30°C)	Rigid muscles; pulse, blood pressure, and respiration continue to decrease
82.4°F (28°C)	Life-threatening heart irregularities

Table 11-4: Progressive physical reactions of hypothermia. As the body's core temperature drops, its ability to function normally decreases. Symptoms include shivering, slurred speech, and other motor and behavioral changes.

Estimated Wind Speed (in mph)	Actual Temperature Reading (°F)											
	50	40	30	20	10	0	−10	−20	−30	−40	−50	−60
	Equivalent Chill Temperature (°F)*											
calm	50	40	30	20	10	0	−10	−20	−30	−40	−50	−60
5	48	37	27	16	6	−5	−15	−26	−36	−47	−57	−68
10	40	28	16	4	−9	−24	−33	−46	−58	−70	−83	−95
15	36	22	9	−5	−18	−32	−45	−58	−72	−85	−99	−112
20	32	18	4	−10	−25	−39	−53	−67	−82	−96	−110	−121
25	30	16	0	−15	−29	−44	−59	−74	−88	−104	−118	−133
30	28	13	−2	−18	−33	−48	−63	−79	−94	−109	−125	−140
35	27	11	−4	−20	−35	−51	−67	−82	−98	−113	−129	−145
40	26	10	−6	−21	−37	−53	−69	−85	−100	−116	−132	−148

(Wind speeds greater than 40 mph have little additional effect.)	LITTLE DANGER			INCREASING DANGER		GREAT DANGER
	In < hr with dry skin. Maximum danger of false sense of security.			Danger from freezing of exposed flesh within one minute.		Flesh may freeze within 30 seconds.

* Developed by U.S. Army Research Institute of Environmental Medicine. Natick, MA.

Equivalent chill temperature requiring dry clothing to maintain core body temperature above 36°C (96.8°F) per cold stress TLV.

Table 11-5: This table illustrates the dramatic cooling effect that moving air can contribute to a worker's environment. Persons who work in cold storage warehouses, outdoors, and in other cold environments are at added risk from cold injury if also exposed to moving air.

available to the worker. For personnel handling cryogenic liquids or exposed to other contact hazards, insulative mittens should be provided and worn.

Insulated clothing and warm-up breaks offer effective methods for preventing injury from cold exposure. Break areas should be insulated and warm, and break periods should be long enough to allow adequate warming of workers. For high-risk workers, shorter work periods and additional insulative clothing may be indicated. Workers whose clothing becomes wet should be required to stop and change into dry clothing. If conditions are very windy, wind-resistant outer clothing is advisable.

Fluid intake is as important in the cold environment as it is in heat stress conditions. Warm drinks such as hot chocolate and soups will provide much-needed fluids and calories, while warming the workers; coffee should be avoided as it has a dehydrating effect.

Finally, working in the buddy system in cold environments allows workers to observe each other for early signs of stress and to react safely. Workers should receive training on recognition of signs of cold stress and understand what steps to take to protect themselves and their co-workers.

Checking Your Understanding

1. Describe five different heat-related illnesses.

2. What correction factor should be applied to the WBGT index value for unacclimatized workers?

3. What are the sources of heat that contribute to the overall heat load of a worker planting trees in a park?

4. Name three factors that can affect a person's ability to work in a hot environment.

5. What measures can be taken for preventing the occurrence of heat-related illnesses?

6. Name two environmental conditions that contribute to potential cold stress conditions.

7. Describe five things that can be done to prevent cold-related injuries.

Summary

Ergonomics, also sometimes called human factors, is the study of human characteristics and the use of those characteristics in designing the work environment. The work environment includes many factors, such as spatial relationships and layout of equipment; the type and number of motions required for performing the task; temperature and lighting; and the physical dimensions of the worker. The repeated use of the body to perform tasks or motions that are unnatural or that the body is poorly adapted to perform can result in injury to the tissues and body parts that are used to perform these motions. Such injuries are called repetitive trauma or repetitive use injuries. Aging workers are at higher risk for these types of injuries, but overuse-related injuries can occur among any member of the working population. Well-known repetitive trauma injuries include carpal tunnel syndrome, Raynaud's, syndrome, and trigger finger, as well as strains and sprains, blisters, and soreness of muscles and other soft tissues.

Ergonomists can make use of anthropometric dimensions to help design workstations that accommodate most workers. This is done through incorporation of adjustable features so that each worker can customize aspects of the work to fit their individual needs. Designs that allow adjustments to accommodate the 5th and 95th percentiles of the population will allow most workers to find a comfortable work setting. Adjustability in workstation design is becoming the rule rather than the exception. Other aspects of workstation design should also be considered, such as lengths of reach; weights and dimensions of objects and tools that must be handled; and the presence of sharp corners and edges. Twisting motions should be avoided during lifting and lowering tasks. The use of pneumatic or hydraulic power tools also requires consideration of the torque and vibration forces experienced by the users of these tools. OSHA's proposed ergonomics standard died in the 1996 Congress. Since then, employers have been cited under the General Duty Clause of the OSHAct for ergonomics-related health and safety violations.

Heat-related illnesses can be serious health threats to workers. It is possible for workers to acclimatize or adjust to working in a hot environment over a period of ten days to two weeks. A high rate of performance, or strenuous work, along with the use of impermeable protective clothing can increase the threat of a heat-related illness. The seriousness of heat-related illnesses varies from heat rash or prickly heat to the life-threatening condition of heat stroke. Factors that influence a worker's susceptibility to heat-related illness include age, degree of physical fitness, and the use of alcohol or drugs. Hot environmental conditions and a demanding work schedule are other possible contributors.

The best treatment for heat-related illnesses is prevention. Adequate acclimatization, appropriate rest periods, and fluid intake are keys to preventing heat stress. Monitoring of the work environment using a WBGT is useful for determining when environmental conditions favor potential heat stress situations. Workers can also be trained to monitor themselves for symptoms of heat-related illness. This may be the most effective method for situations where workers must perform heavy work in impermeable and bulky PPE, such as during investigation and cleanup actions at hazardous waste sites.

Cold injuries may occur among persons who work in cold environments, such as outdoors during cold weather, in cold storage warehouses, or among those who handle cryogenic liquids (liquid nitrogen), or materials such as dry ice. The presence of moving air, whether from wind or from refrigeration units, increases the potential for cold injury. Humid air also increases the hazard. Preventive measures for cold injury include the use of insulating clothing, provision of warm break areas, and drinking warm fluids. As is the case with heat stress, caffeinated beverages should be avoided since dehydration is also a concern in cold environments. Workers can be trained to recognize the early symptoms of cold-related injuries and illness so that they can take appropriate measures before the situation becomes serious.

Critical Thinking Questions

1. Given the following temperatures, compute the WBGT index value:

   Wet bulb: 82°F; dry bulb, 85°F; Globe temperature, 88°F.

2. Assuming the WBGT index computed in the previous question is for use during moderate work in non-permeable PPE, what, if any, recommendations would you make?

3. Explain the importance of work/rest schedules for prevention of heat- and cold-related illnesses. What other similarities can you find between heat and cold stress?

4. What is the effect of wind during work in hot conditions? In a cold environment?

5. What is the equivalent chill temperature for conditions where the ambient temperature is 10°F and the wind speed is 20 mph?

6. Explain acclimatization in terms of its importance for working in a hot environment.

7. What measures should be taken if a worker collapses and loses consciousness during work in a hot environment?

8. Prepare a checklist that you could use to evaluate the potential ergonomic hazards of a workstation. Test it by using it; then incorporate the changes that you find are needed to improve it.

12

Selection and Use of Personal Protective Equipment

Chapter Objectives

Upon completing this chapter, the student will be able to:

1. **Use** a systematic approach for assessing hazards and identifying the appropriate items of personal protective equipment (PPE) for worker use.

2. **Determine** what type of training should be provided to workers assigned to use personal protective equipment.

3. **Locate** and be familiar with OSHA regulatory requirements for personal protective equipment.

4. **List** and briefly describe the basic topics that must be addressed in a respiratory protection program.

5. **Differentiate** between air-purifying and air-supplying respirators and their principles of operation; determine when each is appropriate for use to protect workers' respiratory tracts.

6. **Differentiate** between qualitative and quantitative fit-testing, and describe a method for performing a qualitative fit-test.

7. **Explain** how the characteristics of permeability, resistance to degradation, and compatibility of PPE influence their selection and use.

8. **Use** a logic diagram to select the appropriate respirator for use in a given situation.

Chapter Sections

12-1 Introduction

Personal protective equipment (PPE) is considered by occupational safety and health professionals to be the last resort for worker protection. It is the least desirable choice after the more-preferred methods of engineered and administrative controls. Most of us are familiar with safety glasses, hard hats, and safety shoes, which are the more common PPE items in use. Respirators are not as common, but for some situations, such as removal of asbestos, handling potent sensitizers, protecting health care workers from tuberculosis, and cleanup of hazardous waste sites, they may be the only feasible means for protecting workers against inhaling airborne hazardous materials. Hearing protection, gloves, and chemical-protective clothing round out the choices for PPE. In this chapter, we will discuss how to select and use personal protective equipment, with emphasis on the equipment used to protect workers from hazardous physical and chemical agents.

12-2 General Requirements for Personal Protective Equipment

The regulatory requirements for PPE are found in 29 CFR 1910, Subpart I – Personal Protective Equipment. The specific standards in the subpart address general requirements for the use of specific types of PPE: eye and face protection, respiratory protection, head protection, foot protection, electrical protective devices, and hand protection. We will address the use of various types of PPE for protection against hazardous chemical and physical agents in this chapter. Glove use and selection is addressed in Chapter 8.

General requirements for PPE are found in 29 CFR 1910.132. In this section, OSHA sets forth some basics, namely, that PPE be provided, used, and maintained in a sanitary and reliable condition. If employees provide their own equipment, the employer is responsible for assuring that the equipment will provide adequate protection and that it will be properly maintained. This section also requires employers 1) to perform an assessment of the workplace to determine if hazards are present, and 2) to select the appropriate PPE and have employees use it. The hazard assessments must be certified as complete with a statement that identifies the workplace assessed, the date on which the assessment took place, the name of the person who performed the assessment, and a statement that identifies the document as a hazard assessment. Appendix B to the subpart provides non-mandatory compliance guidelines for performing the hazard assessment. The guidelines in the Appendix are offered by OSHA as an example of procedures that meet the requirements of the standard. The hazard assessment is not required for respirator use or electrical protective devices, since there are specific standards that address the issue of hazard assessment as it pertains to those areas.

Additional topics addressed in the general requirements are a prohibition against use of damaged or defective equipment and a requirement for employee training. The training must ensure that workers understand how to wear and use the PPE, and what its limitations are, before being assigned to perform work while wearing the PPE. Retraining is required if there are changes in workplace conditions or PPE that make the previous training obsolete, or if there are indications that the workers have not retained the necessary knowledge or skills. Additional training requirements for the use of respirators and electrical protective devices are contained in the OSHA standards that pertain to those types of PPE.

Hazard Assessment

OSHA's requirement for a hazard assessment is something new to the PPE standards (it appeared in 1994 revisions to the standards), but it is not a new concept for occupational health and safety professionals. Other terms that refer to a similar process are the job safety analysis, job hazard analysis, and activity hazard analysis. We have discussed hazard assessment in somewhat different forms in other chapters. In this section, we will review a step-by-step process to identify hazards that are present – or have the potential to occur – in a given workplace. Supervisors and experienced workers should participate in this process since they usually know the job and its hazards far more intimately than an observing bystander. For new workers the hazard assessment process will be useful, since it will familiarize them with the hazards involved before they begin the job.

The first step in the process is to perform a walk-through survey of the work area. The walk-through is an opportunity to observe the process, tools, and equipment, and identify potential hazards. As part of the hazard identification process, remember to consider the possible routes of exposure or entry for chemical and physical agents: inhalation, skin absorption, and ingestion. Also, consider the target organs or body parts likely to be affected: lungs, eyes, skin, ears, musculo-skeletal structure. Examples of the types of hazards that might be present are:

—Extreme temperatures both in worker environment and materials;

—Solvents, cleaners, corrosive liquids, and other hazardous chemicals;

—Dusts, mists, vapors, or fumes that are, or might become, airborne;

—Sources of nonionizing radiation: welding, brazing, furnace operations, heat treating, etc.;

—Ergonomic hazards: twisting, repetitive motions; manual material handling, lifting; the layout of the workplace; traffic patterns of personnel and vehicles;

—Sources of ignition; potentially flammable or explosive atmospheres; flammable liquids;

—Oxygen-deficient or enriched atmospheres;

—Confined spaces;

—Handling of sharp objects possibly producing hand injuries;

—Splashing or spraying of chemicals;

—Sources of ionizing radiation;

—Noise levels at or above 85 dB(A).

Once the hazards have been identified, the next step is to look at each one and determine which of the hazards present a risk of injury to the worker. A review of the company's injury and illness record – the OSHA 200 log – will provide additional information about the types and locations of incidents that have occurred in the past.

If it is determined that a risk of injury does exist, that risk should be quantified in terms of how severe the potential consequences could be. For most jobs, identifying the hazards and quantifying their risks can and should be a relatively straightforward task. A simple system can be used: ask yourself how serious the injury would be in relative terms: slight, moderate, or severe? Skin abrasions and small cuts might be considered slight; moderate injuries could include strains and sprains, or irritating dust in the eyes; serious injuries might include acute exposure to irritating or sensitizing chemicals, or traumatic, life- or limb-threatening injuries.

Overexposure to toxic materials should be considered on a case-by-case basis as to seriousness of injury. The occupational health professional must consider whether the material poses an acute or chronic hazard, the relative toxicity of the material, the target organs affected, and the possible dose. For example, an acute exposure at twice the OSHA PEL for lead will probably result in elevated amounts of lead in the worker's body. However, if the situation is recognized and corrected there will probably not be any significant long-term health effects, as-

suming the worker has not been repeatedly exposed at this level. On the other hand, an exposure to two times the PEL for ammonia vapors will have severe acute effects, causing the worker to double over or collapse, with potential for lingering effects such as pulmonary edema or development of chemical pneumonia.

A Process for Selecting PPE

Once the hazards have been identified and ranked as to relative seriousness, PPE selection is the next step in the process. For example, respirators might be needed to protect workers from exposure to irritating or toxic vapors. Let's take a look at some basic selection criteria for different types of PPE that might be used to protect workers from hazardous chemical and physical agents.

Eye and Face Protection

Safety glasses as well as eye and face protective equipment must meet the specifications in 29 CFR 1910.133. Side shields are required on safety glasses for any situations where foreign particles could get into the eyes. A face shield alone does not provide any eye protection; goggles or safety glasses should always be worn under a face shield. Goggles provide protection for the eyes against splash hazards and irritating dusts and mists. Face shields protect the face and neck from splash hazards. Special tinted shields or glasses are required for protection against nonionizing radiation, as discussed in Chapter 10. OSHA provides a chart for guidance in selecting eye and face protection; see Table 12-1.

Hearing Protection

Many brands of hearing protectors are available. Ear muffs and foam plug inserts are among the more common types used. Things to consider when selecting hearing protectors are comfort, hygiene, and the degree of protection required. Foam plug insert-type protectors are the most commonly used since they offer a fair degree of comfort and a high level of protection. However, handling these plugs with dirty hands may soil the plugs, especially if removed and reinserted throughout the day. The hearing protectors that are selected should have a noise reduction rating (NRR) that is adequate to reduce

Eye and Face Protection Selection Chart		
Source	Assessment of Hazard	Protection
IMPACT – Chipping, grinding, machining, masonry work, woodworking, sawing, drilling, chiseling, powered fastening, riveting, and sanding.	Flying fragments, objects, large chips, particles sand, dirt, etc.	Spectacles with side protection, goggles, face shields. See notes 1), 3), 5), 6), & 10). For severe exposure, use faceshield.
HEAT – Furnace operations, pouring, casting, hot dipping, and welding.	Hot sparks	Faceshields, goggles, spectacles with side protection. For severe exposure use faceshield. See notes 1), 2), & 3).
	Splash from molten metals	Faceshields worn over goggles. See notes 1), 2), & 3).
	High temperature exposure	Screen face shields, reflective face shields. See notes 1), 2), & 3).
CHEMICALS – Acid and chemicals handling, degreasing plating.	Splash	Goggles, eyecup, and cover types. For severe exposure, use face shield. See notes 3) & 11).
	Irritating mists	Special-purpose goggles.
DUST – Woodworking, buffing, general dusty conditions.	Nuisance dust	Goggles, eyecup, and cover types. See note 8).
LIGHT and/or RADIATION –		
Welding: Electric arc	Optical radiation	Welding helmets or welding shields. Typical shades: 10-14. See notes 9) & 12).
Welding: Gas	Optical radiation	Welding goggles or welding face shield. Typical shades: gas welding 4-8, cutting 3-6, brazing 3-4. See note 9).
Cutting, Torch Brazing, Torch soldering	Optical radiation	Spectacles or welding face-shield. Typical shades, 1.5-3. See notes 3) & 9).
Glare	Poor vision	Spectacles with shaded or special purpose lenses, as suitable. See notes 9), 10).

Notes to Eye and Face Protection Selection Chart:

1) Care should be taken to recognize the possibility of multiple and simultaneous exposure to a variety of hazards. Adequate protection against the highest level of each of the hazards should be provided. Protective devices do not provide unlimited protection.

2) Operations involving heat may also involve light radiation. As required by the standard, protection from both hazards must be provided.

3) Faceshields should only be worn over primary eye protection (spectacles or goggles).

4) As required by the standard, filter lenses must meet the requirement for shade designations in 1910.1333 (a) (5). Tinted and shaded lenses are not filter lenses unless they are marked or identified as such.

5) As required by the standard, persons whose vision requires the use of prescription (Rx) lenses must wear either protective devices fitted with prescription (Rx) lenses or protective devices designed to be worn over regular prescription (Rx) eyewear.

6) Wearers of contact lenses must also wear appropriate eye and face protection devices in a hazardous environment. It should be recognized that dusty and/or chemical environments may represent an additional hazard to contact lens wearers.

7) Caution should be exercised in the use of metal frame protective devices in electrical hazard areas.

8) Atmospheric conditions and the restricted ventilation of the protector can cause lenses to fog. Frequent cleansing may be necessary.

9) Welding helmets or faceshields should be used only over primary eye protection (spectacles or goggles).

10) Non-sideshield spectacles are available for frontal protection only, but are not acceptable eye protection for the sources and operations listed for "impact."

11) Ventilation should be adequate, but well protected from splash entry. Eye and face protection should be designed and used so that it provides both adequate ventilation and protects the wearer from splash entry.

12) Protection from light radiation is directly related to filter lens density. See note 4). Select the darkest shade that allows task performance.

Table 12-1: Eye and face protection selection chart from 1910.133. The OSHA Standard for Eye and Face Protection requires that PPE that is appropriate to the hazard be provided by the employer. This chart can aid in selecting PPE to protect employees against eye and face hazards that are identified in a hazard assessment.

noise exposure to acceptable levels (See Chapter 9 as well as 29 CFR 1910.95).

Chemical Protective Clothing

Protective clothing, including boots and gloves, is available in a variety of materials that offer a range of protection against different hazards. These materials include cotton and synthetic fibers, natural and synthetic rubber, various plastic films and coatings, and leather. In keeping with OSHA regulations for PPE, the type of clothing must be selected to provide adequate protection against the hazards that workers will encounter. The appropriate clothing material will depend on the contaminants present and the task to be accomplished. For example, leather clothing is appropriate for tasks that involve handling sharp objects or wood, but is not suitable for protection against chemical hazards since they may be absorbed by – and permeate – the leather.

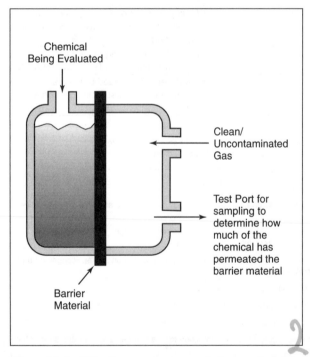

Figure 12-1: This diagram shows a typical test cell that is used for determining a barrier material's permeation rate. Permeation may or may not correspond with a barrier material's resistance to degradation. The permeation rate and the barrier material's resistance to chemical degradation are important characteristics to consider when selecting PPE.

Permeation is the process by which a chemical dissolves in or moves through a barrier material, in this case, protective clothing or gloves. Breakthrough time refers to the amount of time it takes for a chemical to permeate a barrier material. The **permeation rate** is the speed at which the chemical passes through the barrier material; it is reported in units that describe how much chemical passes through a given area of the barrier material per minute, e.g., $mg/cm^2/min$. The permeation rate is usually measured using a standardized testing method, such as ASTM F739. This test can provide information about the barrier material's ability to withstand exposure to the chemical. During the test, a sample of the barrier material is clamped into a test cell so that it divides the cell into two chambers. The chemical being tested is introduced on one side of the barrier material. The gas or air on the other side is monitored to determine when and how much of the chemical has permeated the barrier. This is done by drawing samples of gas from inside the test cell. The samples are analyzed using either a gas chromatograph with a flame ionization detector or a colorimetric method to determine how much of the test chemical is present.

When selecting an item of PPE for use, another important characteristic to consider is the resistance to **degradation**, which is the effect of the chemical on the barrier material. Protective gear such as gloves can become stiff and brittle when exposed to a chemical, or they may swell and soften, which makes them more likely to tear or simply fall apart. A barrier material's ability to resist degradation may or may not be related to permeability. **Penetration** is another characteristic of the clothing that must be considered, that is, the ability of the chemical to leak through seams, zippers, pinholes, and other seemingly invisible openings.

Manufacturers of chemical protective clothing and gloves usually make information regarding their products' permeation, breakthrough, and degradation characteristics available to their customers. Often these are provided in the form of charts or tables that show each product's performance in tests made using different chemicals. The manufacturer may list the actual permeation rate and breakthrough time, or assign relative ratings codes such as excellent, very good, good, and poor. The manufacturer's information should have a key to any codes used in the table; additional information about the testing procedures that were used to evaluate the barrier materials is often included. Each manufacturer of protective clothing and gloves may have proprietary formulations and processes for their products. The gloves and clothing items may have different thicknesses of material or coatings, and the test methods used to evaluate the gear may not be comparable from one manufacturer to another. This means that the permeation and degradation data for one manufacturer does not necessarily apply to other manufacturers; each manufacturer's data must be consulted. Some other issues that must be taken into consideration when selecting PPE include:

— The tests performed for permeability, degradation, and breakthrough are done for one chemical at a time, not for mixtures of chemicals, and not all chemicals are tested by each manufacturer. Most work situations involve more than one chemical; however, since finding a one-material-protects-against-all-chemicals barrier may be impossible, layered PPE or a compromise of materials may provide the best protection.

— The tests on barrier materials are performed at room temperature; the use of protective clothing or gloves for handling heated chemicals will probably influence the way that the barrier material is affected by the chemical. Permeability and degradation will usually be affected negatively. The use of PPE under hot environmental conditions can increase the risk of heat-related illness, and heavy perspiration inside of PPE can alter its permeability and degradation characteristics.

— The conditions under which the PPE will be used must be considered. Loose-fitting clothing and gloves can be caught in moving machinery. Handling abrasive materials or items with sharp edges

will shorten the useful life of gloves and protective clothing made of thin or fragile material. Hot conditions will result in perspiration and subsequent high humidity inside the PPE, possibly changing permeation characteristics or creating additional hazards, such as fogging of respirator facepieces or protective eyewear.

To summarize, the PPE selected must 1) be appropriate for the physical and environmental conditions under which it will be used; 2) it must provide adequate protection against the hazards workers are exposed to; and 3) employees must understand both its capabilities and limitations as well as how to properly use and maintain PPE.

Checking Your Understanding

1. What are the OSHA requirements for training employees to use PPE?

2. Describe, in a series of steps, a process that can be used to help identify and rank hazards.

3. Why must toxic materials be evaluated on a case-by-case basis? What types of issues would such an evaluation address?

4. Explain why eye protection is not provided by a face shield.

5. Define the terms permeation, degradation, and penetration as they pertain to chemical protective clothing.

6. Name and discuss some of the issues that must be considered when selecting PPE for protection against hazardous chemicals.

12-3 Respiratory Protection

The control or elimination of hazardous levels of airborne contaminants should first be attempted by engineered controls such as ventilation or process isolation, or by administrative controls such as changes in work practices. If these steps do not reduce levels of airborne contaminants to acceptable levels, the occupational health professional should consider the use of respirators to protect workers. The following situations may require the use of respiratory protection:

—Engineered and administrative controls are insufficient to reduce levels of airborne contaminants to safe levels;

—Engineered controls are not feasible, or are in the process of being installed;

—An oxygen-deficient atmosphere exists;

—An emergency use, such as firefighting, is required.

Respirators may seem like a good and logical choice for worker protection. However, ask anyone who must wear one while performing the job and you will be told that they are uncomfortable, they're hot, they make it difficult to communicate with others, and they can fog up or otherwise interfere with vision, especially if the user wears corrective lenses. OSHA has issued a standard for the selection and use of respirators, which imposes training, medical surveillance, and other requirements on their use. Add up these issues, and the use of respirators becomes a complicated and expensive prospect.

The OSHA standard for respiratory protection is found at 29 CFR 1910.134. The updated version is effective starting in 1998. The update more closely reflects current knowledge and state-of-the-art practices in respiratory protection and sets forth minimum requirements for establishment of respiratory protection programs. A respiratory protection program should address the following areas:

1. Provisions for selecting the appropriate respirators for use;

2. Medical evaluations for those employees who are required to use respirators;

3. Fit-testing procedures;

4. Procedures for the proper use of respirators in routine and emergency conditions;

5. Procedures and schedules for cleaning, disinfecting, repairing, storing, discarding and otherwise maintaining respirators;

6. Procedures to ensure adequate air quality, quantity, and flow for atmosphere (air)-supplying respirators;

7. Training of employees in the respiratory hazards to which they are potentially exposed during normal and emergency situations;

8. Training of employees in the proper use of respirators, including donning, doffing, limitations on use, and proper maintenance;

9. Procedures for regularly evaluating the effectiveness of the program.

For each program element above, the OSHA standard contains specific minimum requirements that must be met for compliance. Where respirators are required, the employer must provide the respirator, the medical evaluations, and training at no cost to the employee. A qualified person should be appointed as program administrator; often this will be an industrial hygiene professional. OSHA requires that the program administrator be someone who is "qualified by training or experience commensurate with the complexity of the respirator program."

The OSHA standard allows for reduced program requirements in situations where employees wish to wear respirators despite there being no hazards requiring respirator use. Such situations require that the respirators not pose a hazard in themselves; that employees be medically able to use them; and that respirators be maintained (cleaned, used, stored, and repaired) so as not to present a health hazard to the user. The voluntary use provision does not apply to filtering facepieces, otherwise called dust masks. These are the usually white paper-like devices sold in hardware and home improvement stores (see Figure 12-2).

Respiratory protective equipment is tested and certified by NIOSH. Until 1995, the testing and approval of respiratory protection was a joint responsibility of NIOSH and the Mine Safety and Health Administration (MSHA). Because this is a

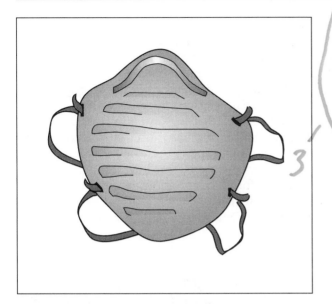

Figure 12-2: Dust masks offer limited protection to the respiratory system. OSHA does not regulate the use of dust masks under the respirator standard.

recent change, there are still many respirators and cartridges in use that bear the old NIOSH/MSHA approvals. These respirators and cartridges may still be used, but new equipment will be issued with the NIOSH approvals. The entire respirator assembly, including the facepiece, cartridges/filters, hoses, and other parts, are included in the testing process. A tested and certified respirator assembly is then issued a number, prefixed with the letters "TC." Many respirators can be assembled in different parts configurations; however, each configuration must be tested, and the approval is good only for the parts assembly as it was tested. Substituting parts from other manufacturers – or even the same one – invalidates the approval if the parts do not belong to an approved assembly. This invalidation applies even if the parts fit together and seem to work. The approvals are listed in the printed information sheet or booklet that comes with the respirator, and may also be on the box label. All corresponding part numbers that were tested will be listed for each approval. An example of an approval certificate is shown in Appendix 7 at the end of this book.

There are three major classes/categories of respirators: 1) **air-purifying** respirators, 2) **air-supplying** respirators; and 3) **self-contained breathing apparatus**, or **SCBAs**, also called **Type C** respirators. Air-purifying respirators, which remove contaminants from the ambient air by means of filtration

or absorption, will be discussed in this section along with some general procedures for all respirators. Air-supplying respirators, also called atmosphere-supplying respirators, provide breathing air from outside of the contaminated work area. SCBAs also provide breathing air, but from a tank that workers carry with them. Air-supplying respirators and SCBAs will be discussed in the next section.

Air-purifying Respirators

Air-purifying respirators generally consist of a tight-fitting facepiece and an air-purifying device. The latter is either a removable component (a cartridge or filter) that snaps or turns onto the facepiece or it is mounted on a belt or body harness and connected to the facepiece by a corrugated breathing hose. **Full-face respirators** cover the entire face, from the hairline to under the chin. **Half-face**, or half-mask respirators, cover half of the face, roughly from under the chin to the bridge of the nose. Because air is drawn through the cartridge or filter by negative pressure that is created inside the respirator facepiece when the user inhales, the term **negative-pressure respirator** is sometimes used to refer to tight-fitting air-purifying respirators.

Powered air purifying respirators (PAPR) are a variation on air-purifying respirators. A PAPR utilizes a battery-powered blower that draws the contaminated air through the cartridge or filter. The cleaned air is then forced through a hose to the facepiece, which may be tight fitting, but may also be a helmet or hood that does not seal tightly against the face of the wearer. PAPRs supply purified air at a positive pressure, which means that if a leak occurs in the facepiece, helmet, or hood, air should move outward. These respirators must deliver at least four cubic feet of air per minute (cfm) to a tight-fitting facepiece and at least six cfm to a loose-fitting helmet or hood. The batteries are designed to operate for eight hours, but it is best to change battery packs every four hours to maintain good airflow. It is possible to **overbreathe** a PAPR: the worker's breathing rate during heavy work may be so high that even with a full-powered battery, the blower cannot supply enough air to keep the facepiece under positive pressure. Under these conditions, it is possible for the worker's breathing to induce a negative pressure in the facepiece, drawing contaminants in. Because of this, tight-fitting PAPRs are assumed to provide the same level of protection as a negative-pressure air-purifying respirator. For the same reason, loose-fitting (hood or helmet-type) PAPRs are prohibited for use in certain applications. Figure 12-3 shows some representative full- and half-face air-purifying respirators.

Because there are limits to the amount of contamination they can remove, air-purifying respirators cannot be used in IDLH (immediately dangerous to life and health) environments. In addition, since they do not supply oxygen, air-purifying respirators are not appropriate for use in oxygen-deficient atmospheres and may be used only when the ambient atmosphere contains at least 19.5 percent

oxygen. Other restrictions and limitations of air-purifying respirators are that they may be prohibited for use as protection against certain substances such as carcinogens and sensitizers.

Air-purifying respirators are not appropriate for use in protection against materials with poor **warning properties**, that is, substances that cannot be detected by the respirator user at safe concentrations. Warning properties rely on the worker's ability to detect them – by smelling, tasting, or feeling them – before concentrations inside the respirator reach unsafe levels. The detection of contaminants inside the facepiece of a respirator is called break-

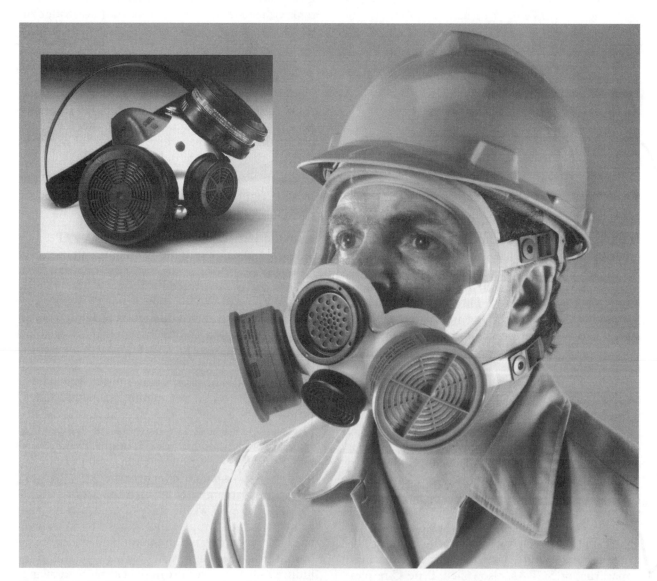

Figure 12-3: Half-face respirators cover less skin and are lighter than full-face respirators. They are more comfortable in some situations, but may be more difficult in terms of fitting well around a worker's nose, chin, and cheeks. Full-face respirators provide a higher level of protection and a better fit; they have the added benefit of built-in eye protection.

through. Because they are somewhat subjective and vary from individual to individual, warning properties are not foolproof. In addition, they are useful only to provide some indication to the worker that the respirator has shifted or has otherwise broken the seal to the face, the cartridges are overloaded, or some other respirator malfunction has occurred. Warning properties should not be relied upon as an indication that the atmosphere outside the respirator is safe. There are some exceptions to this rule; asbestos, for example, is a carcinogen with no warning properties but for which the use of air-purifying respirators is allowed, up to certain concentrations. Radioactive particulates are another potential carcinogen for which air-purifying respirators are adequate, again, up to certain concentrations.

Another qualification for use of air-purifying respirators is that the type and concentration of the air contaminants present in the work area be known and characterized. This requires that some air sampling measurements be performed to determine the maximum and average concentrations that are expected to occur in the workplace. This will assure that the respirator that is selected will provide workers with adequate protection.

Restrictions and limitations on the use of a particular respirator will be found in the NIOSH approvals for the respirator in question. If the respirator, filter or cartridge, and other assembly items are not listed with a NIOSH approval for use as protection against the known or anticipated concentrations of contaminants present, then the respirator may not be used for that situation.

To summarize, conditions that are not suitable for use of air-purifying respirators include:

— Oxygen deficiency;

— IDLH conditions;

— Entry into an unventilated or confined area where the exposure conditions have not been characterized;

— Presence/potential presence of unknown levels of unidentified contaminants;

— Contaminant concentrations exceeding the respirator cartridge's maximum use concentrations;

— Presence of gases or vapors with poor warning properties;

— High relative humidity, rain.

As with other parts of the respirator, the use of any filter or chemical cartridge must be in accordance with the NIOSH approval for the cartridge. The contaminants for which the cartridge or filter is approved will be listed on the label affixed to the cartridge/filter. OSHA requires that – where the technology exists – gas and vapor canisters/cartridges used with air-purifying respirators be equipped with a NIOSH-certified **end-of-service-life indicator (ESLI)**. ESLI is a system – such as an area that changes color – that can alert the respirator user to the fact that the cartridge or canister is approaching the end of its useful life. If the cartridge or canister does not have an ESLI, the respirator program must describe the schedule that will be followed for changing out cartridges/canisters. Car-

Contaminant	Assigned Color
Acid	White
Hydrocyanic acid gas	White with 1/2 inch green stripe
Chlorine gas	White with 1/2 inch yellow stripe
Organic vapors	Black
Ammonia gas	Green
Acid gases and Ammonia gas	Green with 1/2 inch white stripe
Carbon monoxide	Blue
Acid gases and organic vapors	Yellow
Hydrocyanic acid gas and chloropicrin vapor	Yellow with 1/2-inch blue stripe
Acid gases, organic vapors, and ammonia gas	Brown
Radioactive materials, excepting tritium and noble gases	Purple (Magenta)
Particulates – dust, fumes, mists, fogs, or smokes – in combination with any of the above gases/vapors	Canister color for contaminant, as described above, with 1/2 inch gray stripe

Table 12-2: Designated colors for respirator cartridges (OSHA).

tridges are currently assigned a standardized color for each contaminant for which they provide protection (see Table 12-2).

All filters or chemical cartridges have a finite capacity. A filter cartridge is only effective until it is too clogged to breathe through due to loading with particulate. In the case of chemical cartridges, there is only so much contaminant that can be adsorbed or chemically bound to the charcoal or other adsorbent inside the cartridge. The highest concentration that a filter or cartridge has been approved for use in by NIOSH is called the **maximum use con-** centration (MUC). The MUC can be approximated by multiplying the PEL for the contaminant of concern by the **assigned protection factor (APF)** for the respirator being used. The APF is the minimum level of protection that can be expected from a respirator that is properly fitted, worn, and functioning. The APF, like the MUC, is a numerical value assigned by NIOSH. A half-face air-purifying respirator with the appropriate cartridge (chemical or filter) has an APF of 10, meaning it will reduce the amount of contaminant that is present in the outside atmosphere by at least 10 times. Tables 12-3

Respirator	Protection Factor
Single-use or quarter mask respirator	5
Half-mask air-purifying respirator	10
Supplied-air, half-mask, operated in demand mode	10
Power air-purifying, hood or helmet, equipped with particulate filter	25
Supplied-air, helmet or hood, operated in continuous flow mode	25
Full-face, air-purifying equipped with particulate filter	50
Full-face, air-purifying, equipped with HEPA filter	50
Powered air-purifying tight-fitting facepiece with HEPA filter	50
Supplied-air, full facepiece, operated in demand or continuous flow mode	50
Self-contained breathing apparatus (SCBA), full-face, operated in demand mode	50
Supplied-air, half-mask, operated in pressure demand mode	1,000
Supplied-air, full facepiece, operated in pressure demand mode	2,000
Self-contained breathing apparatus, operated in pressure demand mode	10,000
Supplied-air respirator, full facepiece operated in pressure demand equipped with an auxiliary emergency escape bottle	10,000

Table 12-3: NIOSH-assigned respirator protection factors when properly fitted for use and maintained for protection against particulate exposure.

Respirator	Protection Factor
Half-mask, air-purifying, with appropriate cartridges	10
Supplied-air, half-mask, operated in demand mode	10
Powered air-purifying, loose-fitting hood or helmet	25
Supplied-air, continuous flow, equipped with hood or helmet	25
Air-purifying, full facepiece, with appropriate cartridge	50
Powered air-purifying, tight-fitting facepiece, with appropriate cartridges	50
Supplied-air, full facepiece, operated in demand mode	50
Supplied-air, tight-fitting facepiece, operated in continuous flow mode	50
Self-contained breathing apparatus operated in demand mode	50
Supplied-air, half-mask, operated in pressure demand mode	1,000
Supplied-air, full facepiece, operated in pressure demand mode	2,000
Self-contained breathing apparatus, pressure demand mode	10,000
Supplied-air respirator, full facepiece, operated in pressure demand mode with an auxiliary emergency escape bottle	10,000

Table 12-4: NIOSH-assigned respirator protection factors when properly fitted for use and maintained for protection against gas/vapor exposures.

and 12-4 list the NIOSH-assigned protection factors for various types of respirators.

It is not permitted to use a respirator and cartridge in an atmosphere that is above the maximum use concentration for the cartridge. If the filter is overloaded, the worker will have difficulty breathing. If the contaminants are gases or vapors, the cartridge can quickly become saturated, allowing the contaminants to be inhaled by the user.

The maximum use concentration will vary, depending on the contaminant and the type of chemical cartridge used. NIOSH has assigned a maximum use concentration of 1,000 ppm to organic vapor cartridges. For acid gas cartridges, NIOSH has assigned a different maximum use concentration, depending on the gas: 10 ppm for chlorine, 50 ppm for hydrogen chloride, and 50 ppm for sulfur dioxide. In some cases, OSHA has assigned specific respiratory protection for a particular hazardous agent that is different from the NIOSH MUC. The OSHA substance-specific regulation should be consulted to determine the required protection in these cases. Some of the materials for which there are specific OSHA regulations are: formaldehyde, vinyl chloride, inorganic arsenic, lead, benzene, and asbestos.

Fit-testing

OSHA requires that all respirator users have a **fit-test** prior to use. This is a method for evaluating how well the respirator seals against the wearer's face. OSHA allows for one of two types of fit-tests; a **qualitative fit-test** or a **quantitative fit-test**. A qualitative fit-test is one that relies on the wearer's response to a test agent. The more common test agents used in qualitative fit-tests are isoamyl acetate, which smells like bananas, and irritant smoke. The respirator wearer may be asked to perform some physical activities and read aloud a special paragraph called the rainbow passage. The idea is to simulate actual conditions under which the respirator will be worn. If the test agent cannot be detected by the wearer, then that particular brand and size of respirator fits the wearer. If another brand or size of respirator is worn, it must first be fit-tested.

A quantitative fit-test has the same objective as a qualitative fit-test, only for this test an instrument is used that can measure the concentration of the test agent both inside and outside of the respirator. Facepieces used for these types of tests must be modified to accept a small probe for attachment to the instrument. The wearer performs a series of movements (jogging in place, toe touches, turning head from side to side) and reads the rainbow passage during the test. The result of the test is a number called the **fit factor**, which represents how well the respirator seals against the wearer's face. Fit factors can vary dramatically from person to person for the same respirator style and size, since fit is an individual thing. Fit factors may also vary from year to year for the same individual. Generally, a fit factor of at least 100 indicates an acceptable fit, but values in the tens of thousands are not unheard of. An individually determined fit factor will probably not correlate very well with a respirator's assigned protection factor, which is usually much lower.

The rainbow passage is used during respirator fit-testing to simulate variations in facial distortion that might occur while workers perform their tasks and talk to one another. Reading it aloud is a challenge for half-face respirator users being tested with irritant smoke, due to the eye irritation that occurs.

It reads:

"When the sunlight strikes raindrops in the air, they act like a prism and form a rainbow. The rainbow is a division of white light into many beautiful colors. These take the shape of a long round arch, with its path high above and its two ends apparently beyond the horizon. There is, according to legend, a boiling pot of gold at one end. People look, but no one ever finds it. When a man looks for something beyond reach, his friends say he is looking for the pot of gold at the end of the rainbow."

NIOSH recommends – and OSHA requires – that regardless of the pressure of the facepiece under actual use, all respirators must be fit-tested in the negative pressure mode. The fit-testing protocols currently accepted by OSHA are found in Appendix A to 29 CFR 1910.134. A record must be made of the fit-test, containing the name of the employee; the make, model, style, and size of respirator; the type of fit-test performed; the date of the test; and the pass/fail results for qualitative fit-tests, or the fit factor strip charts for quantitative fit-tests.

During any type of fit-testing, the respirator straps must be properly located and as comfortable as possible. Overtightening the straps will sometimes reduce facepiece leakage, but the wearer may be unable to tolerate the respirator during the work period. The facepiece should not cause major discomfort. Although respirator fit-testing is required, users should perform a **seal check** each time they put on the respirator. A seal check may be better known to the reader as a negative/positive pressure fit check. (OSHA uses the term seal check in the respiratory protection standard.) The negative seal check is performed by covering the inlet side of the cartridges and inhaling; the facepiece should collapse slightly and hold a negative pressure for at least 10-15 seconds. The positive seal check is performed similarly, but this time covering the exhalation valve and exhaling slightly. Note that on most respirators, the exhalation valve cover will have to be removed to perform this test. The cover must be replaced after the test. The facepiece should expand and hold a positive pressure without obvious outward leaks. These seal checks should not be confused with the required fit-test.

Fit-tests should not take place if the user's facial hair – sideburns, beard, and moustache – interferes with the seal of the respirator. Changes in hairstyle and/or trimming of beard and moustache may be necessary. Fit-tests must be repeated on an annual basis, or sooner if conditions warrant. If employees experience any changes in physical appearance – such as facial scarring, cosmetic surgery, dental changes, or a change in body weight – they should be fit-tested again. If it is determined that the respirator no longer fits, they must be given the opportunity to select a different respirator and be tested again.

OSHA requires that employees be provided with clean and sanitary respirators. Manufacturer's instructions should be followed for cleaning and sanitizing respirators; the use of solvents, such as alcohol, may not be recommended. Commercial cleaning and sanitizing agents are available for purchase in a variety of solutions and concentrates. Many of the commercially available sanitizers contain quaternary ammonium salts or alcohol, which do not effectively kill all harmful bacteria and viruses. An alternative is to wash the respirator in mild detergent followed by a disinfecting rinse. Reliable, effective disinfectants may be made from readily available household solutions. For example, a hypochlorite solution (50 ppm of chlorine) can be made by adding approximately two ml of hypochlorite (laundry) bleach to one liter of water; a two-minute immersion disinfects the respirator. Or, an aqueous solution of io-

dine (50 ppm of iodine) can be made by adding approximately 0.8 ml tincture of iodine per liter of water. Again, a two-minute immersion is sufficient.

After cleaning and disinfecting, the respirator should be placed on a flat surface and allowed to air dry. The cleaning process provides an opportune time to inspect the respirator and replace any parts that are found defective. Remember that the replacement parts must be from the same manufacturer and made to fit the respirator on which they are used in order to maintain the NIOSH approval.

The methods and procedures for cleaning respirators will vary depending on the size and complexity of the program. Some employers provide the necessary parts, supplies, and equipment but assign responsibility for respirator cleaning and maintenance to employees. Others allow respirators to be traded in for cleaned, inspected, and sanitized respirators with new cartridges/filters. Whatever the method, it must be described in the written respirator program and periodic field checks performed to verify its effectiveness.

The protection factors listed in Tables 12-4 and 12-5 above were compiled by NIOSH. Some state OSHA programs have adopted protection factors derived from ANSI studies that are based on the fit-testing method used. Make sure you are familiar with state, local, or substance-specific regulations that may differ from the NIOSH listed protection factors. If conflicting values are found, always use the protection factor that provides the highest level of protection for the respirator wearer.

Checking Your Understanding

1. Explain the principle of operation of a tight-fitting air-purifying respirator.

2. List at least five conditions where air-purifying respirators would not be appropriate for use.

3. Name three things that will invalidate the NIOSH approval for a respirator.

4. Differentiate between a qualitative and a quantitative fit-test. How does a user seal check relate to respirator fit?

5. What is a fit factor? An assigned protection factor? A maximum use concentration?

6. What is an ESLI?

12-4 Air-supplying Respirators

Air-supplying respirators – also called atmosphere-supplying, supplied-air, and Type C respirators – provide workers with clean breathing air from outside the work area. The air is carried from the clean source through a supply hose that is connected to the worker's facepiece (half- or full-face). Other parts of the system include a compressor, storage tank, air cleaning apparatus, and a reserve air supply. Grade D is the minimum quality of air that can be used in breathing air systems; the specifications for Grade D air are listed below:

Carbon Monoxide (CO)	20 ppm maximum
Carbon Dioxide (CO_2)	1,000 ppm maximum
Condensed Hydrocarbons	5 mg/m³ maximum
Objectionable Odors	None should be detectable
Water vapor	66 ppm minimum (some is needed for CO scrubbers to work)

Compressors used for Type C systems should have a high temperature alarm, a carbon monoxide alarm, appropriate pressure gauges, and safety valves. In addition, a written procedure should be in place that describes how the system is run and maintained.

Perhaps the greatest concern when dealing with Type C systems is carbon monoxide. This odorless, tasteless chemical asphyxiant can be produced by the compressor if it overheats. The overheating causes the lubricating oil to break down releasing carbon monoxide. OSHA requires that oil-lubricated compressors used for breathing air have a high temperature shutoff or carbon monoxide alarm, or both. If a carbon monoxide alarm is used, it must be calibrated at least once a month. If only a high temperature shutoff is used, OSHA requires that the air from the compressor be monitored at intervals sufficient to prevent carbon monoxide from exceeding 10 ppm. The possibility also exists for drawing carbon monoxide from outside into the compressor. To avoid this, the air intake should be positioned some distance away from the compressor itself, and be away from any combustion sources, such as vehicle exhausts, smokestacks, and generators. Sources of other airborne materials that might be drawn into

the system should be identified and appropriate steps taken to prevent these materials from entering the supply. Other contaminants of concern in a Type C system are excess heat and water vapor.

Once the air has been compressed and purified, and an adequate reserve is available for emergencies, it is ready for delivery to the workers. For large-scale activities, or where the project is at a remote location, air lines from a compressor outside the work area are used to feed manifolds inside the work area. In an industrial setting, the worker may connect the air line to a source adjacent to the work area prior to entry. The air supply line from the distribution manifold to a worker cannot exceed 300 feet, the maximum distance approved by NIOSH. The connectors used on the system must be incompatible with other connectors used in the plant. This makes it physically impossible to connect an air supply line to oxygen, welding gases, or any other gas lines in the vicinity other than the approved breathing air supply.

The volume of air needed will depend on many factors, including the type of respirator, number of workers, and auxiliary equipment. As is true of the PAPR system, tight-fitting masks must be supplied with a minimum of four cfm, and hood-type respirators must be provided with at least six cfm. For each of these types of masks, the maximum recommended flow rate is 15 cfm. The use of air cooling devices will require additional airflow according to the specifications of the unit chosen.

Air-supplying respirators can be operated in one of several modes. In **demand mode**, the wearer's inhalation creates a negative pressure inside the facepiece, which causes the regulator to release air into it. The negative pressure condition increases the likelihood of contaminants entering the facepiece. These respirators are not recommended for use and have been largely replaced by the pressure-demand type. In **pressure-demand mode**, the facepiece is maintained under a slight positive pressure at all times, providing a very high level of protection for the wearer. The **continuous flow mode** is one where there is a regulated amount of air supplied to the facepiece at all times. Because of the potential for overbreathing, these have limited applications in welding and grinding.

Air line respirators chosen to protect workers must be approved by NIOSH. As with air-purifying respirators, each is approved as a unit, including facepiece, regulator, and the air line. This means that even if the air line from one manufacturer will work with a respirator made by a different manufacturer, it may not be used unless it has been approved by NIOSH in that particular assembly. Any alteration of the respirator or its subassemblies voids the approval. Alteration means doing anything to the respirator beyond maintenance prescribed by the manufacturer. Replacement parts must be supplied by a manufacturer for their respirators only.

Self-contained Breathing Apparatus (SCBA)

The last major category of respirators is that of the self-contained breathing apparatus. SCBAs are a specialized type of atmosphere-supplying respirator. The SCBA consists of a facepiece, connected by a regulator and hose to a cylinder of compressed Grade D air carried in a harness on the wearer's back.

Figure 12-4: The self-contained breathing apparatus, or SCBA, resembles the more familiar SCUBA gear worn by divers. The primary difference is that the SCBA is designed for use on land.

SCBAs operated in pressure-demand mode offer a high level of protection against airborne contaminants (see Tables 12-3 and 12-4), including IDLH and oxygen-deficient atmospheres. They are recommended for use by firefighters and rescue workers who must enter toxic or oxygen-deficient atmospheres. SCBAs are normally operated as open-circuit systems, where the exhaled air is released into the surrounding atmosphere. The cylinder of air will last anywhere from 30 to 60 minutes; this varies widely depending upon level of activity of the worker, air pressure, and the tank size.

A closed-circuit SCBA uses chemical scrubbers to remove excess CO_2 and water vapor from the wearer's exhaled breath and has a supply of stored oxygen which is added to the air. The air is circulated in the system and breathed again and again rather than being released to the outside as in the open-circuit SCBA. Closed-circuit SCBAs can operate for as long as four hours on a single tank, depending on the model and use conditions.

SCBAs also have wide use as emergency escape respirators. Those that are specially made for escape use generally have a smaller cylinder, designed for up to 20 minutes of use. Emergency escape SCBAs are not approved for use in rescue or emergency entry situations.

Our discussion of SCBA is not complete without mentioning the combination SCBA/air-supplying respirator. These units are primarily operated as an air-supplying respirator, with a backup SCBA, which can be switched to being the primary source of air in an emergency. These types of respirators can be used in an IDLH or oxygen-deficient atmosphere since they have the escape capability. The primary advantage of these types of respirators is that they allow the worker to spend an extended period of time in the area. As with the other air-supplying respirators, operation in the pressure-demand mode is recommended.

Selecting a Respirator

The effectiveness of a respirator depends as much on the selection process as it does on the correct use of the respirator. Without an effective respiratory protection program, workers are not likely to receive adequate protection from the respirator. Training, medical evaluations, fit-testing, proper maintenance, and awareness of the limitations of the

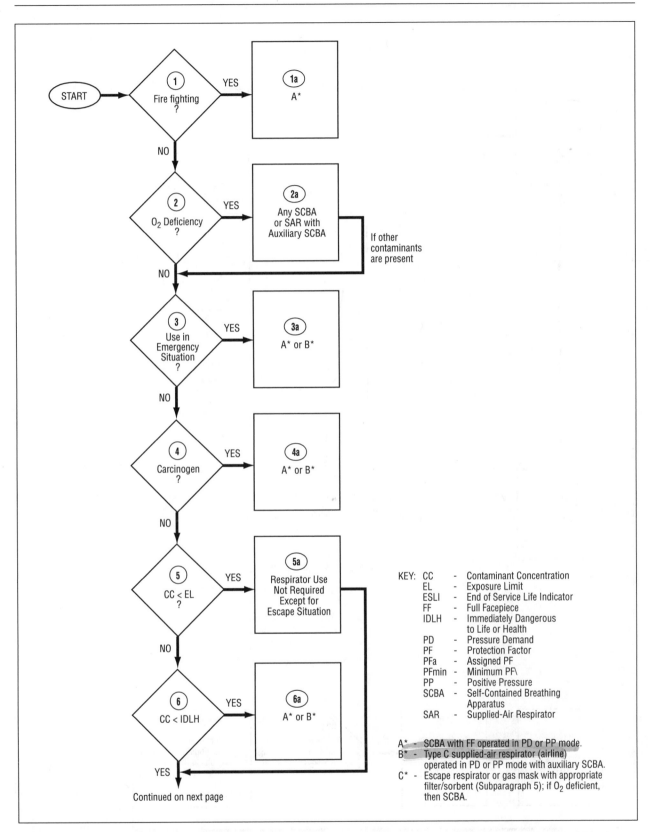

Figure 12-5: The NIOSH respirator decision logic provides guidance for selecting a respirator. The user of the flow diagram must have some knowledge about the anticipated conditions under which the respirator will be used, including the type and concentration of contaminants.

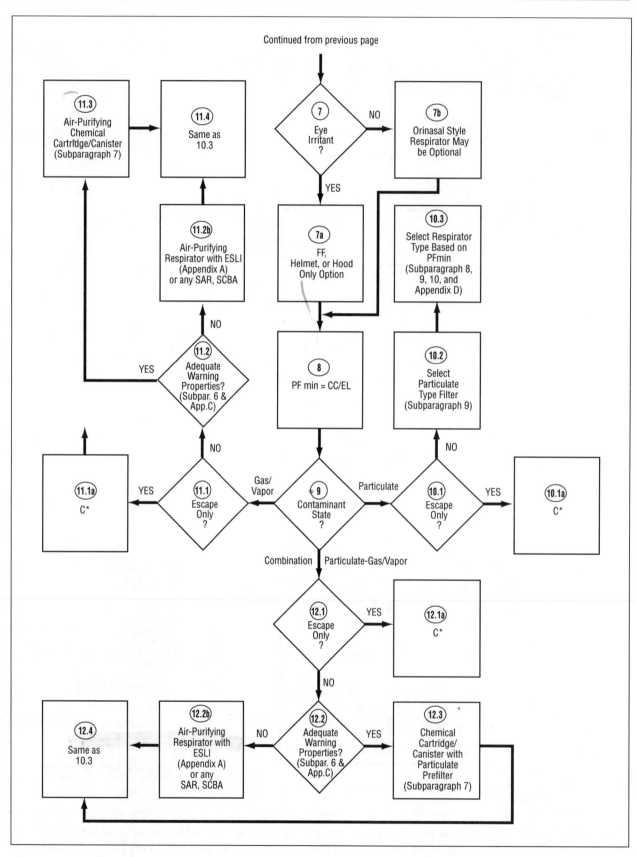

Figure 12-5: Continued.

respirator are essential. If a written program that meets the minimum requirements set forth in the OSHA respirator standard 29 CFR 1910.134 is implemented, it will increase the level of protection that workers realize in the use of respirators.

The selection of the appropriate respirator requires some research and preparation. The occupational health professional must first determine the following information:

1. Is there a written program that covers the use of respirators? If not, one must be prepared and followed.

2. What are the anticipated conditions under which the respirators will be used?

3. What contaminants will be present in the work area?

4. What are the chemical, physical, and toxicological properties of the contaminants?

5. Does the contaminant possess warning properties at levels well below the allowable exposure limit (OSHA PEL, NIOSH REL, or ACGIH TLV)?

6. What is the IDLH concentration of the contaminants? Is there danger of approaching this concentration?

7. Is eye irritation a potential effect of the contaminants?

8. Are cartridges/canisters with ESLIs available (for gas and vapor contaminants)? If not, is any information available about the expected service life of the cartridges/canisters?

Information about the anticipated conditions of use is very important. The job or task description should include details such as the duration of the job; the work location; the physical demands of the work; the types of processes present in the area, other PPE that must be used, and how often the task will be repeated. Information about the environmental conditions – temperature and humidity – can impact the service life of cartridges and canisters; high humidity, for example, has been linked with decreased service life of sorbent-type devices. If more than one contaminant is present, the selection of the appropriate cartridge/canister must include protection against each one at the concentrations that are expected. Numerous combination cartridges and canisters are available that provide protection for several contaminants.

Workers assigned to use respirators must be qualified in accordance with the written program and the OSHA standard. Training, fit-testing, and medical approval for respirator use are required. Respirator use is prohibited in the presence of facial hair that interferes with the respirator seal; corrective lenses must not interfere with the respirator seal.

NIOSH's Respirator Decision Logic contains a flow chart (Figure 12-5) that assists in the selection of the appropriate class of respirator. It walks the user through a series of yes/no questions to determine the respirator that will provide an adequate level of protection. Before using the flow diagram, answers should be obtained for the questions listed at the beginning of this section. For additional information on using the flow diagram, refer to the NIOSH Respirator Decision Logic document.

Checking Your Understanding

1. List all the names used to refer to air-supplying type respirators.

2. What is the minimum quality of air for breathing air systems, and what is its composition?

3. Describe some methods for preventing CO from entering a Type C system.

4. Explain what precaution is required so that breathing air lines are not mistakenly connected to another gas.

5. Name and describe the three modes for operation of supplied-air respirators.

6. What type of respirator is permissible for working in an IDLH atmosphere? For rescue entry into an IDLH atmosphere?

7. What information must be known in order to select a respirator that will provide the required level of protection for the workers?

Summary

The use of PPE is acceptable only if other engineered and administrative controls are not feasible, or while they are being implemented. Several individual components of clothing and equipment may be needed to build an ensemble of PPE to protect workers from all hazards identified in the work area. A hazard assessment process should be used to provide an organized approach to identifying real or potential hazards. Consideration must be given to the potential for worker harm or injury. Toxic materials exposures should be examined on a case-by-case basis in order to properly evaluate their acute and chronic effects. Examples of PPE used to protect workers against hazardous chemical and physical agents include eye and face protection, hearing protection, chemical protective clothing, gloves, and respirators.

Chemical protective clothing and gloves are available in a variety of materials and fabrics, each of which has appropriate uses based on the ability of the barrier material to withstand exposure to the conditions under which it will be used. Permeability, degradation, and penetration are characteristics of the barrier material that must be included for consideration, along with the demands of the task to be performed and the environmental conditions under which it will be used.

The use of respirators requires a written program describing how the respirators will be selected, used, and maintained in routine and emergency conditions. It must also address worker training, medical evaluations, and fit-testing, as well as provisions for periodic evaluations to assure the program's effectiveness. NIOSH tests and certifies respirators, assigns protection factors, and issues approvals for respirators and related equipment. There are three primary types of respirators: air-purifying, air-supplying, and self-contained. Each type is appropriate for specific applications, depending on the contaminants present and the limitations of the respirator. Air-purifying respirators are not appropriate for use in IDLH or oxygen-deficient atmospheres. Air-supplying and SCBA respirators have uses in rescue, emergency escape, and firefighting; combination air supplying/SCBA can be used for working in IDLH atmospheres, if used according to NIOSH approvals.

Selection of the respirator necessary to provide adequate worker protection requires a thorough knowledge of the anticipated conditions of use. Respirators must be used in accordance with the policies and procedures contained in the organization's written respirator program and OSHA 29 CFR 1910.134.

Critical Thinking Questions

1. Describe the steps you would take in determining the appropriate PPE for use by workers assigned to manually clean parts with a solvent.

2. Explain why selecting a single type of gloves for use as protection against a three-solvent solution might be difficult.

3. Compare the three major classes of respirators in terms of their uses and levels of protection.

4. What are the minimum content requirements for a written respiratory protection program according to OSHA?

5. NIOSH tests and certifies respirators. Explain how this affects use, maintenance, and repair of respirators.

6. Using the NIOSH decision flow sequence, determine the type of respirator for the following conditions: (a) handling a potent carcinogen in a laboratory setting; (b) working in an environment where an odorless vapor contaminant is present in a concentration slightly above the PEL; (c) same as (b), only the contaminant's odor is detectable at concentrations well below the PEL.

PRESERVING
THE LEGACY

Appendix 1

Part 1910 – Occupational Safety and Health Standards

Subpart A—General

Sec.
1910.1   Purpose and scope.
1910.2   Definitions.
1910.3   Petitions for the issuance, amendment, or repeal of a standard.
1910.4   Amendments to this part.
1910.5   Applicability of standards.
1910.6   Incorporation by reference.
1910.7   Definition and requirements for a nationally recognized testing laboratory.

Subpart B—Adoption and Extension of Established Federal Standards

Sec.
1910.11   Scope and purpose.
1910.12   Construction work.
1910.13   Ship repairing.
1910.14   Shipbuilding.
1910.15   Shipbreaking.
1910.16   Longshoring and marine terminals.
1910.17   Effective dates.
1910.18   Changes in established Federal standards.
1910.19   Special provisions for air contaminants.

Subpart C —General Safety end Health Provisions

Sec.

1910.20   Access to employee exposure and medical records.

Subpart D—Walking-Working Surfaces

1910.21   Definitions.
1910.22   General requirements.
1910.23   Guarding floor and wall openings and holes.
1910.24   Fixed industrial stairs.
1910.25   Portable wood ladders.
1910.26   Portable metal ladders.
1910.27   Fixed ladders.
1910.28   Safety requirements for scaffolding.
1910.29   Manually propelled mobile ladder stands and scaffolds (towers).
1910.30   Other working surfaces.
1910.31   Sources of standards.
1910.32   Standards organizations.

Subpart E— Means of Egress

1910.35   Definitions.
1910.36   General requirements.
1910.37   Means of egress, general.
1910.38   Employee emergency plans and fire prevention plans.
1910.39   Sources of standards.
1910.40   Standards organizations.

APPENDIX TO SUBPART E–MEANS OF EGRESS

Subpart F—Powered Platforms, Manlifts, and Vehicle-Mounted Work Platforms

1910.66   Powered platforms for building maintenance.
1910.67   Vehicle-mounted elevating and rotating work platforms.
1910.68   Manlifts.
1910.69   Sources of standards.
1910.70   Standards organizations.

Subpart G—Occupational Health and Environmental Control

1910.94   Ventilation.
1910.95   Occupational noise exposure.
1910.96   Ionizing radiation.
1910.97   Nonionizing radiation.
1910.98   Effective dates.
1910.99   Sources of standards.
1910.100   Standards organizations.

Subpart H—Hazardous Materials

1910.101   Compressed gases (general requirements).
1910.102   Acetylene.

Sec.

1910.103   Hydrogen.
1910.104   Oxygen.
1910.105   Nitrous oxide.
1910.106   Flammable and combustible liquids.
1910.107   Spray finishing using flammable and combustible materials.
1910.108   Dip tanks containing flammable or combustible liquids.
1910.109   Explosives and blasting agents.
1910.110   Storage and handling of liquified petroleum gases.
1910.111   Storage and handling of anhydrous ammonia.
1910.112-1910.113 [Reserved]
1910.114   Effective dates.
1910.115   Sources of standards.
1910.116   Standards organizations.
1910.120   Hazardous waste operations and emergency response.

Subpart I—Personal Protective Equipment

1910.132   General requirements.
1910.133   Eye and face protection.
1910.134   Respiratory protection.
1910.135   Occupational head protection.
1910.136   Occupational foot protection.
1910.137   Electrical protective devices.
1910.138   Effective dates.
1910.139   Sources of standards.
1910.140   Standards organizations.

Subpart J—General Environmental Controls

1910.141   Sanitation.
1910.142   Temporary labor camps.
1910.143   Nonwater carriage disposal systems. [Reserved]
1910.144   Safety color code for marking physical hazards.
1910.145   Specifications for accident prevention signs and tags.
1910.146   [Reserved]
1910.147   The control of hazardous energy (lockout/tagout).
1910.148   Standards organizations.
1910.149   Effective dates.
1910.150   Sources of standards.

Subpart K—Medical and First Aid

1910.151   Medical services and first aid.
1910.152   [Reserved]
1910.153   Sources of standards.

Subpart L—Fire Protection

1910.155   Scope, application and definition applicable to this subpart.
1910.156   Fire brigades.

PORTABLE FIRE SUPPRESSION EQUIPMENT

FIXED FIRE SUPPRESSION EQUIPMENT

OTHER FIRE PROTECTION SYSTEMS

APPENDICES TO SUBPART L

Subpart M—Compressed Gas and Compressed Air Equipment

Subpart N—Materials Handling and Storage

Subpart O—Machinery and Machine Guarding

Subpart P—Hand and Portable Powered Tools and Other Hand-Held Equipment

Subpart Q—Welding, Cutting and Brazing

Subpart R—Special Industries

Subpart S—Electrical

GENERAL

DESIGN SAFETY STANDARDS FOR ELECTRICAL SYSTEMS

Subpart T—Commercial Diving Operations

GENERAL

Subparts U-Y [Reserved]

SUBJECT INDEX FOR 29 CFR 1910 – OCCUPATIONAL SAFETY AND HEALTH STANDARDS

SOURCE: 39 FR 23502, June 27, 1974, unless otherwise noted.

Appendix 2

Part 1926 – Construction Standards

TABLE OF CONTENTS

PRESERVING
THE LEGACY

Appendix 3

NIOSH IAQ Questionnaire

INDOOR AIR QUALITY AND WORK ENVIRONMENT SYMPTOMS SURVEY
NIOSH INDOOR ENVIRONMENTAL QUALITY SURVEY

ID Number _____   Location _____   Today's Date ____ / ____ / ____

This survey is being conducted to determine the environmental quality of your office building. This questionnaire asks about your office environment, your work, and your health. Please answer the questions as accurately and completely as you can, regardless of how satisfied or dissatisfied you are with conditions in the office. ALL OF YOUR ANSWERS WILL BE TREATED IN THE STRICTEST CONFIDENCE.

I. WORKPLACE INFORMATION

1.   How long have you worked in this building, to the nearest year?
       ____ Years

       If less than one year, how many months have you worked in this building?
       ____ Months

**INDOOR AIR QUALITY AND WORK ENVIRONMENT SYMPTOMS SURVEY
NIOSH INDOOR ENVIRONMENTAL QUALITY SURVEY (continued)**

2. On average, how many HOURS per WEEK do you work in this building?
    _____ Hours per week

3. During LAST WEEK, how many days did you work in this building?
    _____ Days

4. Which best describes the space in which your current workstation is located?
    _____ Private office (1)
    _____ Shared private office (2)
    _____ Open space with partitions (3)
    _____ Open space without partitions (4)
    _____ Other (specify) _____ (5)

4a. How many people work in the room in which your workstation is located (including yourself)?
    _____ 1   _____ 2-3   _____ 4-7   _____ 8 or more

5. Is there carpeting on most or all of the floor at your workstation?
    _____ Yes (1)   _____ No (2)

6. In general, how clean is your workspace area?
    _____ Very clean (1)
    _____ Reasonably clean (2)
    _____ Somewhat dusty or dirty (3)
    _____ Very dusty or dirty (4)

7. Please rate the lighting at your workstation.
    _____ Much too dim (1)
    _____ A little too dim (2)
    _____ Just right (3)
    _____ A little too bright (4)
    _____ Much too bright (5)

8. Do you experience a reflection or "glare" in your field of vision when at your workstation?
    _____ Rarely   (1)
    _____ Occasionally (2)
    _____ Sometimes (3)
    _____ Fairly often (4)
    _____ Very often (5)

INDOOR AIR QUALITY AND WORK ENVIRONMENT SYMPTOMS SURVEY
NIOSH INDOOR ENVIRONMENTAL QUALITY SURVEY (continued)

9. How comfortable is the current setup of your desk or work table (i.e., height and general arrangement of the table, chair, and equipment you work with)?

_____ Very comfortable (1)

_____ Reasonably comfortable (2)

_____ Somewhat uncomfortable (3)

_____ Very uncomfortable (4)

_____ Don't have one specific desk or work table (5)

10. About how many HOURS per DAY do you work with a computer or word processor, to the nearest hour?

_____ Hours per day   _____ Don't use one (9)

10 a. If you use a computer or word processor, do you usually wear glasses when you use these machines?

_____ Yes (1)   _____ No (2)   _____ Not Applicable (9)

10 b. Do you use a glare screen on your computer?

_____ Yes (1)   _____ No (2)   _____ Not Applicable (9)

11. Which one of the following statements best describes the windows in your work area?

_____ There are no windows in my personal workspace and none in the general area visible from my workspace (when I am either standing or seated). (1)

_____ There are no windows in my personal workspace, but I can see one or more windows in the general area. (2)

_____ There are one or more windows in my personal workspace. (3)

12. If there is a window visible from your workspace, about how far (in feet) is the closest window from your desk chair?

_____ feet   _____ Check here if no window (9)

13. During the PAST THREE MONTHS, have the following changes taken place within 15 feet of your current workstation?

	YES (1)	NO (2)
New carpeting	_____	_____
Walls painted	_____	_____
New furniture	_____	_____
New partitions	_____	_____
New wall covering	_____	_____
Water damage	_____	_____

INDOOR AIR QUALITY AND WORK ENVIRONMENT SYMPTOMS SURVEY
NIOSH INDOOR ENVIRONMENTAL QUALITY SURVEY (continued)

14. How often do you use the following at work? (Check the appropriate area for each item.)

	Several times a day (1)	About once a day (2)	3-4 times a week (3)	Less than 3 times/week (4)	Never (5)
Photocopier	_____	_____	_____	_____	_____
Laser printer	_____	_____	_____	_____	_____
Facsimile (FAX)	_____	_____	_____	_____	_____
Self-copying (carbonless) copy paper	_____	_____	_____	_____	_____
Cleanser, glue, correction fluid, or other strong-smelling chemical	_____	_____	_____	_____	_____

15. Do you presently have any of the following pets at your home?
    Dog: _____ Yes (1) _____ No (2)
    Cat: _____ Yes (1) _____ No (2)
    Bird: _____ Yes (1) _____ No (2)

II. INFORMATION ABOUT HEALTH AND WELL-BEING

1. Have you ever been told by a doctor that you have or had any of the following?

	YES (1)	NO (2)	
Sinus infection	_____	_____	_____ If yes for sinus infection or asthma, in
Asthma	_____	_____	approximately what year was the diagnosis made?
Migraine	_____	_____	
Eczema	_____	_____	
Hay fever	_____	_____	
Allergy to dust	_____	_____	
Allergy to molds	_____	_____	
Allergy to cats	_____	_____	

INDOOR AIR QUALITY AND WORK ENVIRONMENT SYMPTOMS SURVEY
NIOSH INDOOR ENVIRONMENTAL QUALITY SURVEY (continued)

2.   What is your tobacco smoking status?
    _____ Never smoked (1)
    _____ Former smoker (2)
    _____ Current smoker (3)

3.   Do you consider yourself especially sensitive to the presence of tobacco smoke?
    _____ Yes (1)   _____ No (2)

4.   Do you consider yourself especially sensitive to the presence of chemicals in the air of your workspace?
    _____ Yes (1)   _____ No (2)

5.   What type of corrective lenses do you usually wear at work? (Check all that apply)
    _____ None (1)
    _____ Glasses (2)
    _____ Bifocals (3)
    _____ Contact lenses (4)

6.   How old were you on your last birthday?
    _____ under 20   _____ 20-29 years   _____ 30-39 years
    _____ 40-49 years   _____ 50-59 years   _____ over 59 years

7.   Are you:
    _____ Male (1)   _____ Female (2)

INDOOR AIR QUALITY AND WORK ENVIRONMENT SYMPTOMS SURVEY
NIOSH INDOOR ENVIRONMENTAL QUALITY SURVEY (continued)

This page contains questions regarding symptoms you may have experienced while at work during the last 4 weeks.

The following EXAMPLE shows how an employee might fill out this type of questionnaire.

During the LAST FOUR WEEKS YOU WERE AT WORK, how often have you experienced each of the following symptoms while working in this building?

Note: If you answer "Not in Last 4 Weeks" for a symptom, please move down the page to the next symptom.

	Not in last 4 Weeks (1)	1-3 Days in Last 4 Weeks (2)	1-3 Days per Week in Last 4 Weeks (3)	Every or Almost Every Workday (4)
SYMPTOMS				
ringing in ears	X	___	___	___
toothache	X	___	___	___
hiccups	___	X	___	___
leg cramps	___	___	___	X

ANSWER THE FOLLOWING, ONLY IF YOU EXPERIENCED ONE OR MORE OF THE ABOVE SYMPTOMS

During the LAST FOUR WEEKS YOU WERE AT WORK, what happened to this symptom at times WHEN YOU WERE AWAY FROM WORK? (e.g., holidays, weekends)

	Got Worse (1)	Stayed Same (2)	Got Better (3)
SYMPTOMS			
ringing in ears	___	___	___
toothache	___	___	___
hiccups	___	X	___
leg cramps	X	___	___

ANSWER THE FOLLOWING, ONLY IF YOU EXPERIENCED ONE OR MORE OF THE ABOVE SYMPTOMS

While at work TODAY, did you experience this symptom?

	YES (1)	NO (2)
ringing in ears	___	___
toothache	___	___
hiccups	___	X
leg cramps	X	___

The above responses show that DURING THE LAST 4 WEEKS while at work, THIS EMPLOYEE:

1.   DID NOT experience TOOTHACHE or RINGING EARS.
2.   Experienced HICCUPS 1-3 days. HICCUPS STAYED SAME when away from work. NO HICCUPS on day of survey.
3.   Experience LEG CRAMPS almost every day. LEG CRAMPS GOT WORSE when away from work. HAD LEG CRAMPS on day of survey.

(NOTE that the symptoms in this example are for illustration only and ARE NOT the same as those on the following page.)

INDOOR AIR QUALITY AND WORK ENVIRONMENT SYMPTOMS SURVEY
NIOSH INDOOR ENVIRONMENTAL QUALITY SURVEY (continued)

During the LAST FOUR WEEKS YOU WERE AT WORK, how often have you experienced each of the following symptoms while working in this building?

If you answer "Not in Last 4 Weeks" for a symptom, please move down the page to the next symptom.

	Not in last 4 Weeks (1)	1-3 Days in Last 4 Weeks (2)	1-3 Days per Week in Last 4 Weeks (3)	Every or Almost Every Workday (4)
SYMPTOMS				
dry, itching, or irritated eyes	_____	_____	_____	_____
wheezing	_____	_____	_____	_____
headache	_____	_____	_____	_____
sore or dry throat	_____	_____	_____	_____
unusual tiredness, fatigue, or drowsiness	_____	_____	_____	_____
chest tightness	_____	_____	_____	_____
stuffy or runny nose, or sinus congestion	_____	_____	_____	_____
cough	_____	_____	_____	_____
tired or strained eyes	_____	_____	_____	_____
tension, irritability, or nervousness	_____	_____	_____	_____
pain or stiffness in back, shoulders, or neck	_____	_____	_____	_____
sneezing	_____	_____	_____	_____
difficulty remembering things or concentrating	_____	_____	_____	_____
dizziness or lightheaded	_____	_____	_____	_____
feeling depressed	_____	_____	_____	_____
shortness of breath	_____	_____	_____	_____
nausea or upset stomach	_____	_____	_____	_____
dry or itchy skin	_____	_____	_____	_____

INDOOR AIR QUALITY AND WORK ENVIRONMENT SYMPTOMS SURVEY
NIOSH INDOOR ENVIRONMENTAL QUALITY SURVEY (continued)

ANSWER THE FOLLOWING, ONLY IF YOU EXPERIENCED ONE OR MORE OF THE ABOVE SYMPTOMS

During the LAST FOUR WEEKS YOU WERE AT WORK, what happened to this symptom at times WHEN YOU WERE AWAY FROM WORK? (e.g., holidays, weekends)

	Got Worse (1)	Stayed Same (2)	Got Better (3)
SYMPTOMS			
dry, itching, or irritated eyes	_____	_____	_____
wheezing	_____	_____	_____
headache	_____	_____	_____
sore or dry throat	_____	_____	_____
unusual tiredness, fatigue, or drowsiness	_____	_____	_____
chest tightness	_____	_____	_____
stuffy or runny nose, or sinus congestion	_____	_____	_____
cough	_____	_____	_____
tired or strained eyes	_____	_____	_____
tension, irritability, or nervousness	_____	_____	_____
pain or stiffness in back, shoulders, or neck	_____	_____	_____
sneezing	_____	_____	_____
difficulty remembering things or concentrating	_____	_____	_____
dizziness or lightheaded	_____	_____	_____
feeling depressed	_____	_____	_____
shortness of breath	_____	_____	_____
nausea or upset stomach	_____	_____	_____
dry or itchy skin	_____	_____	_____

INDOOR AIR QUALITY AND WORK ENVIRONMENT SYMPTOMS SURVEY
NIOSH INDOOR ENVIRONMENTAL QUALITY SURVEY (continued)

ANSWER THE FOLLOWING, ONLY IF YOU EXPERIENCED ONE OR MORE OF THE ABOVE SYMPTOMS

While at work TODAY, did you experience this symptom?

	YES (1)	NO (2)
SYMPTOMS		
dry, itching, or irritated eyes	_____	_____
wheezing	_____	_____
headache	_____	_____
sore or dry throat	_____	_____
unusual tiredness, fatigue, or drowsiness	_____	_____
chest tightness	_____	_____
stuffy or runny nose, or sinus congestion	_____	_____
cough	_____	_____
tired or strained eyes	_____	_____
tension, irritability, or nervousness	_____	_____
pain or stiffness in back, shoulders, or neck	_____	_____
sneezing	_____	_____
difficulty remembering things or concentrating	_____	_____
dizziness or lightheaded	_____	_____
feeling depressed	_____	_____
shortness of breath	_____	_____
nausea or upset stomach	_____	_____
dry or itchy skin	_____	_____

INDOOR AIR QUALITY AND WORK ENVIRONMENT SYMPTOMS SURVEY
NIOSH INDOOR ENVIRONMENTAL QUALITY SURVEY (continued)

III. DESCRIPTION OF WORKPLACE CONDITIONS

During the LAST FOUR WEEKS YOU WERE AT WORK, how often have you experienced each of the following environmental conditions while working in this building?

If you answer "Not in Last 4 Weeks" for a condition, please move down the page to the next condition.

	Not in last 4 Weeks (1)	1-3 Days in Last 4 Weeks (2)	1-3 Days per Week in Last 4 Weeks (3)	Every or Almost Every Workday (4)
CONDITIONS				
too much air movement	_____	_____	_____	_____
too little air movement	_____	_____	_____	_____
temperature too hot	_____	_____	_____	_____
temperature too cold	_____	_____	_____	_____
air too humid	_____	_____	_____	_____
air too dry	_____	_____	_____	_____
tobacco smoke odors	_____	_____	_____	_____
unpleasant chemical odors	_____	_____	_____	_____
other unpleasant odors (e.g., body odor, food odor, perfume)	_____	_____	_____	_____

ANSWER THE FOLLOWING, ONLY IF YOU EXPERIENCED ONE OR MORE OF THE ABOVE ENVIRONMENTAL CONDITIONS

TODAY, while working at your usual workstation, did you experience this environmental condition?

	YES (1)	NO (2)
CONDITIONS		
too much air movement	_____	_____
too little air movement	_____	_____
temperature too hot	_____	_____
temperature too cold	_____	_____
air too humid	_____	_____
air too dry	_____	_____
tobacco smoke odors	_____	_____
unpleasant chemical odors	_____	_____
other unpleasant odors (e.g., body odor, food odor, perfume)	_____	_____

INDOOR AIR QUALITY AND WORK ENVIRONMENT SYMPTOMS SURVEY
NIOSH INDOOR ENVIRONMENTAL QUALITY SURVEY (continued)

How satisfied are you with the following aspects of your workstation?

A.   Conversational privacy

_____ Very satisfied (1)

_____ Somewhat satisfied (2)

_____ Not too satisfied (3)

_____ Not at all satisfied (4)

B.   Freedom from distracting noise

_____ Very satisfied (1)

_____ Somewhat satisfied (2)

_____ Not too satisfied (3)

_____ Not at all satisfied (4)

IV. CHARACTERISTICS OF YOUR JOB

1.   What is your job category?

_____ Managerial (1)

_____ Professional (2)

_____ Technical (3)

_____ Secretarial or clerical (4)

_____ Other (specify) _____ (5)

2.   All in all, how satisfied are you with your job?

_____ Very satisfied (1)

_____ Somewhat satisfied (2)

_____ Not too satisfied (3)

_____ Not at all satisfied (4)

3.   What is the highest grade you completed in school?

_____ 8th grade or less (1)

_____ Some high school (2)

_____ High school graduate (3)

_____ Some college (4)

_____ College degree (5)

_____ Graduate degree (6)

INDOOR AIR QUALITY AND WORK ENVIRONMENT SYMPTOMS SURVEY
NIOSH INDOOR ENVIRONMENTAL QUALITY SURVEY (continued)

4. Conflicts can occur in any job. For example, someone may ask you to do work in a way that is different from what you think best, or you may find that it is difficult to satisfy everyone. HOW OFTEN do you face problems in your work like the ones listed below? (Check the appropriate blank for each statement.)

	Rarely or Never (1)	Sometimes (2)	Fairly Often (3)	Very Often (4)
Persons equal in rank and authority over you, ask you to do things that conflict.	____	____	____	____
People, in a good position to see if you do what they ask, give you things that conflict with one another.	____	____	____	____
People, whose requests should be met, give you things that conflict with other work you have to do.	____	____	____	____

5. The next series of questions asks HOW OFTEN certain things happen AT YOUR JOB. (Check the appropriate blank for each question.)

	Rarely (1)	Occasionally (2)	Sometimes (3)	Fairly Often (4)	Very Often (5)
How often does your job require you to work very fast?	____	____	____	____	____
How often does your job require you to work very hard?	____	____	____	____	____
How often does your job leave you with little time to get things done?	____	____	____	____	____
How often is there a great deal to be done?	____	____	____	____	____

INDOOR AIR QUALITY AND WORK ENVIRONMENT SYMPTOMS SURVEY
NIOSH INDOOR ENVIRONMENTAL QUALITY SURVEY (continued)

	Rarely (1)	Occasionally (2)	Sometimes (3)	Fairly Often (4)	Very Often (5)
How often are you clear on what your job responsibilities are?	_____	_____	_____	_____	_____
How often can you predict what others will expect of you on the job?	_____	_____	_____	_____	_____
How much of the time are your work objectives well defined?	_____	_____	_____	_____	_____
How often are you clear about what others expect of you on the job?	_____	_____	_____	_____	_____

6. The next series of questions asks about responsibilities OUTSIDE YOUR NORMAL WORKING DAY. Do you currently have the following responsibilities?

	YES (1)	NO (2)
Major responsibility for childcare duties	_____	_____
Major responsibility for housekeeping duties	_____	_____
Major responsibility for care of an elderly or disabled person on a regular basis	_____	_____
Regular commitment of five hours or more per week, paid or unpaid, outside of this job (include educational courses, volunteer work, second job, etc.)	_____	_____

INDOOR AIR QUALITY AND WORK ENVIRONMENT SYMPTOMS SURVEY
NIOSH INDOOR ENVIRONMENTAL QUALITY SURVEY (continued)

PLEASE USE THE REMAINING SPACE TO DISCUSS ANY ASPECTS OF THE BUILDING ENVIRONMENT OR EMPLOYEE HEALTH THAT YOU FEEL APPROPRIATE.

_____

_____

_____

_____

_____

_____

_____

_____

_____

_____

_____

_____

_____

_____

_____

_____

_____

Thank you!!!

Appendix 4

HVAC Checklist

Building: _____ Inspection date: _____

System/zone: _____ Inspected by: _____

System Component	Conditions Observed (distances, contaminant sources, blockage, malfunctions, etc.)	Notes
Outdoor air intake Location:_____ Near cooling tower Near exhaust discharge Near sewer vent/restroom vent Near trash dumpster/storage Loading dock/parking area Bird nest(s) Other sources of contaminants:		

System Component	Conditions Observed (distances, contaminant sources, blockage, malfunctions, etc.)	Notes
Intake vents Blocked/frozen/rusted Debris present		
Exhaust vents Blocked/frozen/rusted Debris present		
Heating coils Clean Dry Debris present		
Cooling coils Clean Dry Debris present		
Humidifier Type: Cold water Steam Water collection pan cleanliness Odors Debris present Algal growth		
Filters Type: Foam Mat/Fibrous Other: Clean Dry Clogged Damaged		

System Component	Conditions Observed (distances, contaminant sources, blockage, malfunctions, etc.)	Notes
Fans/motors Sized per specifications Operating? Wired to power correctly Drive belt condition Motor condition (bearings worn, etc.)		
Evidence of chemicals in system		
Ducts Obstructions/bends Water Wet insulation Diffusers blocked Dirt on ceiling near diffusers		
Temperature controls Accessible Adjustable by occupants Functioning?		
Indoor smoking areas Ventilated to outdoors? Pressure (relative to surroundings) positive negative		
History of flooding, water damage?		
Other sources of odor/contamination (list):		

Appendix 5

Table G-16A of Appendix A of 29 CFR 1910.95, Occupational Noise Exposure and the Hearing Conservation Amendment

This table relates the A-weighted sound level to the corresponding reference duration allowed under the OSHA regulation.

A-weighted sound level, L (decibel)	Reference duration, T (hour)
80	32
81	27.9
82	24.3
83	21.1

A-weighted sound level, L (decibel)	Reference duration, T (hour)
84	18.4
85	16
86	13.9
87	12.1
88	10.6
89	9.2
90	8
91	7.0

A-weighted sound level, L (decibel)	Reference duration, T (hour)	A-weighted sound level, L (decibel)	Reference duration, T (hour)
92	6.1	112	0.38
93	5.3	113	0.33
94	4.6	114	0.29
95	4	115	0.25
96	3.5	116	0.22
97	3.0	117	0.19
98	2.6	118	0.16
99	2.3	119	0.14
100	2	120	0.125
101	1.7	121	0.11
102	1.5	122	0.095
103	1.3	123	0.082
104	1.1	124	0.072
105	1	125	0.063
106	0.87	126	0.054
107	0.76	127	0.047
108	0.66	128	0.041
109	0.57	129	0.036
110	0.5	130	0.031
111	0.44		

Appendix 6

Table A-1 – Conversion from "Percent Noise Exposure" or "Dose" to "8-Hour Time-Weighted Average Sound Level" (TWA)

Table A-1 of Appendix A of OSHA 29 CFR 1910.95, Occupational Noise Exposure and the Hearing Conservation Amendment. This table provides the maximum allowable exposure times for specific levels of noise. The values can be used to determine whether or not an employee's noise exposure exceeds the OSHA PEL for noise.

Dose or percent noise exposure	TWA
10	73.4
15	76.3
20	78.4
25	80.0
30	81.3

Dose or percent noise exposure	TWA
35	82.4
40	83.4
45	84.2
50	85.0
55	85.7
60	86.3
65	86.9
70	87.4
75	87.9
80	88.4
81	88.5

Dose or percent noise exposure	TWA	Dose or percent noise exposure	TWA
82	88.6	170	93.8
83	88.7	175	94.0
84	88.7	180	94.2
85	88.8	185	94.4
86	88.9	190	94.6
87	89.0	195	94.8
88	89.1	200	95.0
89	89.2	210	95.4
90	89.2	220	95.7
91	89.3	230	96.0
92	89.4	240	96.3
93	89.5	250	96.6
94	89.6	260	96.9
95	89.6	270	97.2
96	89.7	280	97.4
97	89.8	290	97.7
98	89.9	300	97.9
99	89.9	310	98.2
100	90.0	320	98.4
101	90.1	330	98.6
102	90.1	340	98.8
103	90.2	350	99.0
104	90.3	360	99.2
105	90.4	370	99.4
106	90.4	380	99.6
107	90.5	390	99.8
108	90.6	400	100.0
109	90.6	410	100.2
110	90.7	420	100.4
111	90.8	430	100.5
112	90.8	440	100.7
113	90.9	450	100.8
114	90.9	460	101.0
115	91.1	470	101.2
116	91.1	480	101.3
117	91.1	490	101.5
118	91.2	500	101.6
119	91.3	510	101.8
120	91.3	520	101.9
125	91.6	530	102.0
130	91.9	540	102.2
135	92.2	550	102.3
140	92.4	560	102.4
145	92.7	570	102.6
150	92.9	580	102.7
155	93.2	590	102.8
160	93.4	600	102.9
165	93.6	610	103.0

Dose or percent noise exposure	TWA
620	103.2
630	103.3
640	103.4
650	103.5
660	103.6
670	103.7
680	103.8
690	103.9
700	104.0
710	104.1
720	104.2
730	104.3
740	104.4
750	104.5
760	104.6
770	104.7
780	104.8
790	104.9
800	105.0
810	105.1

Dose or percent noise exposure	TWA
820	105.2
830	105.3
840	105.4
850	105.4
860	105.5
870	105.6
880	105.7
890	105.8
900	105.8
910	105.9
920	106.0
930	106.1
940	106.2
950	106.2
960	106.3
970	106.4
980	106.5
990	106.5
999	106.6

PRESERVING
THE LEGACY

Appendix 7

NIOSH Approval Certificate

Certificate shown on reverse side of this page.

Mine Safety Appliance Company
Pittsburgh, Pennsylvania, USA
1-800-672-2222

NIOSH

COMFO GMA P100 CARTRIDGE, P/N 814923

THIS CARTRIDGE IS APPROVED ONLY IN THE FOLLOWING CONFIGURATIONS:

CONTROL BOX
815580
SK3094-87
REV.4, 3/14/97
COMFO GMA P100 CARTRIDGE
TAL 5/15/97

RESPIRATOR COMPONENTS

| TC | PROTECTION | ALTERNATE FACEPIECE | CARTRIDGE | REGULATOR / ALT. REGULATOR | HOSE / ALT. | B. TUBE / ALT. | MANIFOLD | SUPPLY UNIT | REG./VALVE | VALVE / ALT. | ALT. BELT ASSY | HARNESS | COVERALL | ALT. ADAPTERS | PLUG | COUP. NUT | OUTSERT | ALT. NOSECUPS | CAUTIONS AND LIMITATIONS |

(Alternate Facepiece part numbers: 449703 COMFO II, 7-201-1 COMFO II MEDIUM, 7-201-2 COMFO II SMALL, 7-201-3 COMFO II LARGE, 7-203-1 ULTRAVUE MED., 7-203-2 ULTRAVUE SMALL, 7-203-3 ULTRAVUE LARGE, 7-204-1 ULTRA TWIN MED., 7-204-2 ULTRA TWIN SMALL, 7-204-3 ULTRA TWIN LARGE, 7-708-1 DUO TWIN MED., 7-708-2 DUO TWIN SMALL, 7-708-3 DUO TWIN LARGE, 7-752-1 DUO TWIN MED., 7-752-2 DUO TWIN SMALL)

CARTRIDGE: 814923 COMFO GMA P100

REGULATOR ASSY: 7-1046-1, ALT: 7-1047-1, 7-1048-1, 7-1427-2

HOSE: 7-664-1 AIRLINE HOSE, PVC; ALT: 7-664-2 AIRLINE HOSE, NEOPRENE

B. TUBE: 456651 BREATHING TUBE; ALT: 457158 BREATHING TUBE, 470734 BREATHING TUBE

MANIFOLD: 7-1064-1

SUPPLY UNIT: 7-1081-1

REG./VALVE ASSEMBLY: 7-679-1

VALVE: 7-862-1; ALT: 5-622-1, 5-713-1

ALT. BELT ASSY: 7-977-1 BELT ASSEMBLY, 1966 BELT ASSEMBLY

HARNESS: 472420

COVERALL: 7-630-1

ALT. ADAPTERS: 69542 UNION ADAPTER, 628232 UNION ADAPTER, 492495 QD ADAPTER, 812839 ADAPTER, 809068 ADAPTER, 484679 ADAPTER, 491766 ADAPTER, 491834 ADAPTER

PLUG: 486566

COUP. NUT: 96547 COUPLING NUT

OUTSERT: 5-939-1

ALT. NOSECUPS: 805125 MEDIUM, 805126 MEDIUM, 805127 SMALL, 805128 SMALL, 805129 LARGE, 805130 LARGE, 471539 SMALL, 471540 MEDIUM, 471541 LARGE

TC	PROTECTION	CAUTIONS AND LIMITATIONS
84A-0153	P100/OV	ABCHJLMNO
84A-0162	P100/OV	ABCHJLMNO
84A-0173	P100/OV	ABCHJLMNO
84A-0183	P100/OV	ABCHJLMNO
84A-0361	P100/OV	ABCHJLMNO
84A-0344	P100/OV	ABCHJLMNO
84A-0388	P100/OV	ABCHJLMNO
84A-0956	P100/OV	ABCHJLMNO
84A-1149	P100/OV/SA	ABCDEGHJLMNOS
84A-1189	P100/OV/SA	ABCDEGHJLMNOS
84A-1208	P100/OV/SA	ABCDEGHJLMNOS
84A-1302	P100/OV/SA	ABCDEGHJLMNOS
84A-1245	P100/OV/SA	ABCDEGHJLMNOS
84A-1263	P100/OV/SA	ABCDEGHJLMNOS

1. PROTECTION

P100- Particulate Filter (99.97% filter efficiency level) effective against all particulate aerosols

OV: Organic Vapor
SA: Supplied Air

2. CAUTIONS AND LIMITATIONS

A. Not for use in atmospheres containing less than 19.5 percent oxygen.
B. Not for use in atmospheres immediately dangerous to life or health.
C. Do not exceed maximum use concentrations established by regulatory standards.
D. Airline respirators can be used only when the respirators are supplied with respirable air meeting the requirements of CGA G-7.1 Grace D or higher quality.
E. Use only the pressure ranges and hose lengths specified in the User's Instructions.
G. If airflow is cut off, switch to filter and/or cartridge and immediately exit to clean air.
H. Do not wear for protection against organic vapors with poor warning properties or those which generate high heats of reaction with sorbent.
J. Failure to properly use and maintain this product could result in injury or death.
L. Follow the manufacturers instructions for changing cartridges and/or filters.
M. All approved respirators shall be selected, fitted, used and maintained in accordance with MSHA, OSHA, and other applicable regulations.
N. Never substitute, modify, add, or omit parts. Use only exact replacement parts in the configuration as specified by the manufacturer.
O. Refer to User's Instructions, and/or maintenance manuals for information on use and maintenance of these respirators.
S. Special or critical user's instructions and/or use limitations apply. Refer to User's Instructions before donning.

Note: This document has been edited to fit format.

Glossary

AA – See Atomic Absorption.

AAIH – See American Academy of Industrial Hygiene.

ABIH – See American Board of Industrial Hygiene.

Absorbance – 1) In the context of solutions, it is the amount of light that is absorbed by a solution and proportional to its concentration. – 2) In the context of sound, it is the inability of a surface to reflect sound energy.

Absorption Spectroscopy – Analytical technique that involves measuring the amount of energy that is absorbed by a compound; it includes atomic absorption.

Acclimatization – The body's gradual adjustment to working in a warm environment; it occurs during the first ten days to two weeks of exposure

ACGIH – See American Conference of Governmental Industrial Hygienist.

Acinus – A minute rounded lobule such as the smallest secreting unit of the liver.

Acoustic Watt – A unit for expressing the energy output of sound.

Action Level – A term used to describe the airborne concentration that triggers certain provisions of a regulation; generally, but not always, it is one-half or 50 percent of the PEL value.

Activated Charcoal – Charcoal that has its air spaces expanded by heating it to 800-900°C in the presence of steam.

Activation – In the context of radioactivity, it is the absorption of a neutron by the nucleus of another atom, thereby forming a different radioactive isotope.

Active Transport – A term used to describe the movement of a molecule across a membrane that would otherwise be impermeable.

Activity – In the context of radioactivity, it is the decay rate of radioactive particles.

Acute Effects – Symptoms of injury or other physical manifestations that follow an acute exposure.

Acute Exposure – Exposure to a high level or concentration for a relatively short duration of time.

Administrative Controls – The use of procedures, work schedules, employee training, and similar methods to reduce worker exposure to health and safety hazards.

AE Filters – See Glass Fiber Filters.

Aerodynamic Diameter – A particle whose diameter gives it the same settling velocity as the contaminant particle.

Aerosol – A suspension of liquid or solid particles (mist, fume, particulate, or dust) that are microscopic in size (0.5 to 10 μm), allowing them to remain airborne for an extended period.

Aerosolize – The process of mixing a gas and a liquid that results in microscopic (0.5 to 10 μm) airborne droplets.

AIHA – See American Industrial Hygiene Association.

Air-purifying Respirator – A respirator that removes airborne contaminants such as particulates, gases, vapors, and fumes from ambient air through filtration, absorption, adsorption, or chemical reactions that take place on the media present in the cartridge or filter.

ALARA – See As Low As Reasonably Achievable.

Albinism – A condition where the person has little or no amounts of melanin in their skin.

Allergic Alveolitis – A response to inhalation of organic particles, which involves the small terminal branches of the bronchioles, just outside the alveolar sacs, causing symptoms such as coughing, increased production of mucus, fever, fatigue, and muscle aches.

Allergic Contact Dermatitis – Skin condition that occurs in response to exposure to a sensitizing material; characterized by redness, swelling and cracking, and sometimes more severe reactions involving the entire immune system.

Allergic Response – The release of antibodies by the immune system in response to recognition of foreign molecules in the body.

Alpha Particles – Positively charged (+2) helium nuclei – composed of two protons and two neutrons – that are spontaneously emitted from the nuclei of radioactive isotopes. Alpha particles are high-energy particles, but can only travel through a few centimeters of air. Simple barriers like paper or skin effectively shield against them. If alpha particles are released internally, however, they transfer enough energy to cause tissue damage.

Alveolar Fraction – Particles with aerodynamic diameters of 0.5-3 mm.

Alveoli – The small, thin sacs at the distal end of the respiratory tract where gas exchange takes place.

American Academy of Industrial Hygiene (AAIH) – A nonprofit professional organization composed of individuals who have successfully met the certification requirements as set forth by the American Board of Industrial Hygiene (ABIH). The purpose and goals of AAIH are to: recruit and train graduates of scientific disciplines; promote recognition of industrial hygiene practices; promote the ABIH Certification as the basic qualification for employment as an industrial hygienist; and establish ethical conduct and practice guidelines for the profession.

American Board of Industrial Hygiene (ABIH) – An independent organization whose membership is composed of AAIH, ACGIH, and AIHA members and that administers the certification program for industrial hygiene professionals. In addition to certification, the ABIH is also responsible for maintaining the re-certification process of practicing industrial hygienists.

American Conference of Governmental Industrial Hygienists (ACGIH) – A professional organization whose mission is to promote excellence in occupational and environmental health. It is composed of 14 technical committees that provide other industrial hygienists with information through various forums. ACGIH publications are considered authoritative works, with one of them establishing the recommended exposure limits (TLVs® and BEIs®) for toxic substances.

American Industrial Hygiene Association (AIHA) – An organization whose membership includes professional industrial hygienists, students, health care professionals, and others with an interest in the area of industrial hygiene. The purpose of the association is to represent and support the member's interests and promote the field of industrial hygiene. The association also publishes the *American Industrial Hygiene Association Journal* and provides input to Congress and Congressional committees on proposed health and safety regulations.

Amorphous Silica – All noncrystalline forms of silica or quartz.

Amosite – Brown asbestos – rich in iron – formerly used as an ingredient in many building materials.

Amphibole – A type of asbestos mineral with a straight, stick-like structure; amosite and crocidolite are amphiboles.

Amplification – The reflection of a sound greater than the amount that originally impacted the surface.

Analyte – The material that is the objective of an analytical procedure; for industrial hygiene samples, it is the airborne contaminant.

Anoxia – The absence of or an abnormally low amount of oxygen in the body tissues. Hypoxia of such severity as to result in permanent tissue damage.

ANS – See Autonomic Nervous System.

Anthropometry – The measurement of humans in terms of heights, depths, breadths, and other distances; values are usually expressed in centimeters (cm).

Antibodies – Specialized proteins released from the immune system; they react with foreign molecules.

Anti-C – Short for anti-contamination clothing; a term commonly used by radiation workers when referring to radiation shielding coveralls. Also called PC, short for protective clothing.

APF – See Assigned Protection Factor.

Area Samples – Samples taken by placing the sampling train in a fixed location in the work area.

As Low As Reasonably Achievable – A widely accepted practice for controlling and reducing worker exposure to radioactive sources. The primary ALARA techniques involve reducing time and increasing distance and shielding.

Asbestos – A group of naturally occurring minerals containing iron, magnesium, and silica dioxide with a tendency to split into fibers. These materials are resistant to heat, chemicals, electricity and mechanical stress.

Asbestosis – Fibrotic scarring of the lungs as the result of inhalation of asbestos fibers.

Asphyxiant – A material that interferes with the uptake of oxygen by the blood.

Assigned Protection Factor (APF) – The minimum level of respiratory protection that a respirator can be expected to provide, assuming it is properly fitted, worn, and functioning. APFs are assigned by NIOSH.

Atomic Absorption – An analytical method where the sample is converted into a vapor by passing it through a flame or other energy source. The absorbance at a particular wavelength is measured and compared with that of a reference substance. The absorbance measured is proportional to the concentration of that substance in the sample.

Atrophy – The deterioration or death of a tissue or an organ from defective nutrition or nerve damage.

Attenuation – The amount of protection or noise reduction that hearing protectors are capable of providing; usually expressed in dB; sometimes noted as NRR.

Audiogram – The report produced by audiometry, showing measured hearing threshold levels at frequencies of 500, 1,000, 2,000, 3,000, 4,000, and 6,000 Hz.

Audiometry – The process by which a person's hearing acuity is measured.

Autonomic Nervous System (ANS) – Peripheral nerve centers that control motor functions and stimulate most of the major internal organs; controls heart rate and breathing.

Axial-Flow Fans – Fans that have blades arranged like propellers and capable of moving large volumes of air, but are incapable of producing the usual pressures necessary to move air through long ducts.

Backward Blades – Centrifugal fan blades that are curved or tilted in the direction opposite rotation; backward blades reach a maximum power requirement regardless of airflow; therefore, their drive motors cannot be overloaded.

Barrier Creams – Topical treatments, such as lotions or creams, which are applied to the skin to provide a layer of protection against chemical skin hazards.

Baseline Audiogram – An initial audiogram that provides a record of hearing acuity before noise exposure.

Basilar Membrane – The thin layer of tissue that lines the interior of the cochlea and supports the hair cells.

Bauxite Lung – See Shaver's Disease.

BEI® – See Biological Exposure Indice.

Benign – Not associated with negative health effects; self-limiting.

Beta Particles – Negatively charged particles – identical in nature to slower moving electrons – that originate in the nucleus of a disintegrating atom.

Bias – Error; may be random or introduced.

Bifurcate – To divide into two branches or split.

Biological Agents – Potential health hazards that are produced directly or indirectly by viruses or organisms such as plants, insects, molds, yeasts, fungi, and bacteria.

Biological Exposure Indices (BEI®) – A set of reference values developed by ACGIH as guidelines for the evaluation of potential health hazards in biological specimens collected from healthy workers who have been exposed to chemicals to the same extent as workers with inhalation exposures to the threshold limit value. The values apply to 8-hour exposures, five days per week.

Biological Half-life – The amount of time required for one-half of the accumulated material in a tissue to be removed.

Blank-Corrected – Describes data that have had trace contamination amounts deducted from the total amount of contaminant detected in the sampling media.

Breakthrough – A condition that exists when the backup section of a sorbent tube is found to contain 20-25 percent of the total amount of contaminant captured in the front section. This is an indication that the front section of the sorbent was completely saturated during the sample collection. It also refers to detection of contaminant inside a respirator facepiece.

Breakthrough Time – In the context of chemical protective clothing, it is the time between initial contact of the chemical on the barrier material surface and the analytical detection of the chemical on the other side of the material.

Breathing Zone – The region defined by a two-foot diameter half-sphere around the front of a worker's head and shoulders.

Bremsstrahlung – A type of x-radiation produced when a beta particle slows and emits energy. The term is a German word that means braking radiation.

Bronchi – The larger air passages of the lungs (bronchus, singular).

Bronchioles – The very small airways that terminate in the alveoli.

Bronchogenic Carcinoma – A lung cancer associated with asbestos exposure.

Byssinosis – Reactive airway disease associated with inhalation of organic textile fibers such as cotton, flax, linen, and hemp.

C – See Ceiling Limit.

Ca – A notation used by NIOSH to indicate that a substance is considered a known or potential occupational carcinogen.

Calibration – Process of verifying the accuracy of a measuring device, such as the flow rate of air through an air-sampling device or sampling train.

Canopy Hood – See Receiving Hood.

Capillaries – Minute blood vessels between the terminations of the arteries and the beginnings of the veins such as those vessels that surround the alveoli.

Capture Hood – A nozzle or hose- type intake into an LEV system; usually placed at the point of contaminant emission.

Capture Velocity – The velocity of air required to move airborne contaminants into the LEV system at the hood opening.

Carboxyhemoglobin – A molecule formed by the combination of carbon monoxide and hemoglobin.

Carcinogen – A substance known to be cancer causing.

Carpal Tunnel Syndrome – The condition in which the median nerve becomes compressed as it passes through the bones in the wrist, known as the carpal tunnel.

Carrier Gas – An inert gas that moves the sample through the column of a gas chromatograph.

Ceiling Limit (C) – A level that should not be exceeded during any part of the working day.

Central Nervous System (CNS) – Composed of the brain and spinal cord; controls important body functions such as behavior, emotion, speech, and memory.

Centrifugal Fan – A fan with blades arranged like the spokes of a wheel; also called squirrel-cage fan. Centrifugal fans with straight blades are capable of moving air that contains particulate without particles accumulating on the blades. These fans are the workhorses of industrial ventilation systems.

Chain of Custody – Documentation necessary to trace sample possession from the time of collection throughout the time of analysis.

Chemical Agents – Potential health hazards that may exist as dusts, mists, fumes, vapors, and gases.

Chemical Asphyxiant – A substance that interferes with the absorption or utilization of oxygen; e.g., carbon monoxide.

Chloracne – A skin irritation caused by chlorine-containing compounds; it resembles adolescent acne.

Cholestasis – Damage to the liver that results in interference with the production of bile and biliary excretion.

Choroid – A vascularized membrane under the retina.

Chronic Exposure – A long-term exposure (several months or more), generally to a relatively low level or dose.

Chronic Toxic Effects – Adverse effects associated with a longer time of exposure, usually several months or more, to a relatively low level or dose of a chemical.

Chrysotile – White asbestos; 90 percent of asbestos used is of this kind.

Cilia – Fine hair-like structures found in the membranes that line the respiratory tract as well as the small intestines.

Cirrhosis – A condition where collagen has been deposited in the liver.

Clearance Samples – Area samples, taken following a lead, asbestos, or other action, which must indicate contaminant concentration at or below a specific level before the area can be released for normal occupation and work activities.

Closed Face – Sampling performed through a small hole in the top of a filter cassette.

CNS – See Central Nervous System.

Coated/Treated Filters – Filters that have been coated with a specific chemical, depending on the contaminant to be collected. The coatings enhance collection by chemically reacting with the contaminant as the air is drawn through the filter.

Cochlea – A snail-shaped, fluid-filled organ of the inner ear. Its inner surface is lined with specialized hair cells that convert sound pressure vibrations into nerve impulses.

Cohort – A group of individuals that share a particular statistical or demographic characteristic, e.g., exposure.

Collagen – A proteinaceous connective tissue; a component of tendons, ligaments, and bones.

Conductive Hearing Loss – Hearing loss caused by blockage or other interference in the path by which sound energy is transferred to the inner ear.

Confined Space – As defined by OSHA, any space that is large enough so that an employee can enter and perform work, has limited or restricted means for entry or exit, and is not designed for human occupancy.

Consensus Standards – Existing standards that are voluntarily being followed by industry; typically, these contain the minimum requirements for materials, procedures, and applications.

Contamination – Radioactive materials present in an unwanted location; also refers to loose radioactive materials that can be easily removed from a surface.

Continuous Flow Mode – The mode of air supply where a regulated amount of air is supplied to the facepiece at all times; has limited applications in welding and grinding.

Cor Pulmonale – A heart condition due to an enlarged heart muscle.

Corium – See Dermis.

Corrosion – A severe irritation-type response characterized by almost immediate – and possibly permanent – changes such as blistering, bleeding, and other severe damage.

Criterion Level – The 8-hour TWA limit for noise exposure that is used for determining the noise dose; OSHA's is 90 dB(A), ACGIH uses 85 dB(A).

Crocidolite – A form of asbestos that is blue.

Cumulative Trauma Disorder – Damage to soft tissues as a result of repetitive tasks or motions that are unnatural.

Cumulative Use Disorder – See Cumulative Trauma Disorder.

Cuvette – Small cylinder resembling a test tube that is used to hold a sample in a photospectrometer.

Cyclone – An air sampling device that separates dust particles according to size. It is typically used for collecting particles in the respirable range.

Daily Noise Dose (DND) – The allowable noise exposure for an 8-hour workday.

dB – See Decibel.

Dead Finger – See Raynaud's Syndrome.

Decay – The spontaneous disintegration of an unstable atomic nucleus and eventual formation of another more stable element or isotope of a lower atomic mass.

Decay Product – The intermediate atomic nuclei formed as an unstable atomic nucleus moves toward a more stable element or isotope of a lower atomic mass.

Decibel (dB) – A dimensionless unit for expressing sound levels. It is based on the logarithm of the ratio between a measured and reference sound level; a quiet room being 40 dB, for example.

Degradation – In the context of PPE, it is the effect that a chemical has on a barrier material; barrier materials may become stiff and brittle, they may swell and soften, or otherwise be affected. This makes them more likely to tear or simply fall apart; a material's ability to resist degradation may or may not be related to permeability.

Demand Mode – The mode of air supply in which inhalation creates a negative pressure inside the facepiece, causing the regulator to release air into the facepiece. Respirators that operate in this mode are not recommended for use and have been largely replaced by respirators operated in the pressure-demand mode.

Depressants – Agents that act to decrease a neuron's sensitivity to nerve impulses.

Dermis – The middle layer of the skin containing connective tissue that supports and strengthens the epidermis, as well as blood vessels, nerves, hair follicles, some muscles, and oil and sweat glands.

Desiccator – A sealed container containing a water-absorbing substance such as silica gel or calcium chloride and used to dry test materials in the laboratory.

Desorbed – Process by which collected compounds are removed from a solid sorbent, using a solvent or heat.

Detector Tube – A direct-reading method for identifying airborne contaminants. Also called length-of-stain tube. It is a convenient tool for detecting and quantifying contaminants in field or emergency situations.

Determinant – A chemical, a metabolic product of the chemical, or a change in the body's chemistry that is induced by the chemical, that can be measured in a biological sample collected from the exposed worker. The level of determinant is then compared to the biological exposure index (BEI®s).

Diffuse Interstitial Fibrosis – A condition characterized by widespread, sometimes isolated, fibrotic lesions in the lung.

Diffusion – The movement of a substance from an area of higher to an area with a lower concentration.

Dilution Ventilation – The use of an added volume of air to dilute contaminants without removing them from the work area atmosphere. This method should be used only when the contaminant is of low toxicity, produced at a steady rate, and an adequate supply of clean air is available.

Directivity – In the context of sound, the characteristic that is associated with sound being energy, in the forms of waves, which moves in a straight line from its source.

Direct-Reading – Sampling approach that provides immediate or very fast feedback in terms of the sample results; examples are meters or colorimetric methods.

Distal Region – Situated away from the point of origin or attachment such as the lower portion of the respiratory tract; lower lungs.

DNA – Deoxyribonucleic acid; the genetic material that contains information for cellular function, metabolism, and growth.

DND – See Daily Noise Dose.

Dose – 1) The level or amount of exposure to a hazardous agent (chemical or physical). – 2) The level or amount of a chemical or ionizing radiation that has been absorbed, usually expressed dose per weight of the exposed organism (e.g., mg/kg).

Dose-Response Relationship – The toxicological concept that states that the toxicity of a substance depends not only on its toxic properties, but also on the amount of exposure, or dose.

Dust – Aerosol composed of particles typically formed by abrasion.

Ear Canal – The opening in the outer ear through which sound energy passes.

Ear Drum – A thin piece of skin that covers the opening to the middle ear.

ED$_{25}$ – The effective dose for 25 percent of the exposed population.

Electrons – Negatively charged particles, with a very small mass, which equal the number of protons and surround the nucleus of an atom.

Electrostatic Attraction – Collection mechanism where the contaminant materials have an electric charge that causes them to be attracted to the filter material.

Emission Spectroscopy – Analytical technique that measures the energy lost by atoms as their electrons return to a ground state.

Enclosing Hood – LEV system intake that surrounds the point of emission or generation, as in a hood around a grinding wheel.

End-of-service-life Indicator (ESLI) – A system that alerts a respirator user that the cartridge or canister is approaching the end of its useful life.

Engineered Control – The removal or reduction of a hazard through implementation of an engineered solution, such as material substitution, process change, or installation of an exhaust ventilation system.

Entrained – Related to substance suspended in a moving air stream.

Entry – In the context of permitted space, the term entry refers to the breaking of the plane, by any part of the body, at the opening of a space requiring a permit.

Entry Loss – The loss of kinetic energy associated with the entry of air into a hood or other inlet.

Epidermis – The outer layer of the skin, which is composed of an outer protective structure known as the stratum corneum or horny layer.

Epigenetic Carcinogen – A substance that causes cancer through a mechanism other than an interaction with the genetic material.

Ergonomic Agent – A potential health hazard and discomfort caused by the interaction of workers with the total work environment, including items such as tools, equipment, the physical arrangement of the work area, uncomfortable or repetitive motions, monotony, and fatigue.

Ergonomics – The science dealing with the application of information on physical and psychological characteristics to the design of the work environment.

Ergonomist – An occupational health professional who specializes in the study and application of human factors or ergonomics.

ESLI – See End-of-service-life Indicator.

Eustachian Tube – The small canal connecting the middle ear and throat.

Evacuated Container – Sampling container that is a sealed vacuum; it is opened for collection.

Exchange Rate – The value of three dB, which represents – because it is logarithmic – a one-half reduction in the amount of sound energy.

Excursion Limit – A time-weighted average, the length of which is specified by OSHA, that cannot be exceeded during the work day; analogous to peak limit.

Experimental Variance – OSHA-granted permission to use an alternative method of worker protection during an approved experiment to demonstrate or validate new safety and health techniques. The variance terminates with the study completion unless another type of variance is applied for and issued by OSHA.

Exposure Limit – Guidelines for worker exposure to physical agents and hazardous chemicals, usually expressed as an allowable time of exposure or an air concentration, below which health hazards are unlikely to occur among most exposed workers.

Face Velocity – The average velocity of air as it enters the hood of an LEV system.

Facilitated Diffusion – Diffusion that occurs across a concentration gradient, but cannot occur unless a specific carrier molecule is present on the cell membrane.

Far – In the context of radiation, a term used to refer to wavelengths that are the greatest distance from visible light, e.g., far infrared.

Ferruginous Bodies – Reddish, iron-containing lesions that form in the lung in response to inhalation of asbestos fibers.

Fetal Exposure – Exposure that occurs to the fetus in the womb.

Fibroblast – Specialized cell that produces fiber-forming material.

Fibrogenic – Associated with a fibrous response.

Fibrosis – A condition where collagen has been deposited in the liver.

FID – See Flame Ionization Detector.

Field Blank – Sample media that are exposed to the same conditions as the media used for the actual sampling, but that are not connected to a sampling pump.

Film Badge – A personal dosimeter containing photographic film that is darkened by ionizing radiation; it is used in comparison to a control film to evaluate the degree of ionizing radiation exposure.

Fit-test – A method for evaluating how well the respirator seals against the wearer's face. OSHA requires that a respirator be found to have a satisfactory fit before it can be used.

Fixed Contamination – In the context of radiation, it is nonremovable contamination.

Flame Ionization Detector (FID) – A nonspecific air sampling instrument, used to identify the amount of a substance by measuring the absorption of electrons resulting from its ionization as it passes through a hydrogen flame. It is generally used for detecting organic compound, and specifically hydrocarbons.

Fluorescence Spectrometry – Analysis in which the intensity and wavelength of the energy that is emitted from excited atoms is used to indicate the presence of certain compounds.

Focal Emphysema – Permanently open areas of tissue in localized areas of the lungs.

Folliculitis – An inflammation of the hair follicles.

Forward Blades – Centrifugal fan blades that are curved or tilted in the direction of rotation. They require more power to overcome air resistance with increased flow, leading to the possibility of motor failure.

Free Radical – A molecular fragment that forms as the result of exposure to ionizing radiation and that has one or more unpaired electrons. Free radicals formed in living tissue may inactivate enzymes, break molecular bonds, affect cell membrane permeability, or cause mutation of genetic material.

Friction Loss – The kinetic energy that is required to overcome the friction between the air and the sides of the duct.

Frit – A porous structure that breaks an air stream into small bubbles thereby maximizing the surface area of air contact with the solution and increasing the amount of contaminant dissolved.

Full-face Respirator – A respiratory protective device that covers the entire face, from the hairline to under the chin.

Fume – Aerosol produced when a material in the gaseous phase condenses to form a solid.

Gamete – Germ cells, such as sperm or eggs.

Gamma Radiation – See Gamma Ray.

Gamma Ray – The shortest wavelength and highest energy type of all electromagnetic radiation. It originates in the nucleus of radioactive isotopes along with alpha particle, beta particle, or neutron emissions.

Gas – One of the three states of matter – along with solids and liquids – that is characterized by a lack of defined shape and volume.

Gas Chromatograph (GC) – An analytical instrument with an internal tube or column that contains a solid sorbent. The sample is injected into the column, which allows some components to pass through more quickly than others, thereby separating the substances in the sample.

GC – See Gas Chromatograph.

GC/MS – The use of a gas chromatograph (GC) followed by a mass spectrometer (MS). The gas chromatograph separates the sample's components, which are then identified by the mass spectrometer. See Gas Chromatography and/or Mass Spectrometer.

General Duty Clause – A clause in the OSHAct that requires the employer to provide a workplace that is free from recognized hazards likely to cause death or serious physical harm.

Genetic Effect – In the context of radiation, a cellular change to eggs and sperm, which can be passed on to the offspring of exposed individuals.

Genotoxic Carcinogen – A substance that causes cancer through a reaction or effect on DNA.

Glass Fiber Filter – Sometimes referred to as AE filter, it is composed of a mat of glass fibers arranged in a haphazard pattern. It is used for collecting particulates and some droplets of contaminants, such as mercury and acid gases.

Glassblower's Cataracts – Cataracts that result from repeated exposures to the near IR radiation emitted by hot glass.

Glomerulus – A bed of capillaries surrounding the tubules of a nephron.

Glove and Stocking Syndrome – A neuropathy characterized by loss of feeling in hands and feet.

Gravimetric Analysis – Analytical technique that involves determining the mass of the contaminant collected, usually on a filter.

Ground State – Atoms with all of their electrons in the most stable energy arrangement.

Hair Cells – Specialized cells that are located between the basilar and tectorial membranes in the cochlea. The bending of the hair cells generates a nerve impulse that travels from the hair cells through nerves to the brain, where it is interpreted as sound.

Half-face Respirator – See Half-mask Respirator.

Half-life – The time it takes for one-half of a radioactive isotope to decay to a stable (nonradioactive) form.

Half-mask Respirator – A respiratory protective device that covers half of the face, roughly from under the chin to the bridge of the nose.

Halogenated Compound – A compound that contains the elements chlorine, bromine, fluorine, or iodine as a part of its structure.

Hanta Virus – An infection caused by inhaling the virus in the dust from nesting materials and droppings of deer mice. The illness produces a series of flu-like respiratory symptoms, which may be mistaken for other diseases, such as flu or bronchitis, sometimes resulting in death.

Hapten – A molecule formed from the combination of a foreign molecule with protein molecules normally present in the body.

Hearing Threshold Level (HTL) – The lowest level at which a person is able to detect a sound or tone. Test frequencies used to establish HTLs include 500, 1,000, 2,000, 3,000, 4,000, and 6,000 Hz.

Heat Cramps – Muscle cramps, usually of the legs and abdomen, caused by heavy sweating that results in an imbalance in the salts and minerals of the muscles.

Heat Rash – A rash that appears as small red bumps on skin that has become wet and remained damp; the bumps are inflamed sweat glands.

Heat Stroke – A heat-induced medical emergency with symptoms including a red, hot face and/or skin; lack of or reduced sweating; erratic behavior, confusion, or dizziness; and collapse or unconsciousness.

Heat Syncope – Fainting that occurs among people who have been standing in one position for a period of time.

HEG – See Homogenous Exposure Group.

Hemoglobin – The protein in red blood cells that bind with and transports oxygen.

HEPA – High efficiency particulate air – an air filtration medium scientifically defined by the Institute of Environmental Science for its 99.97 percent efficiency in capturing particles down to 0.3 microns.

Hertz (Hz) – The frequency with which sound pressure changes; e.g., one pressure oscillation per second is equal to one Hertz.

Histological Change – An observable change in the shape or appearance of cells or an organ.

Histoplasmosis – An infection caused by a fungus found in bird droppings. Also known as bird fanciers disease, it usually affects the lungs and can result in an allergic response, serious illness, or sometimes even death.

Homogenous Exposure Group – A population or group of workers with similar exposures.

Horny Layer – See Stratum Corneum.

HTL – See Hearing Threshold Level.

Human Factors – A term sometimes used synonymously with ergonomics; it may also refer to psychological and sociological aspects of ergonomic issues.

HVAC – The heating, ventilation, and air conditioning system that circulates and delivers filtered, humidified/dehumidified, and cooled or warmed air to the interior of a building.

Hydrophobic – Water resistant or having a lack of affinity for water; usually said of a substance or material that does not absorb moisture.

Hyperpigmentation – A condition where the production of melanin is stimulated, resulting in darkening of the skin.

Hypertension – The condition characterized by an elevated blood pressure.

Hypoxia – A condition characterized by a deficiency of oxygen reaching the tissue.

Hz – See Hertz.

IAQ – See Indoor Air Quality.

IDLH – See Immediately Dangerous to Life and Health.

IH – See Industrial Hygienist.

Immediately Dangerous to Life and Health (IDLH) – A condition that poses a threat of exposure to airborne contaminants that is likely to cause death, or immediate or delayed permanent adverse health effects, or prevent escape.

Immune Cell Infiltrate – The presence of abnormally high numbers of immune cells.

Impaction – Collection mechanism where the contaminants collide with the surfaces in the filter.

Impinger – Sample collector that resembles a graduated cylinder with a long inlet tube fitted into a stopper. The inlet tube extends nearly to the bottom of the outer vial, which holds the solution. The sampling pump is connected so that negative pressure is created inside the impinger, drawing air through the inlet tube, where it exits into the solution, allowing the air to bubble up through the solution.

Impulse Noise – Noise of short duration; three seconds or less; also called impact noise.

Indoor Air Quality (IAQ) – A general term used to refer to the relative acceptability of the air in an occupied building, in terms of noticeable odors, temperature, humidity, and other factors.

Industrial Hygiene – The science and art devoted to the anticipation, recognition, evaluation, and control of those environmental factors or stresses, arising in or from the work place, which cause sickness, impaired health and well-being, or significant discomfort and inefficiency among workers or among the citizens of the community.

Industrial Hygienist (IH) – A college graduate in engineering, chemistry, physics, medicine, or related physical and biological sciences, who has received specialized training in recognition, evaluation, and control of work place stressors and, therefore, achieved competence in industrial hygiene.

Inhalable Fraction – The particles with aerodynamic diameters of 10 μm and greater.

Initiator – In the context of carcinogens, an agent or event that causes a change, making the cell or tissue susceptible to the development of cancer.

Integrated Sampling – Samples taken by drawing air through the sampling medium, which is then analyzed by a laboratory to determine the amount of contaminant present.

Interception – Collection mechanism where the contaminant molecules stick to surfaces as they pass close by.

Interim Order – An official statement issued by OSHA to allow an employer to continue operations under existing conditions while an application for a variance is being considered.

Interstitial – Situated between the cells of a structure or part.

Intrinsically Safe – Attribute of instruments that can be safely operated in a possibly combustible atmosphere.

Ion – A charged particle; formed by an atom gaining or losing one or more electrons or a molecule that has undergone ionization.

Ionization – A chemical change in which a neutral molecule is broken into charged particles, such as water $H_2O \Leftrightarrow H^+ + OH^-$.

Ionization Chamber – A gas-filled container with a positive and a negative electrode. When exposed to ionizing radiation, electrons are displaced from the gas atoms and they are attracted to the positive electrode, thereby allowing for direct measurement of the amount of radiation.

Ionization Potential (IP) – The amount of energy required to remove the most loosely bound electron from an atom or to cause the ionization of a molecule. The ionization potential for each substance is specific.

Ionizing Radiation – Radiation that is capable of causing ionization to occur, either directly or indirectly, through interaction with matter.

IP – See Ionization Potential.

Irritant Contact Dermatitis – A skin condition that occurs in response to irritating industrial chemicals such as solvents, acids, and bases. Similar in appearance and symptoms to allergic contact dermatitis.

Irritation – A reaction of tissues to an injury that results in an inflammation at the site of contact.

Keratin Layer – See Stratum Corneum.

Laboratory Blanks – Sample media that is not sampled on, but is analyzed by the laboratory to detect contamination or other problems associated with preparation and analysis of the samples.

Langerhans Cells – Specialized skin cells that contain sites for chemically binding with haptens that enter the dermis.

LCL – See Lower Confidence Limit.

LD_{50} – The amount of a substance needed to produce death in 50 percent of the treated (exposed) population.

LEL – See Lower Explosive Limit.

Length-of-stain Tube – See Detector Tube.

LEV – See Local Exhaust Ventilation.

LFL – See Lower Flammable Limit.

Limit of Detection (LOD) – The smallest amount of contaminant that can be reliably detected and quantified, using the sampling and analytical techniques in the sampling method.

Limit of Quantitation (LOQ) – The smallest amount of contaminant that can be reliably quantified using a particular analytical technique. See Limit of Detection.

Local Exhaust Ventilation (LEV) – A type of ventilation system designed to capture contaminants at the point of generation.

LOD – See Limit of Detection.

LOQ – See Limit of Quantitation.

Lower Confidence Limit (LCL) – A statistical procedure to estimate whether the true value is lower than the measured value.

Lower Explosive Level – See Lower Explosive Limit.

Lower Explosive Limit – The lowest gas-air mixture that will support combustion.

Lower Flammable Limit – See Lower Explosive Limit.

m/e – The mass to charge ratio determined by dividing the molar mass of the ion by its ionic charge.

Macrophages – White blood cells that phagocytize foreign substances in the alveoli.

Macule – Name given to characteristic lesions that form in the lung in response to deposit of inhaled coal dust.

Manometer – An instrument used for measuring pressure differences. It consists of a U-shaped tube, partially filled with a liquid and constructed so that the displacement of the liquid is proportional to the pressure difference on the two sides.

Marginal Irritant – A material that is capable of causing an irritation response after repeated exposures.

Mass Spec – See Mass Spectrometer.

Mass Spectrometer (MS) – An instrument that identifies substances by causing them to be ionized and subjecting the resulting ions to a strong electromagnetic field. From the curved paths they follow, their mass-to-charge ratio (m/e) can be determined. The degree to which they are deflected, when compared to a standard, can be used to determine their identity.

Mass Spectrum – Printout or plot produced by bombarding a sample with a beam of electrons using an instrument called a mass spectrometer. The plot shows the relative intensities of each ion that is present; the mass spectrum of a specific compound is like a fingerprint and can be used to identify the compound.

Maximum Risk Employees – Workers who are most likely to be exposed to the highest levels of hazardous agents.

Maximum Use Concentration (MUC) – The maximum atmospheric concentration in which a respirator cartridge or filter is recommended for use; can be approximated by multiplying the PEL for the contaminant of concern by the assigned protection factor.

Medical Surveillance Program – The evaluation of an employee's health status that is performed on a regular and periodic basis by a health professional to detect problems associated with exposure to health hazards so that appropriate steps can be taken to prevent permanent or debilitating injury from occurring. Medical surveillance programs may also be used to assure that an employee's health status will allow the continued safe use of personal protective equipment, or the continued safe performance of job duties.

Melanin – The pigment produced by the melanocytes. Melanin imparts color to the skin and protects it from the effects of ultraviolet radiation.

Melanocytes – Cells that produce the pigment or coloration (melanin) of the skin and hair.

Mesothelioma – A relatively rare form of lung cancer that occurs with a higher frequency among persons exposed to amosite asbestos. This cancer can develop in the thin membranes that surround the lungs or line the body cavity.

Metal Fume Fever – Possible response to inhalation of metal fumes characterized by muscle aches, nausea, chills, and fever.

Metal-oxide Semiconductor (MOS) – A solid-state sensor that makes use of the change in electrical conductivity that occurs in relationship to the amount of combustible gas that is adsorbed onto its surface.

Methemoglobin – A brownish compound of oxygen and hemoglobin, with the iron in its +2 oxidation state, that cannot bind with oxygen.

Methemoglobinemia – Asphyxiation resulting from the formation of methemoglobin.

Microscopist – A person who specializes in the use of a microscope, such as a person who analyzes samples for asbestos fibers.

Millirem (mrem) – A measurement of radiation that is one-one thousandth (0.001) of a rem.

Mist – An aerosol consisting of liquid droplets with diameters in the 40–50 micron size range.

Mixing Factor – A number used to express the degree to which mixing occurs in the air that occupies a space.

Mixture Rule – Applies to workers exposed simultaneously to chemicals that act on the same organ or organ system. Levels of exposure must remain at a fraction of the PEL so the sum of the exposures does not exceed unity, or 1. The mixture rule can be illustrated mathematically by the following equation: $PEL_{mixture} = C_1/PEL_1 + C_2/PEL_2 + ... C_n/PEL_n$.

MOS – See Metal-oxide Semiconductor.

mrem – See Millirem.

MS – See Mass Spectrometer.

MUC – See Maximum Use Concentration.

Muco-ciliary Elevator – Mechanism for removal of mucus and particles from the lungs; cilia push the particle-laden mucus upward, toward the larger air passages, where it may be coughed and spit out, or otherwise removed from the respiratory system.

Mucous Membrane – Specialized skin tissue; contains cells that produce mucus.

Mucus – A moist, viscous, and sticky substance produced by lining the nasal, esophageal, and other body cavities.

Mutagenicity – The ability of a substance to produce changes in genetic material (DNA).

Mutations – Changes in the characteristics of an organism produced by modifications in its DNA.

Myelin Sheath – The layer of protective tissue that surrounds nerves.

Nasopharyngeal Region – The portion of the respiratory tract that is comprised of the head, nose and nasal passages, sinuses and mouth, and all associated features such as the tonsils and epiglottis, including the back of the throat.

National Institute of Occupational Safety and Health (NIOSH) – A branch of OSHA that conducts research on occupational safety and health questions and recommends new standards. NIOSH also tests and certifies respirators.

Near – In the context of radiation, a term used to refer to wavelengths that are close to visible light, e.g., near infrared.

Necrosis – The death of one or more cells or a portion of a tissue or organ.

Negative-pressure Respirator – A descriptive term for a tight-fitting air-purifying respirator, where air is drawn through the cartridge or filters by negative pressure that is created inside the respirator facepiece when the user inhales.

NEL – See No Effect Level.

Neoplasia – A term derived from neoplasm meaning new growth; e.g., cancer.

Neoplasm – A term that literally means a new form or a new growth; a cancerous growth, such as a tumor.

Nephron – The filtering and excretory unit of the kidneys, consisting of the glomerulus and tubules.

Neuropathy – Refers to a toxic effect characterized by the dying-back of nerves.

Neurotoxins – Compounds that have a negative effect on the nervous system.

Neutrons – Neutrally charged particles found in the nucleus of most atoms.

NIOSH – See National Institute of Occupational Safety and Health.

NOAEL – See No Observed Adverse Effect Level.

No Effect Level – A term used interchangeably with threshold dose.

No Observed Adverse Effect Level (NOAEL) – Term used interchangeably with threshold dose.

Nodules – Characteristic lesions that may form in response to inhaled silica-containing dusts.

Noise – A general term used to refer to unwanted sound; the terms noise and sound are frequently used interchangeably.

Noise-Induced Hearing Loss – Hearing loss caused by exposure to noise.

Noise-Reduction Rating (NRR) – A numerical value assigned to a hearing protective device; the value, given in dB, represents the amount of attenuation provided by the device under laboratory conditions.

Nonfibrogenic – Not associated with a fibrous response.

Nonspecific – Related to an instrument's response that does not identify the specific contaminant that is present.

Nosacusis – Refers to hearing loss that results from causes other than noise, such as diseases, heredity, drugs, exposure to sudden and severe pressure changes, or traumatic head injuries.

NRR – See Noise-Reduction Rating.

Obstructive – In the context of breathing, a pattern of lung impairment characterized by decreased ability to force air out of the lungs.

Occupational Acne – Red eruptions on the surface of the skin as a result of exposure to oily, greasy, or irritating materials; the condition resembles adolescent acne in appearance.

Occupational Safety and Health Act (OSHAct) – It was enacted December 29, 1970 and became effective April 28, 1971. The act created the Occupational Safety and Health Administration (OSHA) of the Department of Labor, which provides the regulatory vehicle for assuring the safety and health of workers in firms generally employing more than 10 people. Its goal is to set standards of safety that will prevent injury and/or illness among workers.

Occupational Safety and Health Administration (OSHA) – An agency of the U.S. Department of Labor that establishes workplace safety and health regulations. State OSHA programs are also monitored by federal OSHA to ensure that they are at least as effective as the federal OSHA program.

One-Hit Theory – The theory of cancer causation that assumes a single exposure or DNA-damaging event could trigger cancer.

Open Face – Sampling that is performed with the top portion of a filter holder or cassette removed so that the entire filter surface is exposed.

OSHA – See Occupational Safety and Health Administration.

OSHAct – See Occupational Safety and Health Act.

Otitis Media – A condition characterized by fluid in the middle ear; a cause of conductive hearing loss.

Oval Window – The membrane that covers the entrance to the inner ear.

Overbreathe – The condition where the wearer's breathing rate exceeds the ability of the respirator to provide a sufficient volume of air to maintain a positive pressure inside the facepiece.

Oxygen-deficient – An atmosphere containing less than 19.5 percent oxygen.

Oxygen-enriched – An atmosphere containing more than 23 percent oxygen.

PAPRs – See Powered Air Purifying Respirators.

Paracusis – The perception of a sound that does not correspond to the actual sound; musical notes that sound off key, for example.

Partial Pressure – The part of the total pressure of a gas mixture, such as air, that can be attributed to a specific component, such as oxygen.

Passive Sampling – Sampling that relies on diffusion of the contaminant from the air onto a solid sorbent.

Pathogenic – Capable of causing illness or disease; usually refers to microorganisms or viruses.

PC – See Anti-C.

pCi – See Picocurie.

Peak Limit – Refers to a level which exceeds the ceiling limit, but which OSHA allows for a specific limited time during the work shift. Peak limits must never be exceeded and must be compensated for by periods of exposure during which the concentration is low enough so that the employee's cumulative 8-hour TWA is below the PEL.

PEL – See Permissible Exposure Limit.

Penetration – In the context of PPE, the leaking of a contaminant through seams, zippers, pinholes, and other seemingly invisible openings in protective clothing or gloves.

Percent Transmittance – The amount of light that passes through a solution and is proportional to the solution's concentration.

Peripheral Nervous System – Nerve tissues lying outside the brain and spinal cord; functions include the transmittal of sensory information (touch, heat/cold, proprioception, and pain) and motor impulses for movement of the limbs.

Permanent Variance – An alternative method for worker protection, which replaces the OSHA requirement.

Permeation – In the context of PPE, the process by which a chemical dissolves in or moves through a protective clothing material on a molecular level.

Permeation Rate – In the context of PPE, the speed at which a chemical passes through a barrier material; generally reported in $mg/cm^2/min$.

Permissible Exposure Limit (PEL) – A term used to refer to maximum exposure levels allowed by OSHA as 8-hour TWAs. The PELs are enforceable as law.

Permit Space – As defined by OSHA, a confined space that contains a hazardous atmosphere, a material that could engulf an entrant, a configuration that could trap an entrant, or contains any other recognized safety or health hazard.

Personal Protective Equipment (PPE) – Devices and apparel worn by employees to prevent or reduce exposure to health and safety hazards. Examples include respirators, gloves, chemical-resistant coveralls, and safety glasses.

Personal Samples – Samples that are obtained when a worker wears a sampling train for some interval during the work shift.

Phagocytosis – From Greek roots phago, meaning to eat and pino, meaning hungry; the engulfing of a foreign molecule by the cell membrane, or by another cell, such as a white blood cell.

Photoallergic – A response similar to other allergic responses of the skin, except that sunlight or another source of ultraviolet light is required in addition to a sensitizer, to produce the hapten that causes the allergic reaction.

Photoionization Detector (PID) – A nonspecific instrument used to identify the amount of a substance by measuring the absorption of electrons resulting from its ionization by a powerful ultraviolet light source. The energy provided by the ultraviolet lamp may be changed, but it must provide enough energy to reach the substance's ionization potential (IP). The measurement readout is expressed as a ratio of parts of total contaminant per million parts air (ppm).

Photometer – Analytical instrument that contains a light source on one side and a light detector on the opposite side; the amount of light that passes through the sample indicates the result.

Photon – The unit of electromagnetic radiation; photons are discrete concentrations of energy that have no rest mass and move at the speed of light. All electromagnetic radiation – radio waves, IR, visible light, UV, x-rays, and gamma-rays – are composed of photons differing only in their wavelengths.

Phototoxic – A skin response to an irritant, which is made more severe in the presence of sunlight.

Physical Agents – Potential health hazards such as ionizing and nonionizing radiation, noise, vibration, and temperature extremes.

Picocurie (pCi) – A term used to quantify the activity of a radioactive substance; 1 pCi is equal to 0.037 decays per second.

PIDs – See Photoionization Detectors.

Pinna – The outer, cartilaginous, and most visible part of the ear, which amplifies and channels sound into the ear canal.

Pinocytosis – See Phagocytosis.

Pitch – The relative highness or lowness of a sound; higher pitched noises have higher frequencies and lower-pitched sounds have lower frequencies.

Pitot Traverse – A series of measurements of the total and static pressure taken across the area of a duct to determine the air velocity at that point. The sampling distance should be at least 7.5 times the diameter of the duct from any disturbances of air flow.

Pitot Tube – An instrument used to perform a Pitot traverse and named after its inventor, Henri Pitot. It consists of two concentric tubes arranged in such a way as to simultaneously measure both the total and static pressures within a duct.

Plenum – An enclosed space that receives air from occupied areas and returns it to the HVAC unit.

Pneumoconiosis – General term that refers to the response of the lungs to inhaled dust.

Pneumocytes – Specialized skin cells that comprise the walls of the alveoli.

Pneumonitis – Inflammation of the respiratory tract tissue.

Pocket Dosimeter – A tube-like device that contains a charged quartz fiber. Ionizing radiation results in a change in the electric charge on the fiber that is in direct proportion to the amount of radiation exposure.

Polyvinyl Chloride (PVC) Filter – A filter that has good resistance to acids and bases and does not readily absorb water. It is lightweight and commonly used to collect dust samples.

Popliteal – In anthropometry, the distance from the floor to the back of knee.

Pores – Small openings in the surface of a sampling filter.

Positron – A particle identical to a beta particle, except that it carries a charge of plus one (+).

Powered Air Purifying Respirator (PAPR) – A type of air-purifying respirator that utilizes a battery-powered fan to draw contaminated air through the cartridge or filter before entering the facepiece.

PPE – See Personal Protective Equipment.

Preamble – The explanation and reason for implementing an OSHA standard. It typically precedes the text of the standard published in the Federal Register.

Precipitate – An insoluble substance produced by the reaction of two clear solutions; it causes the solution to become cloudy.

Presbycusis – Hearing loss that is attributed to aging.

Pressure-demand Mode – The mode of operation in which the facepiece of a respirator is maintained under a slight positive pressure at all times, providing a very high level of protection for the wearer.

Prickly Heat – See Heat Rash.

Primary Calibration Standard – Calibration standard based on direct measure of a reference value, such as the known volume of a buret.

Progressive Massive Fibrosis – Growth or formation of a large fibrotic mass in response to coal dust inhalation.

Promoter – A chemical substance or specific set of conditions that is favorable to or triggers the development of cancer.

Proprioception – The ability to recognize the relative position of one's body and limbs.

Prospective Epidemiological Study – An epidemiological study that follows a group of exposed individuals over time to see if they develop disease because of an exposure.

Proton – A positively charged subatomic particle found in the nucleus of all atoms.

Psychrometer – An instrument used to measure the relative humidity of air by using two thermometers; the bulb of one being kept moist and ventilated.

PTFE Filter – See Teflon® Filter.

Pulmonary Edema – The collection of fluid in the lungs.

PVC Filter – See Polyvinyl Chloride Filter.

Qualitative Fit-test – A method for evaluating respirator fit that relies on the wearer's response to a test agent. Common test agents are banana oil or irritant smoke. The wearer performs physical activities and reads out loud the rainbow passage, to simulate work conditions. If the wearer fails to detect the test agent under these conditions, then the brand and size of this respirator is considered to fit.

Quantitative Fit-test – A method for evaluating respirator fit that utilizes an instrument to measure the concentration of test agent both inside and outside of the respirator. Facepieces used for these types of tests must be modified to accept a small probe for attachment to the instrument. The wearer performs physical activities and reads the rainbow passage during the test. The result of the test is a numerical expression of fit, called the fit factor, which represents how well the respirator seals against the wearer's face.

R – See Roentgen.

rad – See Roentgen Absorbed Dose.

Radiation – The spontaneous release of energy/matter from an atoms with unstable nuclei as they move toward a more stable nuclear configuration.

Radioactive Materials – Elements that have unstable nuclei that spontaneously disintegrate releasing radiation (mass/energy).

Radon Daughters – The series of isotopes, also with unstable atomic nuclei, that are formed as radon atoms undergo radioactive decay.

Raynaud's Syndrome – A diminished blood flow or loss of blood flow to the digits, due to spasms in blood vessels brought on by vibration. Also known as white finger or dead finger because the fingers turn white and are numb and/or tingling.

Reactive Airway Disease – General term for the response of the lungs to inhalation of organic particles.

Receiving Hood – The entry point into a LEV system that resembles a canopy. It is commonly a hood over a hot process tank or grill in a restaurant. Also called a canopy hood.

Recommended Exposure Limit (REL) – NIOSH-recommended maximum exposure level for hazardous agents. It may be expressed as an 8- or 10-hour TWA.

Reflection – The phenomenon of sound waves impacting a surface, changing direction, and moving back in the direction of the source.

Regulated Areas – An administrative control technique that relies on the establishment of a limited access area to eliminate or reduce employee exposure to hazardous materials. Usually accomplished through the posting of signs and/or the erecting of some type of physical barrier to prevent workers from inadvertently entering the area; OSHA uses the term in many substance-specific standards.

REL – See Recommended Exposure Level.

rem – See Roentgen Equivalent Man.

Removable Contamination – In the context of radiation, contamination that can be removed from surfaces using light pressure; also called smearable.

Repetitive Trauma – A term used to describe injuries from repeated use of the body to perform a task or motion that is unnatural or that the human body is poorly adapted to perform.

Repetitive Use Injuries – See Repetitive Trauma.

Representativeness – The degree to which the sample data accurately represent the situation. It is a qualitative parameter contingent on both proper sampling methods and proper laboratory protocols.

Resonance – The phenomenon of a material vibrating at the same frequency as the emitted sound.

Respirable – Related to particles that are in the size range that allows them to reach the gas exchange region of the lungs.

Response – 1) In the context of a substance, the effect a substance has may be positive, as in the therapeutic dosage of a drug, or negative such as severe irritation of the respiratory tract. – 2) In the context of sound, it is the speed with which the measuring instrument reacts to a sudden increase in sound pressure. A slow response is used to evaluate the average value of the observed sound pressure, while a fast response setting is used for estimating its variability.

Restrictive – Pattern of lung impairment characterized by a decreased ability to inhale air.

Reticulin – A fibrous, thready substance that resembles connective tissue.

Retrospective Epidemiological Study – An epidemiological study that attempts to trace a group of individuals back in time to determine if there was a common exposure that can be attributed to a specific condition or disease.

Roentgen (R) – The amount of x-radiation or gamma-radiation that produces one unit of charge in one cubic centimeter of dry air.

Roentgen Absorbed Dose (rad) – The unit representing an absorbed dose (100 ergs of energy per gram) of ionizing radiation.

Roentgen Equivalent Man (rem) – The dose of any radiation to body tissue in terms of its estimated biological effect, relative to the dose received from an exposure to an ionizing unit of x-rays.

Rotameter – A secondary calibration standard that consists of a clear tube with a metal or plastic float; the position of the float on a graduated scale indicates the flow rate.

Route of Entry – Path by which toxins and other substances may enter the human body. Routes include inhalation, ingestion, and absorption through the skin. Less common routes include injection and absorption through moist surfaces surrounding the eyes and in the ear canal.

SAE – See Sampling and Analytical Error.

Safety Factor – An uncertainty factor that is used in combination with the no-adverse-effect-level data to estimate a safe human dose.

Sample Badge – A small clip-on device that contains solid sorbent and is used for collection of a variety of airborne materials. It is a passive sampler that is typically clipped to the worker's lapel and worn throughout a shift.

Sample Bag – Bag made of an inert polymer, such as Teflon® or Tedlar®, with an attached fitting for connection to an air sampling pump.

Sampling and Analytical Error (SAE) – A numerical factor included as a part of many analytical methods that accounts for uncontrollable errors. This value is used to evaluate whether the sample results indicate exposures within acceptable limits.

Sampling Medium – The device or material through which air is drawn to collect contaminants that are present.

Sampling Train – The assembly of sample medium in its holder, with associated tubing and sample pump.

Secondary Calibration Standards – Calibration devices that are checked against a primary calibration standard; they are often more convenient and practical to carry into the field for use during sampling.

Secondary Ionization – The damage caused to a cell or molecule by free radicals.

Semicircular Canals – The fluid-filled tubes located in the inner ear that provide our sense of balance as well as the perception of our body's position relative to our surroundings.

Sensitization – An adverse reaction that occurs following more than one exposure to a substance.

Sensitizer – A substance that stimulates a response from the immune system.

Sensory Hearing Loss – An irreversible hearing loss resulting from damage to the inner ear tissue that translates sound pressure into nerve impulses.

Serpentine – An asbestos mineral with a wavy appearance; chrysotile is an example.

Shaver's Disease – A disease of the lungs found in workers exposed to fumes or dust containing aluminum oxide. It is a type of pneumoconiosis and results in the formation of interstitial fibrosis and decreased lung function.

Short-Term Exposure Limit – 1) OSHA: a 15-minute TWA exposure that should not be exceeded at any time during a work day, unless another time limit is specified in a notation below the limit, in which case the TWA exposure over the specified time period should not be exceeded at any time during the working day. – 2) ACGIH: a 15-minute TWA exposure that should not be exceeded at any time during a workday, even if workplace concentrations are at or below the TLV-TWA. STEL exposures should not be longer than 15 minutes; they should not occur more than four times each day, and there should be at least 60 minutes between each successive exposure in the STEL range. In some instances, an averaging period of other than 15 minutes is used.

Sick Building Syndrome – A term used to refer to a situation where the indoor building environment is the known or assumed cause of physical complaints and/or symptoms of the building occupants.

Siderosis – A benign reddish discoloration resulting from deposits of iron oxide in the lungs.

Silicosis – Pneumoconiosis resulting from inhalation of crystalline silica (quartz).

Simple Asphyxiant – Substance that displaces air, producing an oxygen-deficient atmosphere.

Size-selective Sampling – Industrial hygiene sampling methods that collect particles having a specific range of aerodynamic diameters.

Skin Notation – The word skin included as part of an exposure limit. It is used for those substances where absorption through the skin is considered to be a significant route of entry into the body.

Sociacusis – Hearing loss that is the result of being exposed to the noises of every day life.

Somatic Effects – In the context of radiation, those effects that are expressed in the cells and tissues of exposed individuals; including cataracts, disruptions of the gastro-intestinal tract and cancers or tissue damage to the thyroid, kidney, spleen, pancreas, and prostate.

Sorbent Tubes – Small glass tubes that contain sampling media such as silica gel or activated charcoal.

Sound Power – The amount of energy carried by the cyclic sound or noise pressure changes.

Sound Pressure – Cyclic pressure changes that are produced by a source and perceived as sound or noise.

Spirometry – A test method used to evaluate lung function; measures volume of (exhaled) air passing through a tube during a given time.

Standard Threshold Shift (STS) – An increase of 10 dB or more in a person's HTL in the 2,000 to 4,000 Hz range; significant because it represents permanent loss of hearing of a significant proportion.

Static Pressure – The pressure exerted in all directions by air moving inside the duct of an LEV system.

Steatosis – The accumulation of fat between liver cells.

STEL – See Short-Term Exposure Limit.

Stimulant – An agent that acts to increase a neuron's sensitivity to nerve impulses.

Stratum Corneum – The outer layer of the epidermis, which is composed of dead keratin-filled cells that have migrated outward from the underlying basal layer. It is also known as the horny layer.

STS – See Standard Threshold Shift.

Subacute Effect – The effect associated with exposures or doses that are below the LD_{50}, but still high enough to cause a toxic response.

Subcutaneous – The deepest layer of the skin, containing fatty and connective tissue, that provides a cushion and insulative base for the skin and also binds the skin to the underlying tissues.

Synergism – The interaction of two or more chemicals or phenomena producing a total effect that is greater than the sum of the individual effects.

Systemic Toxin – A substance that affects target organs or entire organ systems.

Target Organ – Refers to a specific organ where the toxic effect of a substance is manifested.

Tectorial Membrane – The thin layer of tissue that lies on top of the hair cells in the cochlea.

Teflon® Filter – A chemical-resistant and hydrophobic filter composed of polytetraflouroethylene (PTFE) and used for industrial hygiene sampling (e.g., aromatic hydrocarbons).

Temporary Threshold Shift (TTS) – A temporary shift in hearing threshold level that goes away after the person has been in a quiet environment for a few hours; it is confirmed by an audiogram retest – following a suspected STS – after at least 14 hours away from high levels of noise.

Temporary Variance – An OSHA variance issued to an employer who is unable to comply with a standard by its effective date, for reasons beyond their control. The employer must demonstrate a plan for coming into compliance within a period not to exceed one year and provide all available measures to protect the employees in the interim.

Teratogens – Toxins that cause abnormal development or birth defects.

Thermal Conductivity – The ability to conduct heat.

Thermoluminescent Detector (TLD) – Dosimeter used for monitoring beta, gamma, and x-radiation. It contains a small chip of lithium fluoride (LiF) that absorbs the radiation energy. Following exposure, the chip is heated, causing the excited electrons to return to the ground state thereby releasing light energy that is proportional to the amount of energy absorbed.

Thoracic Fraction – Particles with aerodynamic diameters of 5-10 mm.

Threshold Dose – A dosage or exposure level below which the adverse effects of a substance are not realized or expressed by the exposed population.

Threshold Limit Value-Time Weighted Average (TLV®-TWA) – ACGIH term applied to the concentration for a normal 8-hour workday, 40-hour week, to which it is believed nearly all workers may be repeatedly exposed, day after day, with no adverse effect.

Threshold Limit Value (TLV®) – Refers limits of exposure to hazardous agents; represents conditions under which the American Conference of governmental Industrial Hygienists (ACGIH) believes that nearly all workers may be repeatedly exposed, day after day without adverse health effects. TLVs are reviewed and revised annually and are normally based on an assumed 8-hour work day for 40 hours per week.

Tight Building Syndrome – See Sick Building Syndrome.

Time-Weighted Average (TWA) – Averaging a measured exposure over the time period during which the sampling took place.

Tinnitus – The perception of noises that are not there, like ringing, roaring, or hissing sounds.

Titration – An analytical method involving the use of reagents that are added to the absorbing solution until an endpoint is reached as indicated by a color change or formation of a precipitate.

TLD – See Thermoluminescent Detector.

TLV® – See Threshold Limit Value (TLV®).

TLV®-C – See TLV®-Ceiling

TLV®-Ceiling (TLV®-C) – ACGIH term for a concentration that cannot be exceeded during any part of the workday.

TLV®-TWA – See Threshold Limit Value-Time Weighted Average.

Total Pressure – The sum of velocity pressure and static pressure inside the air duct of a LEV system.

Toxicity – The ability of a substance to cause harm or adversely affect an organism.

Toxicology – The science and study of harmful chemical interactions on living tissue.

Tracheobronchial Region – Middle region of the respiratory system; comprised of the trachea or windpipe, and the bronchi.

Trigger Finger – An injury characterized by the inability to bend or straighten a finger; it is caused by a constriction of the tendon, which affects finger movement.

TTS – See Temporary Threshold Shift.

TWA – See Time-Weighted Average.

UCL – See Upper Confidence Limit.

UEL – See Upper Explosive Limit.

UFL – See Upper Flammable Limit.

Upper Confidence Limit (UCL) – A statistical procedure used to estimate whether the true value is higher than the measured value.

Upper Explosive Limit – The highest gas-air mixture that will support combustion.

Upper Flammable Limit – See Upper Explosive Limit.

Urticaria – An allergic reaction to a protein compound produced by a plant or animal; characterized by swelling and reddening, and sometimes small fluid-containing blisters, on the surface of the skin.

Vapor – The term used to describe a substance that is in the gaseous phase, but is a liquid or solid at room temperature and pressure.

Vapor Pressure – The pressure exerted by a solid or liquid when in equilibrium with its own vapor. It is indicative of the evaporation rate of the substance.

Variance – An alternative to an OSHA requirement that ensures that the employer's workplace is as safe as it would be if the employer did comply with the OSHA requirement.

VC – See Vital Capacity.

Vector – A carrier of infectious agents.

Velocity Pressure (VP) – The pressure created by air moving inside such things as a LEV system.

Velometer – An instrument used to measure airflow; provides readings in feet per minute.

Vibration Isolator – A flexible cloth or plastic connection placed between the source of a vibration, for example, a fan housing, and a potential conductor of the vibration, such as ductwork.

Vital Capacity (VC) – The volume of air that can be taken in and pushed out of the lungs.

VP – See Velocity Pressure.

Walk-through Survey – An examination or inspection of a work place involving a review of hazardous materials present and/or used; observing work practices; and talking with individuals to identify all of the actual or potential chemical, physical, biological, and ergonomic hazards that are present.

Warning Properties – The physical/chemical characteristics of a substance that allows it to be tasted or smelled at safe concentration levels.

WBGT – See Wet Bulb Globe Temperature.

Weighting Scale – In the context of noise exposures, the A, B, or C weighting scales were developed to approximate the response of the human ear at different ranges of sound pressure levels. The A weighting scale is specified by OSHA for use in evaluating occupational noise exposures.

Welder's Flash – Corneal damage resulting from exposure to intermediate and far UV wavelengths emitted by welding processes.

Wet Bulb Globe Temperature (WBGT) – An index value computed in °C or °F, which provides information on the potential heat load of the environment.

Wheatstone Bridge Circuit – A circuit designed to measure an unknown resistance or a change in resistance by comparing it to known resistances.

White Finger – See Raynaud's Syndrome.

Wind Chill Cooling Rate – The heat loss due to the combination of ambient temperatures and the movement of the wind across the skin.

x-Rays – Short wavelength electromagnetic energy that originates outside of the nucleus as electrons suddenly move from higher to lower energy levels and must, therefore, give up their energy.

Zone – In the context of air quality control, the area or region of a building serviced by an HVAC unit.

PRESERVING
THE LEGACY

Acknowledgements

Chapter 1

Figure 1-1: The hazards associated with airborne contaminants were already recognized four hundred years ago. These woodcuts from Agricola's *De Re Metallica* (1556) show some of the methods used to provide ventilation in hazardous atmospheres. From *The Mining Magazine*, London, 1912.

Figure 1-2: These woodcuts (also from *De Re Metallica*) illustrate examples of protective equipment used by mine and smelter workers. From *The Mining Magazine*, London, 1912.

Chapter 2

Table 2-1: Approximate LD_{50} values (rats) for The LD_{50} of a material provides some additional information compared to a dose-response curve. Adapted with permission from *Casarett and Doull's Toxicology* by C. D. Klaassen et al. editors, 3rd edition, Macmillan Publishing, New York, 1986.

Table 2-3: Effect of route of administration on the toxicity of various compounds. The relationship between dose and response is affected by the route of exposure. This table illustrates how toxic effect varies depending on the route of exposure. Reprinted from *Essentials of Toxicology* by T. A. Loomis, Lea & Febiger, Philadelphia, 1968.

Figure 2-9: Liver acinus and zonal regions. The liver plays an important role in eliminating toxic materials from the body. It may concentrate, excrete, or biotransform materials to facilitate their removal; because of this role, it is susceptible to damage. Reprinted by permission from *Diseases of the Liver* by L. Schiff editor, 3rd edition, Lippincott Raven, Philadelphia, 1969.

Figure 2-11: This diagram shows significant developmental stages of a human embryo, during which different organs and/or body parts are taking shape. Reprinted by permission from *Principles of Biochemical Toxicology* by J. A. Timbrell, Taylor and Francis, London, 1991.

Figure 2-12: Incidence of malformations in rat embryos exposed to teratogens at different times during gestation. Reprinted by permission from *Environment and Birth Defects* by J. G. Wilson, Academic Press, New York, 1973.

Table 2-9: Many materials have been linked to negative effects on the male reproductive system. This is one area of toxicology, for both males and females, that warrants more investigation. Reprinted by permission from *Casarett and Doull's Toxicology* by C. D. Klaassen et al. editors, 3rd edition, Macmillan Publishing, New York, 1986.

Table 2-10: Effects on the female reproductive system include teratogens and other materials that affect fetal development. Most toxicity studies are limited in the number and type of reproductive effects they are designed to detect. Reprinted by permission from *Casarett and Doull's Toxicology* by C. D. Klaassen et al. editors, 3rd edition, Macmillan Publishing, New York, 1986.

Chapter 4

Figure 4-4: Sizes of various airborne particles. The size of particle has a direct bearing on its ability to enter the respiratory tract. Courtesy of MSA.

Figure 4-5: Photomicrographs showing amphibole and serpentine asbestos. Photos by the United States Geological Survey.

Figure 4-7: Effects of CO concentrations on miners and birds. Based on "Effects of Carbon Monoxide on Man and Canaries" by T. D. Spencer, *Annals of Occupational Hygiene* 5:61, Figure 3.

Table 4-7: Inhalation is the most significant route of entry to occupational airborne hazards. The effects associated with inhalation of the material depend on where in the respiratory tract the inhaled particles are deposited. Reprinted by permission from *Casarett and Doull's Toxicology* by C. D. Klaassen et al. editors, 3rd edition, Macmillan Publishing, New York, 1986.

Chapter 5

Figure 5-1: High-volume (2-15 lpm) air sampling pumps such as this are used for area sampling because they allow a relatively large volume of air to be sampled in a short time. They are also used for clearance sampling following removal of asbestos materials or lead-based paint. Photo Courtesy of Hazco.

Figure 5-3: Personal sampling pumps such as this one are smaller and lighter than pumps commonly used for area sampling. Personal sampling pumps are used by the industrial hygienist to obtain data for evaluation against regulatory and other exposure limits. Photo Courtesy of Hazco.

Figure 5-6: Photomicrograph of surface of air sampling filter. Photo Courtesy of Gelman Sciences.

Figure 5-8: This is a diagrammatic representation of a filter being checked for airborne fibers, the method used for evaluating occupational exposure to asbestos. According to the counting requirements, all structures that are five microns or more in length and three times longer than they are wide are counted. The grid marks superimposed on the slide are from an optical insert – called a Walton-Beckett graticule – in the microscope.

Figure 5-14: Sample bags made out of Teflon® or another inert polymer are useful for obtaining grab samples. The bags are light and easy to use, and are returned to the laboratory for analysis. Photo Courtesy of Hazco.

Figure 5-15: Evacuated containers – usually made of glass or metal – are used for grab samples. Air is drawn inside when they are opened; they are then sealed and returned to the laboratory for analysis. Photo Courtesy of Hazco.

Figure 5-18: Electronic calibrators such as this one operate using the same principle as the bubble meter. They offer the advantage of being faster; also, most can be purchased with options such as computer software, printers, and capabilities for calibrating a wide range of flow rates. Photo Courtesy of Hazco.

Figure 5-28: There are many direct-reading gas detection instruments available to the industrial hygiene professional today. The instrument shown here is build to withstand regular use in a variety of industrial settings. Photo Courtesy of Hazco.

Table 5-4: The response of a direct-reading instrument to a particular gas depends on what gas was used to calibrate the instrument. This chart illustrates how to correctly interpret the response of the instrument shown in Figure 5-28, based on the gas being tested and the gas used for calibration. Courtesy of Industrial Scientific Corp.

Figure 5-30: Photoionization detectors (PIDs) are direct-reading instruments that are used a great deal at hazardous waste sites. They are non-specific, but useful for screening purposes during initial entries at sites where specific contaminants are unknown. Photo Courtesy of Hazco.

Figure 5-31: The flame ionization detector (FID) is another type of nonspecific instrument often used at hazardous waste sites during investigation and cleanup. It is generally better than a PID at detecting hydrocarbon compounds, making them useful for field screening at sites where leaking underground fuel tanks have been removed. Photo Courtesy of Hazco.

Figure 5-32: Detector or length-of-stain tubes are among the most portable and versatile direct-reading methods for chemical detection. Tubes are available for detecting a wide variety of materials within a wide range of concentrations. Photo Courtesy of Hazco.

Chapter 6

Figure 6-3: A velometer is a common instrument for measurement of airflow across the face of a duct or air diffuser. This measurement can be used to calculate the amount of air being supplied to the diffuser by the HVAC unit, which can then be compared to a recommended airflow. Photo Courtesy of Hazco.

Figure 6-4: Map from USGS showing radon potential across the United States. By the United States Geological Survey.

Chapter 8

Figure 8-1: This image shows the depigmented skin of a hospital worker exposed to a phenolic germicidal cleaner. Depigmentation from damage to the melanocytes can be permanent. Photo Courtesy of the National Institute for Occupational Safety and Health.

Figure 8-2: This cross-section of the skin shows the structures found within each of the layers of skin. Many materials used throughout industry damage specific layers or structures found in the skin. Photo Courtesy of the National Institute for Occupational Safety and Health.

Figure 8-3: This allergic contact dermatitis is the result of exposure to phenol-formaldehyde resins. These types of reactions are also called delayed hypersensitivity reactions. Photo Courtesy of the National Institute for Occupational Safety and Health.

Figure 8-4: This acute contact dermatitis resulted from exposure to ethylene oxide, which is a powerful skin irritant. These reactions usually occur fairly soon following exposure to the offending agent. Photo Courtesy of the National Institute for Occupational Safety and Health.

Figure 8-5: The occurrence of folliculitis or occupational acne, is usually associated with exposure to insoluble oils, lubricants, greases, and chlorinated compounds. A case of chloracne is pictured here. Photo Courtesy of the National Institute for Occupational Safety and Health.

Figure 8-6: Chrome holes are ulcerative lesions associated with exposure to Cr^{6+}. The nasal septum can be similarly affected. Photo Courtesy of the National Institute for Occupational Safety and Health.

Figure 8-7: This image shows a photoallergic reaction involving sunlight and exposure to lime juice. Hyperpigmentation can also occur as the result of a photoallergic reaction. Photo Courtesy of the National Institute for Occupational Safety and Health.

Figure 8-8: This photo shows contact urticaria (hives) from exposure to natural latex rubber. The recent rise in the incidence of these types of reactions in the health care industry may be linked to the increased use of latex-containing products such as gloves, respirators, and goggles used for biological substance isolation. Photo Courtesy of the National Institute for Occupational Safety and Health.

Figure 8-9: This is a classic bull's-eye pattern often seen at the site of a tick bite. It is not uncommon for these reactions to take two weeks to go away. Photo Courtesy of the National Institute for Occupational Safety and Health.

Figure 8-10: Cutaneous anthrax is difficult to miss, and the worker will usually obtain prompt treatment for this condition. Pulmonary anthrax is less obvious and may go unnoticed, or be misdiagnosed for some time before treatment is administered. Photo Courtesy of the National Institute for Occupational Safety and Health.

Chapter 9

Figure 9-3: Photomicrograph of cochlea showing healthy (A) and damaged (B) hair cells. The cochlea in B displays significant noise-induced hearing loss. Photos Courtesy of Howard Leight Hearing Protection.

Figure 9-5: The OSHA standard requires that employers make available a variety of hearing protection devices for employees who are required to use them. Photos Courtesy of Howard Leight Hearing Protection.

Chapter 10

Table 10-2: Shielding materials for different radioactive particles. The type and thickness of shielding that is necessary depends upon the type of radiation and its energy. Sometimes a combination of shielding materials is appropriate. Reprinted with permission from *Documentation of the Threshold Limit Values and Biological Exposure Indices*, 6th edition, American Conference of Governmental Industrial Hygienists, 1996.

Table 10-3: Some NCRP recommended limits for exposure to ionizing radiation. The extremities (hands and feet) are able to withstand higher levels of exposure without permanent adverse effects, as compared to the internal organs. Reprinted with permission from *Documentation of the Threshold Limit Values and Biological Exposure Indices*, 6th edition, American Conference of Governmental Industrial Hygienists, 1996.

Chapter 12

Figure 12-3: Half-face respirators cover less skin and are lighter than full-face respirators. They are more comfortable in some situations, but may be more difficult in terms of fitting well around a worker's nose, chin, and cheeks. Full-face respirators provide a higher level of protection and a better fit; they have the added benefit of built-in eye protection. Photo Courtesy of Hazco.

Figure 12-4: The self-contained breathing apparatus, or SCBA, resembles the more familiar SCUBA gear worn by divers. The primary difference is that the SCBA is designed for use on land. Photo Courtesy of Hazco.

Bibliography

Agricola, G., *De Re Metallica*, translated by H. C. Hoover and L. H. Hoover, Dover Press, New York, NY, 1950.

American Board of Industrial Hygiene, "1995 Roster of Diplomats."

American Conference of Governmental Industrial Hygienists, *Documentation of TLVs and BEIs*, sixth edition, Cincinnati, OH, 1991.

American Conference of Governmental Industrial Hygienists, *Industrial Ventilation, A Manual of Recommended Practice*, nineteenth edition, Edwards Brothers Inc., Ann Arbor, MI, 1986.

American Conference of Governmental Industrial Hygienists, *Threshold Limit Values for Chemical Substances and Physical Agents*, Cincinnati, OH, 1996.

American Conference of Governmental Industrial Hygienists, TLV Booklet, Cincinnati, OH, 1997.

American Industrial Hygiene Association, *Who's Who in Industrial Hygiene: The 1995-96 Membership Directory of the American Industrial Hygiene Association*.

American National Standards Institute, *American National Standard for Respiratory Protection, ANSI Z88.2-1992*, New York, NY, 1992.

American Public Health Association, "Control of Communicable Diseases in Man," John D. Lucas Printing Co., 1985.

American Society of Heating, Refrigeration and Air Conditioning Engineers, "ASHRAE 52-1992, Methods of Testing Air Cleaning Devices Used in General Ventilation for Removing Particulate Matter," Atlanta, GA, 1992.

American Society of Heating, Refrigeration and Air Conditioning Engineers, "ASHRAE 55-1992, Thermal Environmental Conditions For Human Occupancy," Atlanta, GA, 1992.

American Society of Heating, Refrigeration, and Air Conditioning Engineers, "ASHRAE 62-1989, Ventilation For Acceptable Air Quality," Atlanta, GA, 1989.

Berger, Elliot, Morrill, J. C., Ward, W. D., Royster, L. H. editors, *Noise and Hearing Conservation*

Manual, American Industrial Hygiene Association, 1986.

Burton, D. Jeff, *Industrial Ventilation Workbook*, third edition, IVE, Inc., 1994.

Klaassen C. D. et al. editors, *Casarett and Doull's Toxicology*, third edition, Macmillan Publishing, New York, NY, 1986.

Clayton & Clayton editors, *Patty's Industrial Hygiene and Toxicology*, fourth edition, Vol. I, Part A, John Wiley & Sons, Inc., New York, NY, 1991.

Clayton, & Clayton editors, *The Industrial Environment, Its Evaluation and Control*, U.S. Department of Health, Education and Welfare, Centers for Disease Control, National Institute for Occupational Safety and Health, Washington, D.C., 1973.

Code of Federal Regulations, 29 CFR 1910, Occupational Safety and Health Standards.

Code of Federal Regulations, 29 CFR 1910, Subpart I, Personal Protective Equipment.

Code of Federal Regulations, 29 CFR 1910.95, Occupational Noise Exposure.

Code of Federal Regulations, 29 CFR 1910.133, Eye and Face Protection.

Code of Federal Regulations, 29 CFR 1910.134, Respiratory Protection, 1998.

Code of Federal Regulations, 29 CFR 1926, OSHA Standards for General Industry and Construction.

Code of Federal Regulations, 29 CFR 1926.1101, Asbestos.

Code of Federal Regulations, 42 CFR 84, Respiratory Protective Devices.

Federal Register, Air Contaminants, Vol. 57, No. 114, Proposed Rule, Friday, June 12, 1992.

Francis, B. Magnus, *Toxic Substances in the Environment*, John Wiley & Sons, Inc., New York, NY, 1994.

Hawkins, Neil, et al., *A Strategy for Occupational Exposure Assessment*, AIHA, 1991.

Hughes, W. William, *Essentials of Environmental Toxicology*, Taylor and Francis, Bristol, PA, 1996.

Key, Marcus M., et al. editors, *Occupational Diseases, A Guide to Their Recognition*, U.S. Department of Health, Education and Welfare, Washington, D.C., 1977.

Kroemer, K., "Cumulative Trauma Disorders," *Applied Ergonomics*, 20:274-280, 1989.

Morrison and Boyd, *Organic Chemistry*, third edition, Allyn and Bacon, 1973.

National Institute for Occupational Safety and Health, *NIOSH Pocket Guide to Chemical Hazards*, Publication No. 94-116, U.S. Government Printing Office, Washington, D.C., 1994.

National Institute for Occupational Safety and Health, *NIOSH Pocket Guide to Chemical Hazards*, Publication No. 94-116, U.S. Government Printing Office, Washington, D.C., 1996.

National Institute for Occupational Safety and Health, *NIOSH Respirator Decision Logic*, U.S. Government Printing Office, Washington, D.C., 1987.

Ness, Shirley, *Air Monitoring for Toxic Exposures*, Van Nostrand Reinhold, New York, NY, 1991.

NIOSH/OSHA/USCG/EPA, *Occupational Safety and Health Guidance Manual for Hazardous Waste Activities*, U. S. Department of Health and Human Services, U. S. Government Printing Office, Washington, D.C., 1985.

Occupational Safety and Health Act of 1970, Public Law 91-596, December 29, 1970.

OSHA Industrial Hygiene Technical Manual, 1984.

Perkins, Jimmy L., *Modern Industrial Hygiene*, Van Nostrand Reinhold, New York, NY, 1997.

Plog, Barbara, Niland, Jill and Quinlan, Patricia J. editors; *Fundamentals of Industrial Hygiene*, fourth edition, National Safety Council, Chicago, IL, 1996.

Stacey, N. H. editor, *Occupational Toxicology*, Taylor and Francis, Bristol, PA, 1993.

Tillotson, Michael R. editor, *Environmental Radiation*, Compilation of Course Material, 1983.

Timbrell, J. A., *Introduction to Toxicology*, Taylor and Francis, London, 1995.

U.S. Department of Health, Education, and Welfare, *NIOSH Occupational Exposure Sampling Strategy Manual*, Publication No. 77-173, U.S. Government Printing Office, Washington, D.C., 1977.

U.S. Environmental Protection Agency, *Managing Asbestos in Place, A Building Owner's Guide to Operations and Maintenance Programs for Asbestos-Containing Materials*, Publication 2OT-2003, Washington, D.C., 1990.

Williams and Burson editors, *Industrial Toxicology*, Van Nostrand Reinhold, New York, NY, 1985.

Index

A

AA. *See* Atomic absorption
AAIH. *See* American Academy of Industrial Hygiene
ABIH. *See* American Board of Industrial Hygiene
Absorbance 122, 221
Absorption spectroscopy 126
Acclimatization 252
ACGIH. *See* American Conference of Governmental Industrial Hygienist
  recommendation for noise exposure 215
ACGIH TLVs 60
  biological exposure indices 61, 143
  categories of carcinogen 62
  comparison with OSHA PELs 60
  guidelines for good practice 60
  heat stress 252
  short-term exposure limit 60
  TLV-ceiling 61
  uncertainty 60
Acinus 34

Acoustic watt 206
Action level 58
Activated charcoal 110
Activation 230
Active transport 32, 36
Activity 166
Activity hazard analysis. *See* Hazard assessment
Acute exposure 23, 24
Administrative controls 10, 171
  for noise 221
  for skin protection 199
Aerodynamic diameter 77
Aerosol 79
Aerosolize 79
AIHA. *See* American Industrial Hygiene Association
Air cleaners 176, 177
Air sampling 101, 140
  calculating results 143
  interpreting results 143
  limitations 141
  reasons for sampling 102
  representativeness 141
  sampling approaches 103
  sources of error 140

Air-purifying respirators 267
  assigned protection factor 270
  conditions not suitable 269
  end-of-service-life indicator 269
  maximum use concentration 270
  warning properties of contaminants 268
Air-supplying respirators 267, 273
  chemical asphyxiant 273
  required air flow 273
Airborne hazards 77, 171
  classifications 77
    aerosol 79
    dust 79
    fumes 79
    gases 79
    mist 79
    particulate 79
    vapors 79
  controls 171
    air cleaners 176
    ventilation systems 173, 183
ALARA. *See* As low as reasonably achievable
Albinism 191
Allergic alveolitis 90